Advances in Postharvest Process Systems

Advances in Postharvest Process Systems

Editors

Daniel I. Onwude
Guangnan Chen

MDPI • Basel • Beijing • Wuhan • Barcelona • Belgrade • Manchester • Tokyo • Cluj • Tianjin

Editors
Daniel I. Onwude
Empa-Swiss Federal
Laboratories of Material Science
and Technology
Switzerland

Guangnan Chen
University of Southern
Queensland
Australia

Editorial Office
MDPI
St. Alban-Anlage 66
4052 Basel, Switzerland

This is a reprint of articles from the Special Issue published online in the open access journal *Processes* (ISSN 2227-9717) (available at: http://www.mdpi.com).

For citation purposes, cite each article independently as indicated on the article page online and as indicated below:

LastName, A.A.; LastName, B.B.; LastName, C.C. Article Title. *Journal Name* **Year**, *Volume Number*, Page Range.

ISBN 978-3-0365-2373-6 (Hbk)
ISBN 978-3-0365-2374-3 (PDF)

Contents

About the Editors

Dr. Daniel I. Onwude is currently a Research Scientist at the Swiss Federal Laboratories For Material Science and Technology, Empa, ETHz domain, Switzerland. He is also a Lecturer in the Department of Agricultural and Food Engineering, University of Uyo, Nigeria. Dr. Onwude received his Ph.D. degree in Agricultural Process Engineering from Universiti Putra Malaysia in 2018. In 2019, he received a post-doctoral scholarship at ETH Zurich, Switzerland, with joint affiliation with Empa, under the Swiss Government Excellence Scholarship.

Dr. Onwude has published three book chapters and over 50 articles in international peer-reviewed journals, which were cited over 1000 times, and resulted in a Scopus h-index of 14. He has been a principal investigator in several national and international research projects. He is currently an Associate Editor for the *Journal of Thermal Science and Engineering Progress* (TSEP-Elsevier), and an Editorial Board Member in the *Journal of Agricultural and Food Engineering* (JAFE). He is a member of international Food and Agricultural organizations including the American Society of Agricultural and Biological Engineers (ASABE), Nigerian Institution of Agricultural Engineers (NIAE), Institute of Food Technologist (IFT), International Society for Horticultural Science (ISHS), and Institution of Engineering and Technology (IET). His research interests include food preservation and processing, drying technologies, computational fluid dynamic, and mathematical modelling and computer simulations.

Dr Guangnan Chen is currently an Associate Professor and Discipline Leader in agricultural engineering at the University of Southern Queensland, Australia. He graduated from the University of Sydney with a PhD degree in 1994. Before joining the University of Southern Queensland in early 2002, he worked for two years as a post-doctoral fellow and for more than five years in a private consulting company based in New Zealand.

Dr Chen teaches and researches the subjects of agricultural machinery, agricultural materials and post-harvest technologies, and sustainable agriculture. He has published more than 140 papers in various international journals and conferences, including three edited books published by the CRC Press, Taylor & Francis Books and 10 invited book chapters. He is currently also the Secretary of Board of Technical Section IV (Energy in Agriculture), Commission Internationale du Genie Rural (CIGR), which is one of the world's top professional bodies in agricultural and biosystems engineering.

processes

Editorial

Special Issue "Advances in Postharvest Process Systems"

Daniel I. Onwude [1,2,*] and Guangnan Chen [3,*]

1 Empa-Swiss Federal Laboratories of Material Science and Technology, Überland Str. 129, 8600 Dübendorf, Switzerland
2 Department of Agricultural and Food Engineering, Faculty of Engineering, University of Uyo, Uyo 52021, Nigeria
3 Faculty of Health, Engineering and Sciences, University of Southern Queensland, Toowoomba, QLD 4350, Australia
* Correspondence: daniel.onwude@empa.ch (D.I.O.); Guangnan.chen@usq.edu.au (G.C.)

The world population is predicted to increase from the present 7.7 billion to 9.7 billion in 2050, demanding a significant increase in food supply and production. However, around 25–30% of food is wasted worldwide every year due to poor postharvest supply chain design and management in different stages of the food supply chain, including postharvest handling, processing, and storage systems.

This special issue presents state-of-the-art information on the important innovations and research in the agricultural and food industry. Different novel technologies and their implementation to optimize postharvest processes and reduce losses are reviewed and explored. In particular, it examines a range of recently developed and improved technologies and systems to help the industry and growers to manage and minimize postharvest losses, enhance reliability and sustainability in the postharvest food value chain, and generate high-quality products that are both healthy and appealing to consumers.

This special issue consists of three sections, focusing on food storage and preservation technologies [1–4], food processing technologies [5–8], and the applications of advanced mathematical modeling and computer simulations [9–11]. We wish to acknowledge the expert contributions of all authors here. We also wish to acknowledge and thank MDPI staff for their professional assistance in editing the published articles. We sincerely hope that this special issue will assist all readers and stakeholders working in or are associated with the fields of agriculture, agri-food chain, and technology development and promotion. After all, efficient postharvest technology is an essential and key factor underlying future global food security, and ultimately human survival and development.

Citation: Onwude, D.I.; Chen, G. Special Issue "Advances in Postharvest Process Systems". *Processes* **2021**, *9*, 1426. https://doi.org/10.3390/pr9081426

Received: 2 August 2021
Accepted: 2 August 2021
Published: 18 August 2021

1. Food Storage and Preservation Technologies

There are many ways to store and preserve food, each with its benefits and limitations. Different food and crops also need to be stored and preserved in particular ways to maintain their quality best. Onwude et al. [1] reviews the recent advances in technologies to reduce the quality loss of fresh agricultural produce in the supply chain, including the applications of imaging technology, spectroscopy, multi-sensors, electronic nose, radio frequency identification, printed sensors, acoustic impulse response, and mathematical models. The Internet of Things (IoT) and virtual representation models of a particular fresh produce (digital twins) are particularly identified as emerging technologies that can help monitor and control the quality during postharvest.

Ndukwu et al. [2] investigates the effectiveness of different household storage strategies and also the plant-based preservatives for dehulled and sun-dried breadfruit seeds. The obtained results revealed the high potential of alligator pepper (*Zingiberaceaeaframomum melegueta*) as a botanical insecticide in preventing insect infestation and mold growth. An aluminum silo bin with alligator pepper powder is subsequently recommended to store dried and dehulled breadfruit seeds as a baseline for other tropical crops.

Tomlins et al. [3] evaluates the various factors contributing to the storage shelf-life of fresh Cassava roots. In this study, three different types of bag materials were tested. Microclimate related to temperature, humidity, and carbon dioxide (CO_2) was monitored. The results showed that fresh cassava roots could be stored for 8 days, with minimal postharvest physiological deterioration and starch loss. It was also shown that carbon dioxide concentration in the stores was significantly correlated with the starch loss in fresh cassava roots and is therefore proposed as a possible method for continuously and remotely monitoring starch loss in large-scale commercial operations and reducing postharvest losses.

Esua et al. [4] reviews the applications of ultraviolet-C radiation and ultrasound technology for postharvest preservation and handling of fruits and vegetables. It was identified that scale-up and optimization of the process would provide considerable opportunities for industry-led collaborations and chart effective commercialization paths.

2. Food Processing Technologies

Food processing technologies include the applications of different physical, chemical, or microbiological methods and techniques to enhance the food quality and also to ensure more diversity for the increasing demand of different consumers. Raut et al. [5] discusses the impact of process parameters and bulk properties on the quality of dried hops. The results showed that it is important to define and consider optimum bulk and process parameters in order to optimize the hop drying process to improve the process efficiency as well the product quality.

Gurdil et al. [6] evaluates the postharvest processing of hazelnut kernel oil extraction using uniaxial pressure and organic solvent. It was found that the increased speed caused a serration effect on the force–deformation curve, resulting in lower oil yield. Lower and upper oil point forces were also observed to be useful for predicting the pressure for maximum output oil. It was further found that the peroxide value and free fatty acid content of kernel oil decreased with increasing temperature. In designing new presses, it is suggested that there is a need to consider compression and relaxation processes to reduce the residual kernel cake oil.

Tan et al. [7] explores the physicochemical changes, microbiological properties, and storage shelf life of cow and goat milk from industrial high-pressure processing (HPP). No significant changes were found in the physicochemical properties of the treated milk except for pH. HPP-treated cow and goat milk both achieved microbial shelf life of 22 days at 8 °C storage temperature.

Kwofie et al. [8] investigates the relationship between the classification and force deformation characteristics of common beans and the influence of bean softeners on cooking time. This work classified ten bean cultivars as either easy-to-cook (ETC) or hard-to-cook (HTC) based on a traditional subjective finger pressing test and a scientific objective hardness test. The results showed that a modified three-parameter non-linear regression model could accurately predict the rate of bean softening. The influence of bean softeners, such as potassium carbonate (K_2CO_3) and sodium chloride (NaCl), to reduce cooking time was also investigated in this paper.

3. Application of Mathematical Modeling and Computer Simulations in Postharvest Technology

With increased computing power, it is now very attractive to investigate the applications of mathematical modeling and computer simulations in the postharvest processes. Many models have now been successfully developed and employed in food and crop monitoring, grading, and classification, predicting and modeling quality properties, and forecasting chemical, physical, and nutrient characteristics during processing and postharvest storage. Khaled et al. [9] demonstrates the application of computational intelligence in describing the drying kinetics of persimmon fruit during vacuum and hot air-drying processes. Kinetic models were developed using selected thin layer models and computational intelligence methods, including multi-layer feed-forward artificial neural network (ANN), support vector machine (SVM), and k-nearest neighbors (kNN). The validation results indicated good agreement could be obtained from the computational intelligence methods between the predicted values and the experimental data.

Azman et al. [10] studies the physical properties and mass modeling of pepper berries at different maturity levels. It was found that the quadratic model was best fitted for mass prediction at all mass maturity levels (immature, mature, and ripe). The results showed that mass modeling based on the actual volume of pepper berries was more applicable compared to other properties. The findings would be potentially useful in developing improved grading, handling, and packaging systems.

Due to the numerous side effects of synthetic pesticides, including environmental pollution, threats to human health, harmful effects on non-target organisms, and pest resistance, the use of alternative healthy, available, and efficient agents in pest management strategies is attracting increased attention. Ebadollahi et al. [11] examines the optimization and modeling of susceptibility of Tribolium castaneum (Coleoptera: Tenebrionidae) to the fumigation of two essential Satureja oils. The insecticidal properties of the essential oils were modeled and optimized using response surface methodology. It was suggested that the essential oils of S. hortensis and S. intermedia could be offered as promising pesticidal agents against Tribolium castaneum in the management of such pests instead of detrimental synthetic pesticides.

Funding: This research received no external funding.

Conflicts of Interest: The authors declare no conflict of interest.

References

1. Onwude, D.I.; Chen, G.; Eke-Emezie, N.; Kabutey, A.; Khaled, A.Y.; Sturm, B. Recent advances in reducing food losses in the supply chain of fresh agricultural produce. *Processes* **2020**, *8*, 1431. [CrossRef]
2. Ndukwu, M.C.; Onwude, D.I.; Ehiem, J.; Abada, U.C.; Ekop, I.E.; Chen, G. The effectiveness of different household storage strategies and plant-based preservatives for dehulled and sun-dried breadfruit seeds. *Processes* **2021**, *9*, 380. [CrossRef]
3. Tomlins, K.; Parmar, A.; Omohimi, C.I.; Sanni, L.O.; Adegoke, A.F.; Adebowale, A.-R.A.; Bennett, B. Enhancing the shelf-life of fresh cassava roots: A field evaluation of simple storage bags. *Processes* **2021**, *9*, 577. [CrossRef]
4. Esua, O.J.; Chin, N.L.; Yusof, Y.A.; Sukor, R. A Review on individual and combination technologies of UV-C radiation and ultrasound in postharvest handling of fruits and vegetables. *Processes* **2020**, *8*, 1433. [CrossRef]
5. Raut, S.; von Gersdorff, G.J.E.; Münsterer, J.; Kammhuber, K.; Hensel, O.; Sturm, B. Impact of process parameters and bulk properties on quality of dried hops. *Processes* **2020**, *8*, 1507. [CrossRef]
6. Gürdil, G.A.K.; Kabutey, A.; Selvi, K.Ç.; Mizera, Č.; Herák, D.; Fraňková, A. Evaluation of postharvest processing of hazelnut kernel oil extraction using uniaxial pressure and organic solvent. *Processes* **2020**, *8*, 957. [CrossRef]
7. Tan, S.F.; Chin, N.L.; Tee, T.P.; Chooi, S.K. Physico-chemical changes, microbiological properties, and storage shelf life of cow and goat milk from industrial high-pressure processing. *Processes* **2020**, *8*, 697. [CrossRef]
8. Kwofie, E.M.; Mba, O.I.; Ngadi, M. Classification, force deformation characteristics and cooking kinetics of common beans. *Processes* **2020**, *8*, 1227. [CrossRef]
9. Khaled, A.Y.; Kabutey, A.; Selvi, K.Ç.; Mizera, Č.; Hrabe, P.; Herák, D. Application of computational intelligence in describing the drying kinetics of persimmon fruit (Diospyros kaki) during vacuum and hot air drying process. *Processes* **2020**, *8*, 544. [CrossRef]
10. Megat Ahmad Azman, P.N.; Shamsudin, R.; Che Man, H.; Ya'acob, M.E. Some physical properties and mass modelling of pepper berries (*Piper nigrum* L.), variety kuching, at different maturity levels. *Processes* **2020**, *8*, 1314. [CrossRef]
11. Ebadollahi, A.; Taghinezhad, E.; Setzer, W.N.; Chen, G. Susceptibility of tribolium castaneum (Coleoptera: Tenebrionidae) to the fumigation of two essential satureja oils: Optimization and modeling. *Processes* **2021**, *9*, 1243. [CrossRef]

processes

MDPI

Review

Recent Advances in Reducing Food Losses in the Supply Chain of Fresh Agricultural Produce

Daniel I. Onwude [1,2,*], Guangnan Chen [3], Nnanna Eke-emezie [4], Abraham Kabutey [5], Alfadhl Yahya Khaled [6] and Barbara Sturm [7]

1 Empa, Swiss Federal Laboratories for Materials Science and Technology, Laboratory for Biomimetic Membranes and Textiles, Lerchenfeldstrasse 5, CH-9014 St. Gallen, Switzerland
2 Department of Agricultural and Food Engineering, Faculty of Engineering, University of Uyo, Uyo 52021, Nigeria
3 Faculty of Health, Engineering and Sciences, University of Southern Queensland, Toowoomba, QLD 4350, Australia; Guangnan.Chen@usq.edu.au
4 Department of Chemical and Petroleum Engineering, Faculty of Engineering, University of Uyo, Uyo 52021, Nigeria; nnanna.chimaobim@gmail.com
5 Department of Mechanical Engineering, Faculty of Engineering, Czech University of Life Sciences Prague, Kamýcká 129, 165 00 Prague, Czech Republic; kabutey@tf.czu.cz
6 Department of Biosystems & Agricultural Engineering, College of Agriculture, Food and Environment, University of Kentucky, Lexington, KY 40503, USA; Alfadhl.Alkhaled@uky.edu
7 Process and Systems Engineering in Agriculture Group, Department of Agricultural and Biosystems Engineering, University of Kassel, DE-37213 Witzenhausen, Germany; Barbara.sturm@uni-kassel.de
* Correspondence: daniel.onwude@empa.ch

Received: 12 September 2020; Accepted: 5 November 2020; Published: 9 November 2020

Abstract: Fruits and vegetables are highly nutritious agricultural produce with tremendous human health benefits. They are also highly perishable and as such are easily susceptible to spoilage, leading to a reduction in quality attributes and induced food loss. Cold chain technologies have over the years been employed to reduce the quality loss of fruits and vegetables from farm to fork. However, a high amount of losses (≈50%) still occur during the packaging, pre-cooling, transportation, and storage of these fresh agricultural produce. This study highlights the current state-of-the-art of various advanced tools employed to reducing the quality loss of fruits and vegetables during the packaging, storage, and transportation cold chain operations, including the application of imaging technology, spectroscopy, multi-sensors, electronic nose, radio frequency identification, printed sensors, acoustic impulse response, and mathematical models. It is shown that computer vision, hyperspectral imaging, multispectral imaging, spectroscopy, X-ray imaging, and mathematical models are well established in monitoring and optimizing process parameters that affect food quality attributes during cold chain operations. We also identified the Internet of Things (IoT) and virtual representation models of a particular fresh produce (digital twins) as emerging technologies that can help monitor and control the uncharted quality evolution during its postharvest life. These advances can help diagnose and take measures against potential problems affecting the quality of fresh produce in the supply chains. Plausible future pathways to further develop these emerging technologies and help in the significant reduction of food losses in the supply chain of fresh produce are discussed. Future research should be directed towards integrating IoT and digital twins for multiple shipments in order to intensify real-time monitoring of the cold chain environmental conditions, and the eventual optimization of the postharvest supply chains. This study gives promising insight towards the use of advanced technologies in reducing losses in the postharvest supply chain of fruits and vegetables.

Keywords: food security; food quality; agricultural production; crop storage and processing; food distribution; smart digital technology; industry 4.0; refrigeration

1. Introduction

Food losses in the postharvest supply chain amount to a great loss of investments in the packaging, transportation, and storage operations. About 25–30% of global food produced is lost between on-farm food production and its storage at a retail facility, largely as a result of poor chain management and spoilage [1,2]. Food losses occur due to a reduction in quality and safety standards driven by consumer preferences, particularly in developed countries [3]. A high amount of losses (up to 30% per year) is often experienced during the postharvest handling of fresh agricultural produce, such as fruits and vegetables [4]. Advanced technologies are required to reduce the losses of fruits and vegetables in the postharvest supply chain. The reduction of these losses would increase the number of fresh produce available for consumption.

Fruits and vegetables are important sources of nutrients such as vitamins, minerals, and bioactive compounds, which provide many health benefits [5–8]. However, they are highly perishable goods that need to be appropriately preserved, to reduce the degradation of macro and micro-nutrients and extend shelf life [7,9,10]. As a result, fruits and vegetables are often packaged and kept in a desired low-temperature range using various refrigeration systems during the transportation and storage postharvest handling processes. This process delays or reduces microbial growth and enzymatic reaction, thereby improving overall quality, reducing mass loss, and extending shelf-life. The succession of refrigeration steps along these chains can be referred to as a postharvest cold chain of fruits and vegetables [11]. A description of the postharvest cold chain of fruits and vegetables is shown in Figure 1.

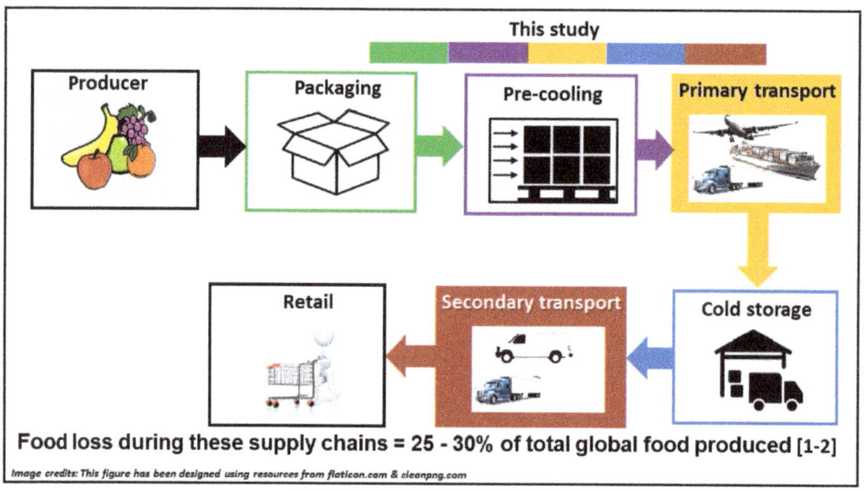

Figure 1. Postharvest refrigerated supply chain of fruits and vegetables.

Refrigeration is a key element in enhancing the quality of fresh produce and extending the shelf-life, thereby enabling their adequate supply to an increasingly urbanized world [11–13]. However, more than 90% of perishable goods are still not refrigerated [1,14]. Inadequate refrigeration infrastructure or access to energy accounts for more than 20% losses of perishable goods [15]. These losses also encompass a huge amount of energy and water losses, but also carbon dioxide emissions [15,16]. Therefore, sustainable cold chain technologies in terms of being more resource-efficient, improving product quality retention, and reducing induced food losses become indispensable.

Several studies have been conducted on the postharvest cold chain of fruits and vegetables with a view to gain more insight on ways to address these technological and developmental challenges. The losses in the mass and nutritional qualities of strawberries, raspberries, red currants, drupes, cherries, and sour cherries were reduced using refrigerated containers at 4 °C when compared to storage at room temperature [17]. Packaging methods such as edible coating, active modified atmosphere packaging (MAP), nano-composite based packaging (NCP), and polypropylene/polyethylene bags have been used to reduce quality losses of cherry tomato, kiwifruits, guava, mushroom, cucumber, and berries during cold chain processes [18–23]. Recently, active packaging such as oxygen scavengers, ethylene absorbers, moisture regulators, and intelligent packaging including the use of chemical sensors, temperature, freshness and gas indicators, barcodes and radio frequency identification devices (RFID), have been developed to maintain the quality and improve the safety of fresh produce [24–27]. These different packaging methods are simple to design, easily affordable and can help to extend product shelf life [15]. However, retailers in the food supply chain are increasingly looking for ways to minimize or eliminate the use of packaging, to project sustainable eco-friendly products [28]. Consequently, the negative impact of most packaging materials (e.g., plastic packaging) is largely overestimated by consumers in comparison to other actions with much higher impacts [29–31]. As an example, the controversy between paper bags versus plastic packaging comes to mind. Paper bags hold a much higher environmental impact, due to its higher weight [32], but are often perceived to be more eco-friendly by the consumer. In a similar manner, a life cycle analysis of a commonly consumed fruit or vegetable with and without packaging will show that the environmental impact of plastic packaging, for example, is by far smaller than the impact of the food losses [33,34]. In addition, plastic packaging presently reduces food losses by up to 4.8% at retail and also reduces induced food losses at households as a result of prolonged shelf life [35]. Despite all these improvements and awareness from peer-reviewed literature, the question of why a significant amount of food losses in the postharvest supply chain (see above) arises, suggesting that more insight and advances into cold chain technology are required to further reduce food losses by preventing excessive quality loss of fresh produce.

Key drivers that accelerate food losses during the postharvest supply chain of fruits and vegetables include lack of innovative packaging materials, inadequate monitoring technology, variations in the temperature, approach air velocity and relative humidity in cold chain systems, rate of metabolism, long shipment duration and the heterogeneity of fruits and vegetables. During shipments, there is often a wide variation of temperature and relative humidity at different locations in a cold chain system. Great variations in the approach airspeed of different fruits and vegetables are often observed as a result of the heterogeneous nature of refrigeration equipment, food properties, and packaging container. These variations can affect the final mass loss, overall quality, and the remaining shelf life of fresh produce [36–38].

Understanding the physics behind different phenomena that occur during the different postharvest supply chains and linking these phenomena to measurable output using sensors that provide actionable data may be the key in optimizing the design of packaging, storage, and transport processes for fruits and vegetables. Unfortunately, studies on these advances are limited.

This paper aims to explore ways on how food losses can further be reduced in the postharvest supply chain of fruits and vegetables. Particularly, we discuss the current state-of-the-art in monitoring and optimizing cold chain systems for a reduction in quality loss during the packaging, transportation, and storage of fruits and vegetables. We also analyze the potential of applying emerging technologies such as the Internet of Things (IoT) and digital twins for reducing food losses. We then put forward how the future should look towards reducing food losses during the packaging, storage, and transportation supply chain.

2. The Need to Reduce Food Losses in the Postharvest Supply Chain of Fruits and Vegetables

Food losses can be referred to as "the reduction in the amount of fresh fruits and vegetables that was originally meant for human consumption" [39–41]. Globally, one third to half of all food produced is lost or wasted along postharvest supply chains, with packaging, storage, and transportation value chains the most impacted [42,43]. Losses of fruits and vegetables worldwide are between 40% and 50% of which 54% occur in stages of production, postharvest handling, and storage [3,44,45].

During packaging, transportation, and storage of fresh agricultural produce, food losses are often induced as a result of a reduction in the quality (e.g., color, texture, mass) of the produce. These postharvest handling operations affect the nutritional and sensory quality of the agricultural produce, the mass of the fresh produce as well as the quantity of fresh produce available to the consumers. The quality of fresh agricultural produce can be referred to as the excellent characteristics of such products that are acceptable to a consumer [46]. Consumers typically purchase fresh agricultural produce based on their biochemical characteristics such as appearance, texture, flavor, and nutritive value [46,47]. Fresh agricultural produce such as fruits and vegetables provide an essential part of human nutrition, as they are important sources of vitamins, dietary fibers, minerals, and other biochemical (e.g., carbohydrate, protein, etc.) with tremendous health benefits [48]. Adequate in-transit monitoring of environmental conditions and changes in the quality attributes of fresh produce during transport and storage will help reduce food losses and ensure the availability and accessibility of fresh fruits and vegetables with high nutritional density to the consumers [49,50]. Therefore, emerging technologies are needed to help reduce the overall quality loss of fresh agricultural produce, thereby reducing food losses in the postharvest supply chain.

3. Advanced Technologies for Quality Assessment in Postharvest Supply Chain: State-Of-The-Art

In recent decades, several modern food quality techniques have been applied to monitor, control, and predict the quality evolution of various fruits and vegetables in postharvest supply chains. These techniques include imaging systems, spectroscopy, multi-sensors, electronic nose (E-nose), acoustic impulse response (AIR), radio frequency identification (RFID), printed sensors (PTS), and mathematical modeling. In this section, we analyze the application of these techniques in advancing cold chain operations and process optimization during the packaging, storage, and transportation of fruits and vegetables within the past 10 years.

3.1. Application of Imaging Technology, Spectroscopy, Multi-Sensors, E-Nose, AIR, RFID and PTS in the Postharvest Supply Chain of Fruits and Vegetables

Imaging technology is an advanced method used by the food and agro-allied industries to monitor changes in food quality [51]. This technology includes computer vision (CV), hyperspectral imaging (HSI), multispectral imaging (MSI), thermography, and X-ray imaging. Image technology is particularly useful in detecting and evaluating the external quality attributes (color, geometrical, size, appearance, and surface structure) [52,53], and in some cases, the internal structures (X-rays and hyperspectral imaging) of fruits and vegetables. This technology involves collecting and analyzing spatial information gained from captured images of products, such as color, geometrical, size, appearance, and surface structure. The application of imaging technology in postharvest supply chains is mainly limited to surface detection. The surface properties of an object can be detected due to the interaction of light. A typical imaging system consists of a CCD camera, a light source, a computer, and related software (Figure 2). The camera captures the images of the product based on the region of interest. The captured images are then processed to evaluate and quantify the quality changes that have occurred during a particular postharvest operation. The image processing steps often consist of image acquisition, segmentation, feature extraction and recognition, classification, and interpretation [54–57].

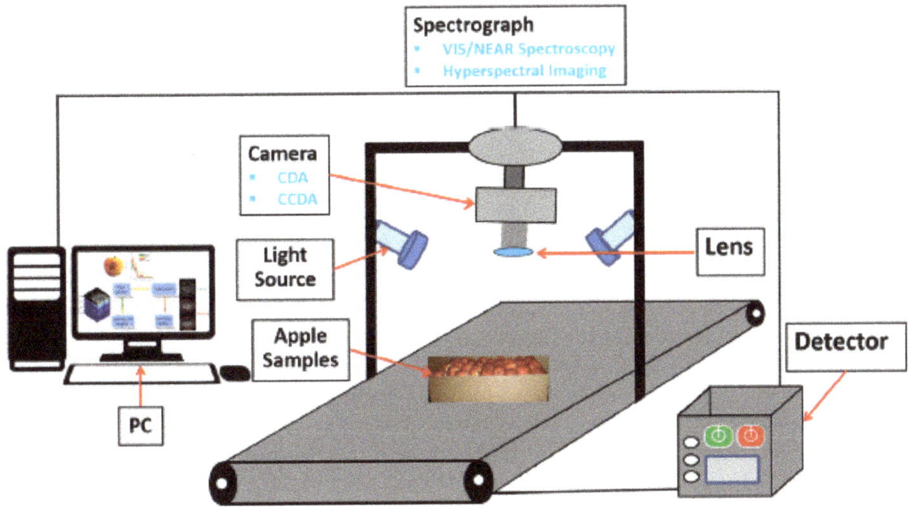

Figure 2. Typical set-up of an imaging system for monitoring the quality of fresh agricultural produce.

In order to adequately discriminate and analyze the captured numerous images, chemometrics and deep learning methods are often employed. These methods have already been found reliable in quantifying the accuracy of processed images and associated quality changes of fruits and vegetables in the postharvest supply chains (Table 1). They include Savitsky–Golay (SG), Standard Normal Variate (SNV), Principal Component Analysis (PCA), Partial Least Squares Regression (PLSR), Multiple Scatter Correction (MSC), Partial Least Squares Discriminant Analysis (PLS-DA), Artificial Neural Network (ANN), Convolutional Neural Networks (CNN), Linear Discriminant Analysis (LDA), k-Nearest Neighbors (kNN), Correlation-based Feature Subset Selection (CFS), Gini Impurity Algorithm (GIA), Sequential Forward Selection (SFS), Backpropagation Neural Network (BPNN), Extreme Learning Machine (ELM), Sparse Logistic Regression (SLR), Support Vector Machine(SVM), Radial Basis Function (RBF), Neural Networks (NN), Genetic Algorithm (GA), Support Vector Regression (SVR), Student–Newman–Keuls (SNK), Least Squares Support Vector Machines (LS-SVM), and Random Forest (RF).

The applications of different imaging and smart digital technologies in monitoring the quality of fruits and vegetables in postharvest cold chains are summarized in Table 1. These technologies include computer vision (CV), hyperspectral imaging (HSI), multispectral imaging (MSI), X-ray imaging, spectroscopy, multi-sensors, E-nose, and acoustic impulse response.

From Table 1, the majority of the study was done using CV [58–66]. This involves the capturing of images of a product using a digital camera, and the ability of computers to understand the processed image data using computational intelligence tools (e.g., chemometrics, deep learning) [54,55]. This imaging method is rapid, reliable, and consistent. However, this technique has some limitations such as the use of artificial lighting during image capturing and the inability to detect internal attributes. Table 1 further shows that the bulk of the studies using CV was on cold storage, mostly to monitor and detect spoilage, chilling injury, and shelf-life of grapes, lettuce, tomato, zucchini, banana, strawberry, oranges, and mango [58–64,66]. CV with several chemometric and statistical analytic approaches was able to quantify quality losses with 75–92% accuracy (Table 1). Only a single study explored the application of CV in quantifying the quality losses of lettuce as a result of packaging material used, with 86% accuracy (Table 1) [65].

Hyperspectral and multispectral imaging (HSI and MSI) are advancements of computer vision, which involves the capturing of image data at a different wavelength (e.g., continuous 400–1700 nm in steps of 1 nm for HSI and targeted 400–1100 nm in steps of 20 nm for MSI) across the electromagnetic spectrum [67–70]. Hyperspectral imaging particularly integrates both imaging and spectroscopy features to simultaneously gather spectral and spatial information from a product, thus making it a more powerful imaging technology compare to CV and multispectral imaging [71]. Both hyperspectral and multispectral imaging technologies can detect internal and external quality attributes of fresh produce. However, they also require artificial lightning, are very sensitive to environmental conditions, have limited penetration depth, and are very expensive to use. Closely following CV, several studies have been conducted on the application of hyperspectral imaging in monitoring the quality of spinach, mushrooms, cucumber, mango, apples, and citrus fruit during cold storage and packaging (Table 1) [72–75]. This imaging system coupled with PCA, CFS, GIA, SFS, SLR, LDA, kNN, and NN was able to give 89–98% accuracy (Table 1). However, not so much for multispectral imaging, as only two studies quantified the quality losses of mangoes during the cold storage, with a classification rate of ≈92% using PLS-DA (Table 1) [70,76].

Spectroscopy, which is the study of the interaction of electromagnetic waves, including ultraviolet, visible, and infrared spectra, has been applied to monitor and optimize the cold storage process of peach and mango (Table 1). Although only a few studies have been carried out (Table 1) [77,78], this spectral approach can however give an accurate prediction (≈96%) of the total soluble solids, and phenolic content of peaches during cold storage. This technique gives the advantage of repeatable spectral data and provides high resolution of spectra. Additionally, this method is toxic-free. Nevertheless, the spectra data often contains redundant information due to hundreds of spectral variables, limited sensitivity to minor components, and complicated analysis.

X-ray imaging was applied to detect both the internal disorder and external changes (firmness) of pears and kiwifruit during cold storage, respectively (Table 1) [11,79]. This technology involves the production of electromagnetic radiation by an X-ray tube when passed through a product to absorb part of X-ray beam energy [80]. Using SVM, FEA, NN, and LDA, X-ray adequately quantified the columella firmness of kiwifruit and discriminated healthy pear from defective ones, with accuracy ranging from 90% to 95% (Table 1).

Other techniques used to monitor the quality of fruits and vegetables in postharvest cold chains are multi-sensors, electronic nose (E-nose), acoustic impulse response, radio frequency identification (RFID), and printed sensors (PTS). In this study, multi-sensors involves the use of numerous sensors placed at different locations on the produce and in the cold chain equipment (storage container or transport vehicle) to capture important quality attributes (e.g., color, firmness) and food losses (weight loss, temperature, time) metrics [81]. Data from the sensors are processed using sensor fusion (soft sensors). Soft sensors are virtual software code to process multiple sensor information for identified quality classifiers and for the development of warning systems (e.g., quality decline in fruits) [82]. They can be developed using different methods including mechanistic modeling based on physics of specific measured quality and food loss metrics, statistical modeling based on low-level representations in the feature space, and chemometrics or deep learning-based sparse representation techniques for multi-modal event modeling. Figure 3 depicts the use of multiple sensors (e.g., temperature, humidity) coupled with imaging technology during the cold storage of fruits and vegetables. From Table 1, only two studies applied multi-sensors to improve the accuracy of continuous sensor data acquisition in order to enhance transparency and traceability of the cold storage and transportation logistics of pear [83,84]. The multi-sensors monitored critical parameters that affect the quality attributes of fresh produce, including temperature and relative humidity (using portable low-energy-demanding temperature and humidity sensors). The detection of these parameters during shipment allows for effective control of safety and quality changes of the pear. Similarly, a study was reported on the use of an electronic nose (an instrument used to detect volatile organic compounds) [85] and acoustic impulse response to evaluate the quality of tomato and apple during cold storage, respectively (Table 1). The accuracy of the measured quality attribute (firmness) and mass loss data using ANN was ≈85%. This value is lower than those obtained using the aforementioned imaging technologies. This could be because of the complex nature of the method, which is based on the measurement of the sound emitted by fruit as it vibrates in response to a gentle tap with a small pendulum [86].

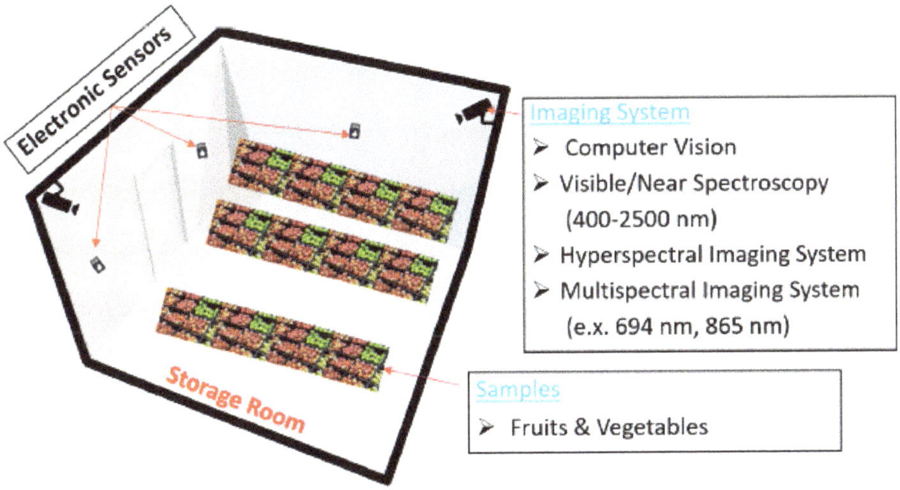

Figure 3. Schematics of multi-sensors coupled with imaging system for monitoring the quality of fruits and vegetables during cold storage.

RFID has also been applied as an advanced tool for identifying internal and external changes in the physical, biochemical, and physiological processes of packaged food [87–89]. This non-contact identification communication technology can automatically identify multiple objects moving at high-speed, and therefore can be applied in the transport cold chain, specifically as an IoT enabler [90]. Similarly, PTS which uses the printing process, such as inkjet printing, nanoimprinting, screen printing, etc., to prepare electronic circuits on a flexible substrate enables the monitoring of temperature, moisture, pressure, and motion of fresh produce [91]. This technology has the advantage of flexibility when printed on substrates, ease of distribution, and low cost especially when compared to RFID [91]. However, their application as tools in monitoring and optimizing cold chain processes is scarce.

Furthermore, there is no study on the application of imaging technology and smart digital technologies to monitor food quality losses during the transportation of fruits and vegetables. This is surprising considering that food losses in the transportation stage of the food supply chain can be as high as 30% as in the case of Poland, for example [92]. For this reason, future studies on the application of imaging technology, spectroscopy, multi-sensors, electronic nose, acoustic impulse response, RFID, PTS in the postharvest cold chain of fruits and vegetables should focus on the transportation chain.

Table 1. Published articles on the application of imaging technology, spectroscopy, multi-sensors, E-nose, and acoustic impulse response in assessing the quality of fruits and vegetables in the postharvest cold supply chains.

S/N	Technology	Cold Chain Operation	Purpose	Food Type	Statistical Approach	Significant Results	Reference
1.	Imaging technology hyperspectral Imaging	Packaging and storage	To study the image characteristics of vegetables acquired during packaging and storage	Spinach	SG, SNV, PCA	The arterial images were able to sense the aging of the leaves	[72]
2.	Imaging technology hyperspectral Imaging	Packaging	To study the refrigeration conditions of vegetables during packaging	Mushrooms	PLS, PLSR, MSC, PLS-DA	This method suggested an effective packaging solution to extend shelf life and prevent food losses of mushrooms during storage.	[73]
3.	Imaging technology multispectral imaging technology	Storage	To monitor and evaluate agro-food spoilage during storage	-	PLS-D	The spoiled food was predicted with an overall classification rate of 91.8%	[76]
4.	Imaging technology hyperspectral image	Storage	To explore the potential for the detection of chilling induced damage in fruits and vegetables	Cucumber	PCA	An overall classification rate of 90%	[74]
5.	Imaging technology computer vision system	Storage	To evaluate the effect of hydration degree during storage	Strawberry	PCA	Lighter appearance up to 75%	[58]
6.	Imaging technology machine vision systems	Storage	To monitor the quality change of food during storage	Orange	ANN, CART, CNN, LDA, kNN	An overall classification rate of 91.5%	[59]
7.	Imaging technology NIR hyperspectral image	Storage	To study the mechanical damage in fruits	Mango	CFS, GIA, SFS, SLR, LDA, kNN	Accuracy of 97.9%	[75]
8.	Imaging technology X-ray CT	Storage	To preserve the quality of fresh fruit during the supply chain and long-term storage	Pear	SVM, FEA	The X-ray computed tomography successfully detected the internal disorder severity of pear fruit with classification accuracies ranging between 90% and 95%	[79]
9.	Imaging technology computer vision	Storage	To develop a shelf life prediction model for postharvest handling of fruits and vegetables	Grape	RBF, NN	The method found the prediction accuracy of R^2 0.91	[60]
10.	Imaging technology computer vision	Storage	To study changes in color features of fruits during storage and to evaluate the use of image analysis technique as a rapid and nondestructive method	Banana	RBF, SVR, ANN	The computer vision technology with SVR of color parameters provided a useful model for prediction of the quality indices of bananas	[61]
11.	Imaging technology computer vision	Storage	To provide an application designed for embedded devices such as mobile Android smartphones to objectivize the measurements using machine vision	Tomato and zucchini	HA	The proposed method successfully achieved by predicting the quality characteristics of the products during cold storage	[62]
12.	Imaging technology computer vision system	Storage	To develop a computer vision system to predict the quality levels of vegetables during storage	Lettuce	SNK	The color information measured by the computer vision system achieved nondestructively to evaluate the quality level of iceberg lettuce with R^2 of 0.77	[63]
13	Imaging technology computer vision	Storage	To demonstrate the applicability of Random Forests (RF) for estimating the internal qualities of fruits based on peel color	Mango	RF	The relationship between peel color and fruit quality was strongly found in different storage temperatures with a correlation coefficient up to 0.98	[64]
14.	Imaging technology X-ray	Storage	To design a methodology for sorting fruits with X-ray image processing and pattern recognition techniques	Kiwifruits	NN, LDA	The model built with LDA predicted the columella firmness in kiwifruit with a 94.6%	[11]
15.	Imaging technology thermography	Packaging	To study the temperature distribution on a pallet of fruits during plastic boxes and cardboard packaging	Apples	ANN	Thermal imaging showed the cardboard boxes to be a better packaging material for apples compared to plastic boxes	[93]

Table 1. *Cont.*

S/N	Technology	Cold Chain Operation	Purpose	Food Type	Statistical Approach	Significant Results	Reference
16.	Imaging technology hyperspectral imaging systems	Storage	To develop a feature selection technique in classifying problems for detecting rottenness in tropical fruits	Citrus fruits	ROC, NN	Hyperspectral images found the classification success rate of around 89% for detecting the rottenness in citrus fruits	[75]
17.	Imaging technology multispectral Imaging	Storage	To evaluate chilling injury in fruits during storage using multispectral imaging	Mangoes	LS-SVM, PCA	The statistical results demonstrated significant changes in the reference quality properties of samples before and after storage	[70]
18.	Imaging technology computer vision systems	Packaging	To evaluate the quality of vegetables nondestructively using computer vision during packaging	Iceberg lettuce	CNN	The CNNs method was able to identify the lettuce quality with an accuracy of 86%	[65]
19.	Imaging technology computer vision systems	Storage	To evaluate fruit quality nondestructively by computer vision	Grapes	RF	The system achieved a cross-validation classification accuracy up to 92% which support its capability of powerfully, flexibly, and continuously monitoring the quality of the complete production along the whole supply chain	[66]
20.	Visible/shortwave near-infrared spectroscopy	Storage	To establish optimal spectral models for the assessment of fruits in actual production and therefore obtain early warning information of mechanical injuries during storage	Peach	GA	The optimal spectral model through the GA-PLS method found prediction accuracy ranges from 0.89 to 0.91 of the mechanical injuries of peaches during storage	[77]
21.	Handheld spectroscopy	Storage	To develop a nondestructive assessment of fruit quality using handheld micro NIR spectroscopic device	Mango	SVM, PLS	The proposed method was able to detect the mango fruit quality during the storage with prediction accuracy up to 96%	[78]
22.	Multi-sensor technology	Transportation and storage	To provide decision support to quality change and control	Pear	BPNN	The results indicated that this method could improve the accuracy of continuous sensor data acquisition	[83]
23.	Multi-sensor technology	Storage	To monitor and improve the data accuracy of sensory and physiological quality attributes of fruits during cold storage	Korla fragrant pear	BPNN	The multi-sensors technology including temperature, relative humidity, O_2, CO_2, and ethylene sensors improved the accuracy of data acquisition for gas content, pH, firmness, and total soluble solids	[84]
24.	Electronic nose	Storage	To investigate the reliability and validity of using the electronic nose to evaluate the quality and freshness of vegetables after high-pressure argon treatments	Cherry tomatoes	ELM, PLS	The results demonstrated E-nose technology combined with ELM provided a reliable and valid method for evaluating the quality and freshness of cherry tomatoes during cold storage with fitting correlation coefficients ($R^2 > 0.95$)	[85]
25.	Acoustic impulse response	Storage	To apply for apple classification nondestructively	Apple	ANN	The accuracy was 84.9% and 84.7% for Golden Delicious and Red Delicious, respectively	[86]

3.2. Application of Mathematical Modeling Techniques in the Postharvest Supply Chain of Fruits and Vegetables

Modeling is the act of representing phenomena or processes in such a way as to explicitly describe an observed system and to predict or optimize different behaviors, parameters, and conditions [94]. Mathematical modeling is essential for efficient engineering design and optimization. With adequate mathematical models, undesirable effects that significantly causes food losses such as weight loss, or quality changes can be predicted, thus cold chain logistics can be optimized or controlled.

Mathematical modeling techniques are becoming increasingly popular as an alternative to expensive and difficult experiments of postharvest cold chain operations as a result of the sophistication and reliability of computers as well as the affordability and availability of modeling software [95–99]. Agricultural and food engineers, and other researchers have over the years developed different mathematical models for postharvest supply chains. Depending on the complexity, these different modeling techniques have been developed to predict heat and mass transfer, fluid flow, and quality changes in and around fresh produce. Gas exchange and in-depth understanding of migration from packaging material to fresh produce have been described using mathematical models. Additionally, several deterministic, stochastic and kinetic models have also been developed to predict the overall quality of fresh produce, mass loss, fluid flow, and heat and mass transfer during the transportation and storage of fruits and vegetables [98,100–106].

In this study, mathematical models used to enhance cold chain operations chain can be separated into six different types based on their specific process application, namely: migration models (MM), membrane gas separation (MGS) model, heat and mass transfer (HMT) model, structural behavior models (SBM), stochastic models (SM), and kinetics rate models (KRM) (Table 2). MM are often used to study the migration of organic compounds such as Benzophenone, Diisobutyl phthalate, and Phenanthrene from packaging material to fresh produce. On the other hand, MGS models are often used in a modified atmosphere (e.g., modified atmosphere storage or modified atmosphere packaging) to study the lifespan of fresh produce by reducing the respiration rate through the adequate regulation of atmospheric conditions (e.g., CO_2, O_2). Additionally, to abstract the packaging of different fruits and vegetables [107]. This type of model works on four ideal flow patterns for a mixed gas module including co-current flow, cross flow, counter-current flow, and perfect mixing [108–110].

HMT modeling also called hygrothermal modeling involves using a numerical physics-based method such as computational fluid dynamic (CFD) to solve the governing partial differential equations of heat and mass transport phenomena in a system, often using finite element analysis (FEA) [100,105,111,112]. HMT models describe the underlining physics inside fresh produce and how they are affected by the surrounding conditions (Figure 4). They generally are independent of experimental calibration and validation. HMT models can also be used to investigate the impact of different packaging designs on the convective heat transfer rate of fruits and the surrounding [105]. By reducing heat transfer from the outside environment, effective packaging can help to shield the product from temperature variation in the storage and transport process. In addition, increasing the packaging heat transfer resistance can also ensure temperature stability of the fresh agricultural produce [113,114]. There is therefore a need for simulation tools and numerical models that analyze all factors affecting optimum packaging design.

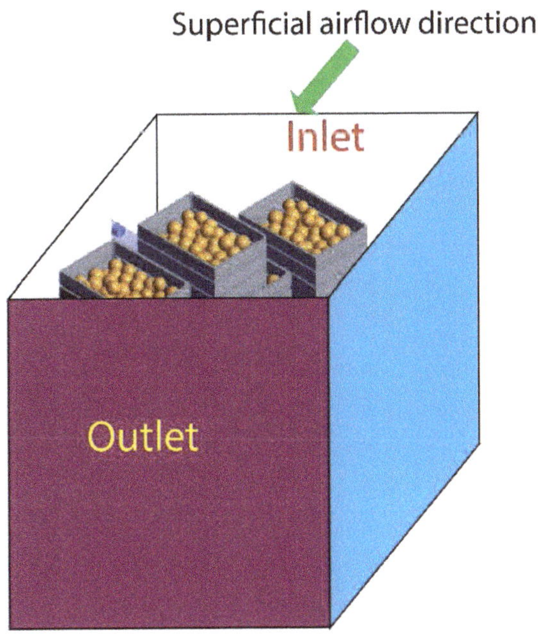

Figure 4. Simulation domain showing loaded fruits in stacked cartons in a virtual wind tunnel system with internal and external surrounding conditions [101].

SBM involves the study of structural and mechanical properties of packaging materials for fresh produce using FEA (Table 2). This modeling approach encompasses a geometric representation, material representation, and boundary conditions (loading and restraints).

SM involves predicting the variability of certain generated data following their probability distribution, and then evaluating of results statistics until the minimum error becomes constant [115]. These models are often used to analyze the effect of biological variability on food quality and losses during the supply chain, and also to quantify the efficiency of the cold chain technology [102,103,116]. Although they do not provide a fundamental understanding of the underlying physics, they are however very reliable and flexible.

KRM are temperature-dependent and are frequently used to study the combination of the rate of reaction with the material balance to predict the behavior of a particular system. Nutritional and sensory qualities of fresh agricultural produce in the postharvest supply chain can be quantified based on kinetics, such as zero-order, first-order, second-order, mixed order, or higher-order reactions [117].

From Table 2, most modeling studies were conducted on packaging followed by storage with very little modeling studies on the transportation of fruits and vegetables. Over 25 modeling studies on the packaging of fruits and vegetables during the postharvest cold chain were conducted within the last decade (Table 2). The products studied were apple, tomato, carrots, strawberries, capsicum, citrus, avocado, grape, feijoa fruits, and pears [100,101,105,107,112,116,118–137]. The bulk of the studies used HMT models to describe the cooling process in a packaged material during storage [100,101,104,105,112,119,120,130,131,137–140]. Seven studies applied MGS models to modify the atmospheric conditions of fresh agricultural produce in packaging material (Table 2) [118,121,125–127,135]. While, four studies examined the migration of chemical compounds (e.g., Benzophenone, Diisobutyl phthalate, and Phenanthrene) from packaging material to fresh produce using MM (Table 2) [122,124,134,136]. Several researchers also reported the application of KRM, SBM, and SM during the packaging of fruits and vegetables (Table 2) [116,123,128,129,132,141].

Table 2. Summary of recent literature on the application of mathematical modeling in monitoring quality loss of fruits and vegetables in postharvest supply chain fruits and vegetables.

S/N	Technology	Cold Chain Operation	Purpose	Food Type	Significant Results	Reference
1.	Membrane gas separation modeling	Packaging	To predict changes in fresh produce on molecular level based on changes in environmental conditions; to reduce losses	Apple	A membrane-based model can be used to abstract packaging for different fruit and vegetables; respiration of the process is considered important in modeling, and also environmental biological and technical factors; quality (taste, texture, color, appearance) is based on some subjective consumer evaluation	[107]
2.	Migration modeling	Packaging—paper and board	Deepened understanding of migrants from paper and board into foodstuffs	Tomato	Migration rate from paper and board to food at low temperatures is small compared to plastic material. For modeling, paper and board can be regarded as a two-layer system	[134]
3.	Membrane gas separation modeling	Packaging	To study the effect of external turbulence on the gas exchange rate; to develop a mathematical model to predict the effect of tube dimensions	Carrots	Different hydrodynamic condition affects the gas exchange rate in PM-MAP (perforation-mediated modified atmosphere packaging); the effect of temp, air velocity, and tube diameter on O_2 and CO_2	[135]
4.	Kinetic modeling	Storage	To predict the remaining shelf-life after storage; to predict dipp loss (zero-order), vitamin C (first-order), color (zero and first order) food quality	Broccoli	The model only looked at temperature effect on shelf life and adequately predicted the shelf life based on a 50% threshold of vitamin C loss at −18°C	[144]
5.	Migration modeling	Packaging migration controlled by diffusion	To use the Weibull distribution model to quantify migration in food packaging systems	-	Migration depends on the type of contact, food stimulant, type of paper, chemical nature of migrant in a paper, temperature and time of contact; migration is faster in paper than in plastics and involves the simultaneous transfer into food and also to the atmosphere	[136]
6.	Heat and mass transfer modeling	Packaging; storage	To quantify the impact of ventilation vent on temperature distribution of product	-	Decreasing the number of vents increased the cooling uniformity; the model, based on velocity and temperature simulation can be used as a design tool to provide homogenous temperature distribution to reduce food losses	[112]
7.	Heat and mass transfer modeling	Packaging	To predict O_2, CO_2, N_2, and H_2O concentrate in perforation-mediated polymeric packages; transport of O_2, CO_2, N_2, and H_2O was modeled using Maxwell Stefan equation for gas and Fick's law for diffusion through the micro-perforated package	Strawberries	The model result suggests an improvement in material properties, especially with regard to the permeability of polymeric packaging film; the model predicted a packaging with 30 µm thickness, 6 micro-perforation of 50 µm diameter each as the most suitable	[137]
8.	Heat and mass transfer modeling—based on compartments	Transport; storage	To predict temperature distribution during transport and storage of fruits and vegetables	Spinach, apricots, and peaches	The model adequately predicted the maximum and minimal load temperature distribution with lower computational time compared to CFD simulation	[104]
9.	Membrane gas separation modeling (Michaelis–Menten kinetic model + mass transfer model)	Packaging; storage	To describe the evolution of MAP of capsicum using a mathematical model; to quantify the performance of different packaging under the dynamic condition of use	Capsicums	Temperature and perforation have a significant effect on MAP conditions of capsicum; the combined model adequately predicted O_2 and CO_2 under different storage conditions	[118]
10.	Heat and mass transfer modeling	Packaging; storage	To assess the sensitivity of produce cooling uniformity and cooling time with respect to the packaging vent design	-	Increase in the number of vents increased cooling uniformity and reduce cooling time	[119]
11	Stochastic modeling (Monte Carlo simulations) + kinetic modeling	Storage; transportation	To estimate the expected fraction of perished products	Peaches	The simulation study predicted a fraction of 8.00% perished products based on a 100.00% quality threshold. The model quantifies thereof.	[143]
12.	Heat and mass transfer modeling	Packaging	To develop a 3D heat and mass transfer model of a fresh food produce packaging system to predict airflow and heat transfer characteristics	Citrus	Airflow, temperature, and turbulence property distribution inside a single product are nonuniform; good correlation between air velocity and temperature; the accuracy of the model depends on geometry, thermal, physical, and chemical properties of the package, the cooling air, and the produce	[120]
13.	Membrane gas separation modeling	Packaging; storage; transportation	To develop a mathematical model to investigate the effect of stage cut on the gas separation performance of hollow fiber membrane modules	Avocado	The model utilizes pure gas permeances of membrane material to predict the mixed gas separation performance; the performance of gas separation declines over time as they age with exposure to pressure, temperature, and contaminant loading	[121]

Table 2. *Cont.*

S/N	Technology	Cold Chain Operation	Purpose	Food Type	Significant Results	Reference
14.	Kinetic modeling	Storage	To evaluate the impact of storage conditions on vegetable color, firmness, weight loss, and phenolic content; to investigate the kinetics of quality parameters alterations of stored vegetables	Tomato	Fractional kinetic model fitted adequately well with experimental data; Arrhenius model describes well the temperature effect on all factors studied	[142]
15.	Migration modeling	Packaging; storage	To study the influence of box material and plastic cover on the distribution of 1-MCP (methyl cyclopropane) in cold storage to delay ripening of fruit	Apple	Diffusion, convection, and adsorption were modeled to simulate the temporal distributions of 1-MCP inside a storage container, boxes, and fruits; the plastic cover does not affect the adsorption of 1-MCP; wooden boxes notably adsorbed 1-MCP from the treatment atmosphere and may reduce the efficacy	[122]
16.	Kinetic modeling	Packaging	To investigate the impact of temperature and relative humidity on fruit transpiration rate (TR); to develop a prediction model for quantifying TR; to integrate TR model into engineering packaging design and quantify	Strawberries	Temperature and relative humidity have a significant impact on the transpiration rate of strawberries; increase in relative humidity increases TR; decreasing temperature decreases TR; the model predicted the water vapor barrier properties required for maintaining optimal relative humidity inside package	[123]
17.	Heat and mass transfer modeling	Storage	To model the airflow and temperature distribution in a natural connection thermal energy storage refrigerator; to determine the performance of the refrigerator with different phase change material (PCM) vertical-horizontal	-	Horizontal PCM configuration produces lower compartment temperatures than a vertical configuration; combining horizontal and vertical configuration gives better design performance	[140]
18.	Migration modeling	Packaging; storage	To study the diffusion, convection and adsorption of 1-MCP gas in cold stores; to understand the mechanism of 3D distribution of 1-MCP	Apple	The model demonstrated the absence of significant spatial variation of 1-MCP gas in a container; diffusion–convection in air and diffusion–adsorption in the product	[124]
19.	Heat and mass transfer modeling	Packaging	To develop a porous medium model was develop on volume averaging of transport equations of momentum and 1-MCP in air and product	Apple	The velocity field in and around the stack was well reproduced by the porous medium model; the porosity, skin mass transfer coefficient, and specific surface area strongly affected the simulation process	[100]
20.	Heat and mass transfer modeling	Packaging (bunch carry bag and plastic liners)	To determine the effect of the packaging component and box stacking on airflow, heat, and mass transfer rate	Grape	The use of carry bag resulted in an increase in the cooling time; the addition of plastic liner over the bunch carry bag increased cooling time; moisture loss was most prevented using nonperforated liners; CFD simulation determined optimum table grape packaging and costing procedure	[101]
21.	Heat and mass transfer modeling	Storage; packaging	To evaluate the performance of corrugated fiberboard, Supervent, Ecopack re-usable plastic container; to check the influence of airflow rate and cooling	Citrus	With respect to cooling, Eco-pack showed lower convective heat transfer rate but cooled in a uniform way, which improves fruit quality	[105]
22.	Membrane gas separation modeling	Packaging	To predict the shelf life of MAP systems	-	The model was able to predict the mass transfer phenomena for O_2 and CO_2 and also the microbial growth in the food system	[125]
23.	Kinetic model	Storage	To describe the product time–temperature history along the cold chain; the model considered front and rear air circulation in the cold room	Apples	The model adequately predicted cooling rate, the temperature at different positions, and weight loss; the model has a short CPU computational time (<1 s) when compared to CFD models. This enabled a rapid evaluation of input parameters such as air temperature	[106]
24.	Stochastic modeling	Storage	To evaluate the quality of perishable foods using a generic algorithm + center of gravity model; to estimate the environmental level	Fruits	The algorithm adequately predicted temperature and humidity levels; the algorithm was integrated to gauge the use of RFID (radio frequency identification) and sensors for real-time information gathering	[102]
25.	Stochastic modeling	Transport; storage	To estimate the heat generation and also the cooling efficiency during cold transport and storage chain	Banana	10% accuracy for the heat of respiration and cooling efficiency was reached after 4–7.5 days of transport	[103]
26.	Membrane gas separation modeling	Packaging; storage	To describe product respiration and gas exchange through package using Michaelis–Menten kinetics + Fick's equation; taking into account diffusive gas permeation through packaging film and perforation respiration rate and storage temperature	Tomato	The model adequately predicted the required package surface area and perforation diameter to achieve a specific O_2 concentration in the headspace; the model can be used to set a specific equilibrium concentration of O_2 and CO_2 by modifying the configuration of the package	[126]

Table 2. *Cont.*

S/N	Technology	Cold Chain Operation	Purpose	Food Type	Significant Results	Reference
27.	Membrane gas separation modeling	Packaging	To develop describe the evolution of water vapor, O_2, and CO_2 in packaging headspace, weight loss, and condensation of water in a MAP system	Feijoa fruits	The model adequately predicted weight loss and relative humidity in the MAP system	[127]
28.	Structural behavior modeling	Packaging	To develop a validated structural behavior model to predict the compression strength of a ventilated paperboard carton	-	The model adequately predicted the compression strength of a ventilated corrugated paperboard (VCP) packaging; the effect of vent area, vent height, and buckling load on the performance of VCP were adequately quantified by the model	[128]
29.	Heat and mass transfer modeling	Storage	To model the airflow, heat, and mass transfer in the storage chamber of Chinese cabbage; to predict the velocity, temperature, and relative humidity distribution	Cabbage	The model gave qualitative insight into the flow patterns in the cold room; the model adequately predicted temperature in the bulk and relative humidity of the air	[139]
30.	Stochastic modeling	Packaging; storage	To model the gas exchange in pear fruit taking the effect of biological variability	Pears	The model predicted that O_2 and CO_2 gas profiles inside the fruit were highly impacted by diffusivity, maximal respiration rate, and morphology of fruit; the model was used to analyze the incidence of fermentation at reduced O_2 levels during controlled atmosphere storage	[116]
31.	Structural behavior modeling	Packaging	To simulate the compression of paper and paperboard packaging material for food using finite element analysis (FEA)	-	The developed FEA model accurately predicted the incident buckling load of the corrugated paperboard; the modulus of elasticity was observed to be sensitive to the environmental conditions; the model can adequately be used to optimize corrugated paperboard packages	[129]
32.	Heat and mass transfer modeling	Packaging	To develop a more accurate model for describing the cooling process of freshly harvested apples and pears	Apples; pears	The model was able to describe the cooling behavior and uniformity of fruits in fiberboard boxes; there was large variability in convective heat transfer coefficients from the apples and pear filling; the fruit shape affects the model accuracy	[130]
33.	Heat and mass transfer modeling	Packaging	To develop a 3D HMT model to quantify cooling behavior of 10 different carton designs based on cooling rate, energy consumption, uniformity, weight loss, and chilling injury of apples	Apples	The model adequately quantified the effect of airflow, and packaging design on the product quality; vent area, shape and number of vent have less impact on the fruit cooling; homogeneity and symmetry of packaging vent positions have more impact on the fruit cooling rate; the model proposed airflow velocity between 0.4 and 1.0 m/s	[131]
34.	Structural behavior modeling	Packaging	To develop a validated structural behavior model to study the structural behavior of VCP packages by considering the geometrical nonlinearities of the packages	Fruits and vegetables	The model accurately predicted the compression strength of the corrugated paperboard, control package, and standard vent package; compression strength of the standard vent packages was found to be linearly affected by paperboard liner thickness; increasing and decreasing the baseline liner thickness of the standard vent package by 80% resulted in an increase and decrease in compression strength by about 15% and 19%, respectively; from the contact FEA model, maximum Von Mises stress was produced at the corners of the package; Von Mises stress was significantly affected by the coefficient of friction	[132]
35.	Heat and mass transfer modeling	Storage; transport	To investigate the airflow distribution inside two types of refrigerated shipping containers (T-bar floor and flat floor) used for transporting fresh fruit handling		The airflow distribution in the two container designs was markedly different. Good agreement was found between measured and predicted values of air velocities. The reefer with T-bar floor design exhibited a noticeable reduction of air recirculation zone and enhanced uniform vertical air movement compared to the reefer with flat floor design	[138]

Table 2. *Cont.*

S/N	Technology	Cold Chain Operation	Purpose	Food Type	Significant Results	Reference
36.	Kinetic modeling	Storage	To examine the effect of relative humidity (RH) conditions on the shelf life of strawberries including both the sensory and nutritional quality; to study the kinetics of sensory and chemical changes occurring in strawberries during storage by comparing three kinetic models; to examine and predict through modeling the waste that would occur depending on the storage conditions	Strawberries	Weight loss significantly increased when storage RH decreased; the weight loss was correlated with the changes that occurred in visual appearance and chemical properties. Overall appearance (i.e., the average score of color and shriveling), was modeled with a zero-order kinetics model for the various RH conditions; lower RH increased the rate of appearance deterioration thereby reducing the remaining shelf life of strawberries. The Weibull model adequately fitted the chemical properties data and it was found to be an important tool in describing the changes that occur with varying storage conditions	[98]
37.	Stochastic modeling	Whole chain	To develop an Agro-Chain Greenhouse gas emissions calculator (ACGE calculator) to calculate the percentage of food losses per chain stage	Cut vegetables	The ACGE calculator can be applied for understanding the impacts of different operations along a postharvest cold chain and for analyzing chain configuration scenarios: such as weighing impacts of the operations/impacts along the chain, comparison between various options for supplying a specific food component, and comparing a reference scenario with an 'improved scenario'; low temperature in the cold chain may result in extended retail shelf life and a lower percentage of losses, but will cost more energy; application of protective packaging leads to a reduction of losses, but at the cost of the packaging	[141]
38.	Heat and mass transfer modeling	Transportation; storage; packaging	To obtain more insight into the cooling process and quality evolution of fruits parked in ventilated cartons in a pallet	Citrus	Fruits packed in downstream cartons exhibited lower cooling heterogeneity compared with those in upstream cartons. Precooling reduced quality loss by 23%	[133]

With respect to storage, several researchers developed various models to predict quality loss during the storage of carrots, strawberries, spinach, apricots, peaches, capsicums, banana, cabbage, pears, citrus, and broccoli, however not all together (Table 2) [98,102–106,116,118,119,121,122,124,126,133,138–140,142–144]. A bulk of the models used were based on heat and mass transfer simulations of the weight loss, and temperature distribution; KRM for quality decay and shelf-life prediction (Table 2) [98,104–106,112,119,133,138–140,142,144].

Concerning the transportation supply chain, only a few authors have applied SM, HMT models, and KRM to estimate heat generation, cooling efficiency, temperature distribution, and the expected fraction of perishable products (Table 2). These products include spinach, peaches, and banana [103,104,143].

The above analysis shows that mathematical models have been widely developed and applied in the packaging and storage of fruits and vegetables with the view to improve quality and reduce food losses. However, not so much modeling study for the transportation supply chain of fruits and vegetables. Future mathematical modeling studies should focus on the transportation supply chain taking into account the shipment time, the varying environmental conditions (e.g., temperature, humidity, and airflow), packaging, and vehicular movement. In addition, the KRM could be integrated with HMT models to give more insight into what extent the quality attributes of fruits and vegetables are preserved better and also to quantify the effect of other drivers for decay processes (e.g., relative humidity, light) in the fresh produce supply chain. This can be achieved by developing a digital twin of the product (see Section 4.2).

4. Emerging Opportunities in Reducing Food Losses in the Postharvest Supply Chain

The application of IoT and digital twins in optimizing shelf-life and reducing food losses during an entire shipment has gained significant interest in recent years. This section analyzes the potential of applying the Internet of Things (IoT) and digital twins in reducing food losses in the postharvest supply chains of fruits and vegetables.

4.1. Application of IoT in the Postharvest Supply Chain of Fruits and Vegetables

IoT has emerged in different fields such as e-commerce [145,146], manufacturing [147,148], education [149–151], medicine and healthcare [152–158], and agriculture [159,160]. This is because of the enormous number of devices connected to the Internet, as well as the widely available internet and data storage service providers [148,161,162]. Basically, IoT allows humans, objects, and things to connect and communicate at any time and anywhere. The European Commission Information Society defined IoT as different things exhibiting identical and virtual personalities, connecting and communicating in a smart space using intelligent interfaces within social, economical, and user contexts [163].

The IoT system consists of networks of physical objects that contain embedded technology to sense, communicate, and interact with their internal states or the external environment [164]. The key enablers for a typical IoT system include RFID, printed sensors, web service, machine-to-machine communication (M2M), WSN, imaging system, multi-sensors, cloud, blockchain, among others, but not necessarily altogether [87,91,147,161,165,166].

The application of IoT is well-established in various agricultural production sectors such as controlled environment agriculture, open-field agriculture, and livestock applications [167,168]. In recent years, the use of IoT has gained significant interest in the food industry for product tracking, traceability and the monitoring of environmental conditions (e.g., temperature, humidity), weight loss, and the overall quality loss in the postharvest supply chain [87,161,165,169–171]. This technology has also received significant attention in developing intelligent packaging in the food sector [172–174].

Intelligent packaging involves the use of sensors (biosensor, printed, chemical, and gas sensor) and indicators (time-temperature indicators, freshness indicators, gas indicators, and integrity indicators) to detect biological, chemical, or gaseous changes from packaged fresh produce [24,25,27,91,170,174–176]. The sensor-based RFID tags as an example can detect hygrothermal and chemical changes (e.g., temperature, CO_2, light exposure, pH, etc.) of the fresh produce in the post-harvest supply chain [25,87,175]. The timely information obtained within the package system can be used to inform stakeholders in the supply chain of an event that may damage the packaging material or the fresh produce itself.

Generally, the application of IoT in different cold chain processes results in a large amount of real-time data which can pave the way for new computational approaches such as artificial intelligence and big data analytics [161,177]. This data will help various stakeholders in the supply chain control and optimize the cold chain technology to reduce quality loss and also make informed decisions regarding food safety. However, the application of IoT in controlling cold chain technologies in order to reduce food losses in the supply chain of fruits and vegetables is still inadequate.

Table 3 shows that several studies applied IoT in tracking and tracing temperature changes and food quality during the shipment of fruits and vegetables in the past decade [20,159,165,178,179]. Two studies applied IoT on the packaging of fruits and vegetables, as well as during cold storage [159,178]. The application of IoT in the shipment of fruits and vegetables is accompanied by multi-sensors such as temperature and humidity sensors, light exposure sensors, and global positioning system (GPS) sensors (Figure 5) during shipment of fruits and vegetables (Table 3). These sensors are installed in the food containers to monitor the changes in the environmental cold chain conditions such as air temperature, airspeed, light exposure, and relative humidity using a sensor data fusion (soft sensors). They are connected to a wireless network and computers to communicate with control stations, producers, or other stakeholders in the supply chain. The collected data can then serve as input data in analyzing the changes in the food attributes (e.g., weight loss, shelf life, nutritional, or sensory qualities), using a mechanistic physics-based model or a digital twin. It is worth mentioning that multi-sensors (e.g., chemical sensors, biosensors, etc.), imaging systems (see Section 3.1), E-nose (see Section 3.1), spectroscopy (see Section 3.1), and AIR (see Section 3.1) can also be used to directly measure changes in some quality attributes of fresh produce in the postharvest supply chain. IoT has become a very important tool in monitoring and controlling the process conditions of food, allowing the controllers to implement proper decisions. All of these can help to significantly reduce food losses. More so, the reduced cost of software and hardware wireless devices [180], digital sensors, accompanied by IoT technology in food transportation, packaging, and/or storage already increases the potential of IoT as a veritable and sustainable tool for reducing food losses.

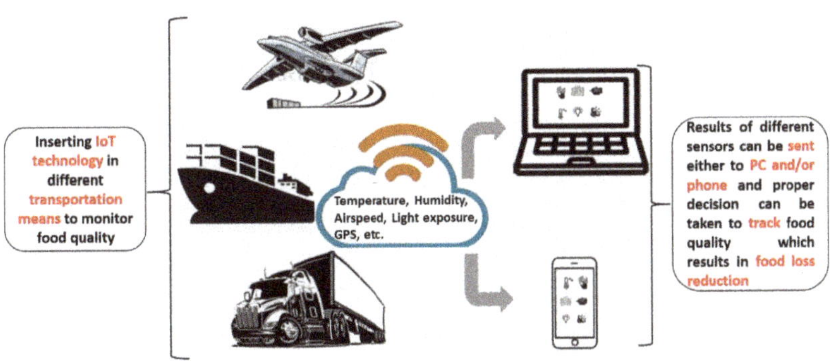

Figure 5. The implementation of Internet of Things (IoT) during shipment of fresh fruits and vegetables.

4.2. *Digital Twin as an Advanced Tool in Reducing Food Losses in the Postharvest Supply Chain of Fruits and Vegetables*

Digital twins have recently gained significant interest in postharvest engineering, as a way of expanding mathematical models and computer simulations by linking input data to the solutions implemented after the simulation study [181]. Simply, a digital twin of a product can be defined as a virtual model of the product's real-life representation containing all realistic characteristics.

The virtual model contains all essential elements, including geometrical components and material properties, and accurately and realistically simulates all relevant physics and their kinetics throughout the product's life-cycle. Digital twins can be mechanistic (physics-based), statistical (empirical-based), and intelligent (e.g., machine learning, deep learning) in nature. However, only the physics-based mechanistic digital twins can adequately evaluate the processes that cause quality loss in fresh produce. This involves linking measured sensor data of the environmental conditions (e.g., the air temperature around the fruit), as input data to the currently still uncharted product's quality evolution of fresh produce (Figure 6), preferable in a real-time update, using a physics-based model. In this way, the digital replica reacts hygrothermally and biologically in the same way as its physical counterpart (a real fresh fruit or vegetable).

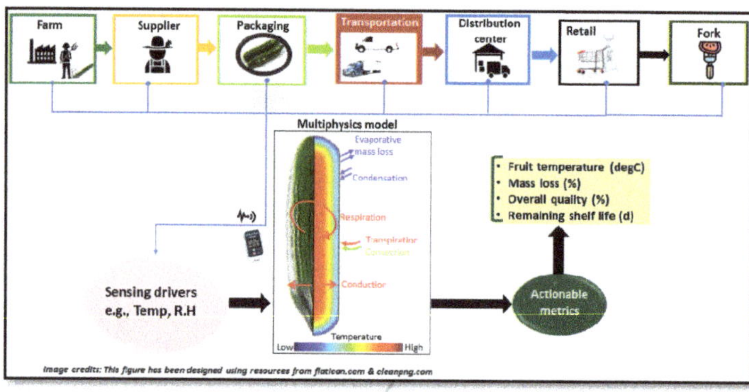

Figure 6. A schematic of a simple digital twin for fruit during the postharvest supply chain.

By enriching current real-time monitoring capabilities using sensors, digital twins can be used to diagnose and predict potential problems in the supply chain that will increase food losses. These problems can be caused by physiological (e.g., chilling injury), hygrothermal (e.g., mass loss), biotic (e.g., phytosanitary pests, pathogens), and mechanical (e.g., puncture injury, bruising) effects. This unique attribute shows that the digital twin has the potential of incorporating several physics-based thermal, physiological, mechanical, biological, and decay models for corresponding quality and shelf-life metrics. This insight can then help remotely analyze the quality performance of the fresh produce in each shipment and also predict the remaining shelf life days. Based on the analysis, a proactive preventive measure can be taken early to reduce quality losses throughout the cold chain. Such measures can also help predict and optimize future product quality and process design.

As a next step, digital twins can be implemented in real-time with actual multiple shipments. This is expedited with the integration of the already available big data technologies (e.g., IoT devices, blockchain devices, soft sensors, cloud systems, etc.) [182,183]. However, such a system is not yet in place, to the best of our knowledge (Table 3). From Table 3, only two studies developed a digital twin for mango. The mechanistic models developed for these studies included HMT models, as well as KMR for various quality attributes such as firmness, soluble solids content, and vitamin content [184] [185]. The air temperature data of the actual mango cold chain, collected as input from a temperature sensor was linked to these models to create a digital twin of a virtual mango fruit [184,185]. With the digital twins, the fruit quality evolution was quantified for multiple overseas shipments. However, the twin did not use other significant input data history such as the humidity of the products at the different supply chains. The temperature data collected was not in real-time, but offline, so a-posteriori. In addition, the digital twin did not integrate models to estimate the mass loss, chilling injury, and other biochemical models which are important for quantifying food losses in the entire cold chain (Table 2).

Table 3. Summary of recent literature on the application of IoT and digital twins in reducing food quality losses in the postharvest supply chain of fruits and vegetables in the last 10 years.

S/N	Technology	Cold Chain Operation	Purpose	Food Type	Significant Results	Reference
1.	Internet of Things	Transportation	To build an intelligent model for food quality monitoring and control in a multi-temperature distribution center	Food	The food spoilage rate of food reduced during transportation due to the real-time food quality monitoring and control using an intelligent model	[165]
2.	Internet of Things	Transportation	To improve transport efficiency in order to save the fruits from spoilage	Fruits	IoT system in a truck refrigerator adequately used to monitor the quality condition of fruits during transport	[20]
3.	Internet of Things	Packaging, storage, and transportation	To monitor food quality and safety	Fruits and vegetables	IoT obtain real-time food traceability and monitoring data to control the logistic and process parameters causing quality loss	[178]
4.	Internet of Things	Packaging, storage, and transportation	To reduce the food losses during the food chains since 50% is lost	Agro-food	IoT used to automate the packaging system with proper tracking and monitor the temperature of the produce at cold storage and during transportation	[159]
5.	Internet of Things	Transportation	To monitor the temperature changes and inefficient management during transportation as the insufficient temperature can pose a high risk to food quality	Fruit and vegetables	Application of IoT in food chains facilitated safety, intelligence, and deliver quick decisions	[179]
6.	Digital twin	Storage; transport	To develop a digital twin for the cold chain shipment of fruits	Mango	Based on measured environmental conditions, the impact of shipment duration, heat of respiration, airspeed and delivery air temperature history on quality of mango for different transport pathways was easily quantified using a digital twin	[184]
7.	Digital twin	All chain	To gain a better insight into how fruits behave under convective cooling	Mango	At low speeds, a more uniform cooling can be achieved and thereby a more homogeneous quality decay within the mango; digital twin was able to evaluate the heterogeneity of the temperature field and identified the zones with the highest temperature inside the product, which can be valuable information for the placement of temperature probes	[185]

5. Future Opportunities to Reduce Food Losses in the Postharvest Supply Chain of Fruits and Vegetables

With the gradual depletion of resources, there is a need to look at sustainable ways of achieving food security by reducing food losses in the postharvest supply chain. Looking ahead, the major challenges that cause food losses in the postharvest supply chain have to be addressed.

One emerging field is the development of intelligent packaging systems to reduce food losses, especially fresh agricultural produce. Intelligent packaging systems through the use of internal and external monitors (sensors, nanosensors, and indicators) provide valuable information on the interaction of food with the packaging material and the environment at different phases of processing, transportation, and storage. It also takes into consideration the ergonomic features of the packaging to reduce inconvenience in the transportation, storage, use, and eventual disposal of the packaging material [174,186]. With the recent interest and development in intelligent packaging, there is a need to integrate the sensors, indicators, and data carriers technologically to provide real-time information about fresh food in different cold chain logistics through the use of IoT based technologies and digital twin.

Although the potential of a digital twin in minimizing quality losses and increasing the shelf-life of fresh produce has been demonstrated (Table 3), the holistic implementation of digital twins in the entire value chain (from planting-fork) and for a wide range of fresh produce is yet to be demonstrated. The existing digital twins (Table 3) should be improved upon to include other relevant models that simulate thermal, physiological, mechanical, and biological damages that cause food losses in the postharvest supply chain. For example, a mass loss model can be included to quantify the salable weight at the end of the chain. This can help quantify the market value of fresh produce due to the subjective acceptable consumer product appearance. Additionally, tropical fruits such as banana or papaya experience chilling injury due to low-temperature storage and long cold chain process (Table 1). Therefore, thermal damage models predicting chilling injury during cold chain processes should be included. The potential of linking pathogens with decay severity should also be a future focus. Future digital twins should also capture the biological variability of fresh produce in order to give more realistic actionable metrics as multiple fresh produce have different individual pre-harvest and postharvest history. This can be achieved by integrating stochastic simulations (e.g., Monte Carlo simulations) with the existing digital twin physics-based models.

An additional future focus is to integrate IoT systems (including soft sensors) in real-time with digital twins. This real-time coupling will enable stakeholders to monitor and control each supply chain shipment at all times and take dynamically corrective measures to reduce quality loss and increase the remaining shelf-life days. Furthermore, by adding more "intelligence" to the coupled IoT and digital twins system, the cold chain technology (e.g., a refrigerated container) can independently optimize its process parameters to increase the shelf life of fresh produce and reduce food losses of the entire shipment. This added value can be easily quantified especially in this current time, considering the COVID-19 situation. Due to COVID-19, food producers have seen a decrease in the timely distribution of fresh produce to supply chain retailers. This is attributed to the decrease in transport labor, and longer shipment time because of shipment re-routing. This development has led to increased food losses. As a consequence, for example, about 5 billion US dollar worth of fresh fruits and vegetables were lost in the USA alone during the COVID-19 peak period from March 2020 to June 2020 [187]. With intelligent coupled IoT and digital twins, different processes and cold chain technology can be optimized to reduce the dependency on human labor, faster shipment duration, and possible damages caused by physiological, hygrothermal, mechanical, and biotic factors. A reduction in damages on the fresh produce implies a reduction in food losses.

6. Conclusions

Fruits and vegetables are important sources of nutrients such as vitamins, minerals, and bioactive compounds, which provide many health benefits. However, due to poor postharvest management processes, large quantities of fruits and vegetables perish before they reach the consumer. Of all the techniques for extending the shelf life of perishable produce, storage at low temperatures is by far the most effective.

This study, therefore, provides in-depth insight on the application of advanced technology in improving food security, by reducing food losses during postharvest cold chain processes for fruits and vegetables. It has been found that:

- Computer vision, hyperspectral imaging, multispectral imaging, spectroscopy, and X-ray imaging are already widely used in monitoring and optimizing the cold chain processes of fresh agricultural produce.
- The application of MM, MGS models, HMT models, SBM, SM, and KRM in improving the cold chain processes and evaluating the quality losses of fresh produce is well established. These models can help control and optimize the packaging and storage operations of fruits and vegetables in order to reduce quality losses.
- IoT is widely applied in monitoring and controlling the process conditions of fresh produce, enabling various stakeholders to implement proper decisions.
- Digital twins are significant in quantifying the quality evolution of fresh produce in each shipment and also predict the remaining shelf life days.
- There is a very huge potential for coupling digital twins with big data technologies (IoT devices, printed sensors, RFID, multi-sensors, soft sensors) to monitor, optimize, and make significant changes that will reduce food losses in the postharvest supply chain of fresh produce. However, such a system does not exist.

The augmented insight on the application of emerging technologies can serve as a roadmap for future cold chain studies on fresh agricultural produce.

Author Contributions: Conceptualization, D.I.O. and G.C.; methodology, D.I.O.; investigation, D.I.O., N.E.-e., A.K., A.Y.K., B.S.; data curation, D.I.O., N.E.-e., A.K., A.Y.K.; writing—original draft preparation, D.I.O.; writing—review and editing, D.I.O., N.E.-e., A.K., A.Y.K., G.C., B.S. All authors have read and agreed to the published version of the manuscript.

Funding: This research received no external funding.

Conflicts of Interest: The authors declare no conflict of interest.

Nomenclature

FAO	Food and Agriculture Organization
AIR	Acoustic Impulse Response
CV	Computer Vision
HSI	Hyperspectral Imaging
MSI	Multispectral Imaging
CCD	Charge-Couple Device
SG	Savitsky–Golay
SNV	Standard Normal Variate
PCA	Principal Component Analysis
PLSR	Partial Least Squares Regression
MSC	Multiple Scatter Correction
PLS-DA	Partial Least Squares Discriminant Analysis
ANN	Artificial Neural Network
CNN	Convolutional Neural Networks
LDA	Linear Discriminant Analysis
kNN	k-Nearest Neighbors
CFS	Correlation-based Feature Subset Selection
GIA	Gini Impurity Algorithm
SFS	Sequential Forward Selection
BPNN	Backpropagation Neural Network

ELM	Extreme Learning Machine
SLR	Sparse Logistic Regression
CMS	Central Monitoring System
RTS	Real-Time analytic System
RFID	Radio Frequency Identification Tags
MSA	Multi-Sensors Analysis
SVM	Support Vector Machine
RBF	Radial Basis Function
LCA	Life Cycle Assessment
PTS	Printed Sensors
GA	Genetic Algorithm
SVR	Support Vector Regression
SNK	Student–Newman–Keuls
LS-SVM	Least Squares Support Vector Machines
RF	Random Forest
LW	Local Order
WSNs	Wireless Sensor Networks
ROC	Receiver Operating Characteristic
MM	Migration Models
MGS	Membrane Gas Separation
HMT	Heat and Mass Transfer
SBM	Structural Behavior Models
SM	Stochastic Models
KRM	Kinetics Rate Models
CO_2	Carbon Dioxide
O_2	Oxygen
N_2	Nitrogen
H_2O	Water
FEA	Finite Element Analysis
MAP	Modified Atmosphere Packaging
1-MCP	Methyl Cyclopropane
TR	Transpiration Rate
CFD	Computational Fluid Dynamic
VCP	Ventilated Corrugated Paperboard
IoT	Internet of Things
GPS	Global Positioning System
NN	Neural Networks
NCP	Nano-Composite based Packaging
CFD	Computational Fluid Dynamics

References

1. Coulomb, D. Refrigeration and cold chain serving the global food industry and creating a better future: Two key IIR challenges for improved health and environment. *Trends Food Sci. Technol.* **2008**, *19*, 413–417. [CrossRef]
2. Shafiee-Jood, M.; Cai, X. Reducing Food Loss and Waste to Enhance Food Security and Environmental Sustainability. *Environ. Sci. Technol.* **2016**, *50*, 8432–8443. [CrossRef] [PubMed]
3. Gustavsson, J.; Stage, J. Retail waste of horticultural products in Sweden. *Resour. Conserv. Recycl.* **2011**, *55*, 554–556. [CrossRef]
4. Yahaya, S.; Ay, M. Review of Post-Harvest Losses of Fruits and Vegetables. *Biomed. J. Sci. Tech. Res.* **2019**, *13*, 10192–10200. [CrossRef]
5. Onwude, D.; Hashim, N.; Janius, R.; Nawi, N.; Khalina, A. Modeling the thin-layer drying of fruits and vegetables—A review. *Compr. Rev. Food Sci. Food Saf.* **2016**, *15*, 599–618. [CrossRef]
6. Salehi, F.; Aghajanzadeh, S. Effect of dried fruits and vegetables powder on cakes quality: A review. *Trends Food Sci. Technol.* **2020**, *95*, 162–172. [CrossRef]

7. Jiang, Q.; Zhang, M.; Xu, B. Application of ultrasonic technology in postharvested fruits and vegetables storage: A review. *Ultrason. Sonochem.* **2020**, *69*, 105261. [CrossRef]
8. Cömert, E.D.; Mogol, B.A.; Gökmen, V. Relationship between color and antioxidant capacity of fruits and vegetables. *Curr. Res. Food Sci.* **2020**, *2*, 1–10. [CrossRef] [PubMed]
9. Huang, L.L.; Zhang, M. Trends in Development of Dried Vegetable Products as Snacks. *Dry. Technol.* **2012**, *30*, 448–461. [CrossRef]
10. Liu, D.K.; Xu, C.C.; Guo, C.X.; Zhang, X.X. Sub-zero temperature preservation of fruits and vegetables: A review. *J. Food Eng.* **2019**, *275*, 109881. [CrossRef]
11. Mercier, S.; Villeneuve, S.; Mondor, M.; Uysal, I. Time–Temperature Management along the Food Cold Chain: A Review of Recent Developments. *Compr. Rev. Food Sci. Food Saf.* **2017**, *16*, 647–667. [CrossRef]
12. Arah, I.K.; Ahorbo, G.K.; Anku, E.K.; Kumah, E.K.; Amaglo, H. Postharvest Handling Practices and Treatment Methods for Tomato Handlers in Developing Countries: A Mini Review. *Adv. Agric.* **2016**, *2016*, 6436945. [CrossRef]
13. Raut, R.D.; Gardas, B.B.; Narwane, V.S.; Narkhede, B.E. Improvement in the food losses in fruits and vegetable supply chain—A perspective of cold third-party logistics approach. *Oper. Res. Perspect.* **2019**, *6*, 100117. [CrossRef]
14. James, S.J.; James, C. The food cold-chain and climate change. *Food Res. Int.* **2010**, *43*, 1944–1956. [CrossRef]
15. Defraeye, T.; Cronje, P.J.R.; Berry, T.; Opara, U.L.; East, A.; Hertog, M.L.A.T.M.; Verboven, P.; Nicolai, B.M. Towards integrated performance evaluation of future packaging for fresh produce in the cold chain. *Trends Food Sci. Technol.* **2015**, *44*, 201–225. [CrossRef]
16. Thompson, J.F.; Mejia, D.C.; Singh, R.P. Energy Use of Commercial Forced-Air Coolers for Fruit. *Appl. Eng. Agric.* **2010**, *26*, 919–924. [CrossRef]
17. Piljac-Žegarac, J.; Šamec, D. Antioxidant stability of small fruits in postharvest storage at room and refrigerator temperatures. *Food Res. Int.* **2011**, *44*, 345–350. [CrossRef]
18. Chu, W.; Gao, H.; Chen, H.; Wu, W.-J.; Fang, X. Changes in Cuticular Wax Composition of Two Blueberry Cultivars during Fruit Ripening and Postharvest Cold Storage. *J. Agric. Food Chem.* **2018**, *66*, 2870–2876. [CrossRef]
19. Fagundes, C.; Moraes, K.; Pérez-Gago, M.; Palou, L.; Maraschin, M.; Monteiro, A. Effect of active modified atmosphere and cold storage on the postharvest quality of cherry tomatoes. *Postharvest Biol. Technol.* **2015**, *109*, 73–81. [CrossRef]
20. Hong, K.; Xie, J.; Zhang, L.; Sun, D.; Gong, D. Effects of chitosan coating on postharvest life and quality of guava (*Psidium guajava* L.) fruit during cold storage. *Sci. Hortic.* **2012**, *144*, 172–178. [CrossRef]
21. Hu, Q.; Fang, Y.; Yang, Y.; Ma, N.; Zhao, L. Effect of nanocomposite-based packaging on postharvest quality of ethylene-treated kiwifruit (*Actinidia deliciosa*) during cold storage. *Food Res. Int.* **2011**, *44*, 1589–1596. [CrossRef]
22. Jiang, T.; Feng, L.; Li, J. Changes in microbial and postharvest quality of shiitake mushroom (*Lentinus edodes*) treated with chitosan–glucose complex coating under cold storage. *Food Chem.* **2012**, *131*, 780–786. [CrossRef]
23. Zhang, Y.; Zhang, M.; Yang, H. Postharvest chitosan-g-salicylic acid application alleviates chilling injury and preserves cucumber fruit quality during cold storage. *Food Chem.* **2015**, *174*, 558–563. [CrossRef]
24. Ghaani, M.; Cozzolino, C.A.; Castelli, G.; Farris, S. An overview of the intelligent packaging technologies in the food sector. *Trends Food Sci. Technol.* **2016**, *51*, 1–11. [CrossRef]
25. Müller, P.; Schmid, M. Intelligent Packaging in the Food Sector: A Brief Overview. *Foods* **2019**, *8*, 16. [CrossRef] [PubMed]
26. Haass, R.; Dittmer, P.; Veigt, M.; Lütjen, M. Reducing food losses and carbon emission by using autonomous control—A simulation study of the intelligent container. *Int. J. Prod. Econ.* **2015**, *164*, 400–408. [CrossRef]
27. Janjarasskul, T.; Suppakul, P. Active and intelligent packaging: The indication of quality and safety. *Crit. Rev. Food Sci. Nutr.* **2017**, *58*, 808–831. [CrossRef]
28. Williams, H.; Wikström, F. Environmental impact of packaging and food losses in a life cycle perspective: A comparative analysis of five food items. *J. Clean. Prod.* **2011**, *19*, 43–48. [CrossRef]
29. White, A.; Lockyer, S. Removing plastic packaging from fresh produce—What's the impact? *Nutr. Bull.* **2020**, *45*, 35–50. [CrossRef]
30. Bilstein, F. What Reduces Our Personal CO_2 Footprint? We Have no Clue! 2019. Available online: https://www.linkedin.com/pulse/what-reduces-our-personal-co2-footprint-we-have-clue-frank-bilstein (accessed on 19 October 2020).

31. Wynes, S.; A Nicholas, K. The climate mitigation gap: Education and government recommendations miss the most effective individual actions. *Environ. Res. Lett.* **2017**, *12*, 074024. [CrossRef]
32. Hischier, R. Schwierige Suche Nach Plastikalternativen. *Migros Mag.* **2019**. Available online: https://www.migrosmagazin.ch/schwierige-suche-nach-plastikalternativen (accessed on 19 October 2020).
33. Conte, A.; Cappelletti, G.; Nicoletti, G.; Russo, C.; Del Nobile, M. Environmental implications of food loss probability in packaging design. *Food Res. Int.* **2015**, *78*, 11–17. [CrossRef]
34. Gutierrez, M.M.; Meleddu, M.; Piga, A. Food losses, shelf life extension and environmental impact of a packaged cheesecake: A life cycle assessment. *Food Res. Int.* **2017**, *91*, 124–132. [CrossRef]
35. Denkstatt. How Packaging Contributes to Food Waste Prevention. 2017. Available online: www.denkstatt.at (accessed on 19 October 2020).
36. Duan, Y.; Wang, G.-B.; Fawole, O.A.; Verboven, P.; Zhang, X.-R.; Wu, D.; Opara, U.L.; Nicolai, B.; Chen, K. Postharvest precooling of fruit and vegetables: A review. *Trends Food Sci. Technol.* **2020**, *100*, 278–291. [CrossRef]
37. Praeger, U.; Jedermann, R.; Sellwig, M.; Neuwald, D.A.; Hartgenbusch, N.; Borysov, M.; Truppel, I.; Scaar, H.; Geyer, M. Airflow distribution in an apple storage room. *J. Food Eng.* **2020**, *269*, 109746. [CrossRef]
38. Badia-Melis, R.; Carthy, U.M.; Ruiz-Garcia, L.; Garcia-Hierro, J.; Villalba, J.I.R. New trends in cold chain monitoring applications—A review. *Food Control* **2018**, *86*, 170–182. [CrossRef]
39. Gustavsson, J.; Cederberg, C.; Sonesson, U.; van Otterdijk, R.; Meybeck, A. *Global Food Losses and Food Waste*; FAO: Rome, Italy, 2011.
40. Morris, K.J.; Kamarulzaman, N.H.; I Morris, K. Small-scale postharvest practices among plantain farmers and traders: A potential for reducing losses in rivers state, Nigeria. *Sci. Afr.* **2019**, *4*, e00086. [CrossRef]
41. Parfitt, J.; Barthel, M.; Macnaughton, S. Food waste within food supply chains: Quantification and potential for change to 2050. *Philos. Trans. R. Soc. B Biol. Sci.* **2010**, *365*, 3065–3081. [CrossRef]
42. Lundqvist, J.; de Fraiture, C.; Molden, D. *Saving Water: From Field to Fork: Curbing Losses and Wastage in the Food Chain*; Stockholm International Water Institute: Stockholm, Sweden, 2008.
43. Bustos, C.A.; Moors, E.H. Reducing post-harvest food losses through innovative collaboration: Insights from the Colombian and Mexican avocado supply chains. *J. Clean. Prod.* **2018**, *199*, 1020–1034. [CrossRef]
44. Santos, S.F.; Cardoso, R.D.C.V.; Borges, Í.M.P.; Almeida, E.A.C.; Andrade, E.S.; Dos Santos, S.F.; Ramos, L.D.C. Post-harvest losses of fruits and vegetables in supply centers in Salvador, Brazil: Analysis of determinants, volumes and reduction strategies. *Waste Manag.* **2019**, *101*, 161–170. [CrossRef]
45. FAO. *Food Wastage Footprint—Impacts on Natural Resources: Summary Report*; Food and Agriculture Organization of the United Nations: Rome, Italy, 2013.
46. Sivakumar, D.; Wall, M.M. Papaya Fruit Quality Management during the Postharvest Supply Chain. *Food Rev. Int.* **2013**, *29*, 24–48. [CrossRef]
47. Mampholo, B.M.; Sivakumar, D.; Thompson, A.K. Maintaining overall quality of fresh traditional leafy vegetables of Southern Africa during the postharvest chain. *Food Rev. Int.* **2015**, *32*, 400–416. [CrossRef]
48. Bradford, K.J.; Dahal, P.; Van Asbrouck, J.; Kunusoth, K.; Bello, P.; Thompson, J.; Wu, F. The dry chain: Reducing postharvest losses and improving food safety in humid climates. *Trends Food Sci. Technol.* **2018**, *71*, 84–93. [CrossRef]
49. Gordon, A.; Gordon, D. Food safety and quality systems implementation along value chains. In *Food Safety and Quality Systems in Developing Countries*; Elsevier BV: Amsterdam, The Netherlands, 2020; pp. 81–124.
50. Canadian Food Inspection Agency. *Food Safety Enhancement Program Manual*; Canadian Food Inspection Agency: Ottawa, ON, Canada, 2014.
51. Siripatrawan, U. Hyperspectral imaging for rapid evaluation and visualization of quality deterioration index of vacuum packaged dry-cured sausages. *Sensors Actuators B Chem.* **2018**, *254*, 1025–1032. [CrossRef]
52. Dowlati, M.; De La Guardia, M.; Mohtasebi, S.S. Application of machine-vision techniques to fish-quality assessment. *TrAC Trends Anal. Chem.* **2012**, *40*, 168–179. [CrossRef]
53. Feng, Y.Z.; Sun, D.-W. Determination of total viable count (TVC) in chicken breast fillets by near-infrared hyperspectral imaging and spectroscopic transforms. *Talanta* **2013**, *105*, 244–249. [CrossRef]
54. Aghbashlo, M.; Hosseinpour, S.; Ghasemi-Varnamkhasti, M. Computer vision technology for real-time food quality assurance during drying process. *Trends Food Sci. Technol.* **2014**, *39*, 76–84. [CrossRef]
55. Martynenko, A. Computer Vision for Real-Time Control in Drying. *Food Eng. Rev.* **2017**, *9*, 91–111. [CrossRef]

56. Zhang, H.; Li, D. Applications of computer vision techniques to cotton foreign matter inspection: A review. *Comput. Electron. Agric.* **2014**, *109*, 59–70. [CrossRef]
57. Zhang, B.; Huang, W.; Li, J.; Zhao, C.; Fan, S.; Wu, J.; Liu, C. Principles, developments and applications of computer vision for external quality inspection of fruits and vegetables: A review. *Food Res. Int.* **2014**, *62*, 326–343. [CrossRef]
58. Agudelo-Laverde, L.M.; Schebor, C.; Buera, M. Water content effect on the chromatic attributes of dehydrated strawberries during storage, as evaluated by image analysis. *LWT-Food Sci. Technol.* **2013**, *52*, 157–162. [CrossRef]
59. Cubero, S.; Aleixos, N.; Moltó, E.; Gómez-Sanchis, J.; Blasco, J. Advances in Machine Vision Applications for Automatic Inspection and Quality Evaluation of Fruits and Vegetables. *Food Bioprocess Technol.* **2010**, *4*, 487–504. [CrossRef]
60. Li, Y.; Tang, X.; Shen, Z.; Dong, J. Prediction of total volatile basic nitrogen (TVB-N) content of chilled beef for freshness evaluation by using viscoelasticity based on airflow and laser technique. *Food Chem.* **2019**, *287*, 126–132. [CrossRef]
61. Sanaeifar, A.; Zakidizaji, H.; Jafari, A.; De La Guardia, M. Early detection of contamination and defect in foodstuffs by electronic nose: A review. *TrAC Trends Anal. Chem.* **2017**, *97*, 257–271. [CrossRef]
62. Emana, B.; Afari-Sefa, V.; Nenguwo, N.; Ayana, A.; Kebede, D.; Mohammed, H. Characterization of pre- and postharvest losses of tomato supply chain in Ethiopia. *Agric. Food Secur.* **2017**, *6*, 87. [CrossRef]
63. Pace, B.; Cefola, M.; Da Pelo, P.; Renna, F.; Attolico, G. Non-destructive evaluation of quality and ammonia content in whole and fresh-cut lettuce by computer vision system. *Food Res. Int.* **2014**, *64*, 647–655. [CrossRef] [PubMed]
64. Fukuda, S.; Yasunaga, E.; Nagle, M.; Yuge, K.; Sardsud, V.; Spreer, W.; Müller, J. Modelling the relationship between peel colour and the quality of fresh mango fruit using Random Forests. *J. Food Eng.* **2014**, *131*, 7–17. [CrossRef]
65. Cavallo, D.P.; Cefola, M.; Pace, B.; Logrieco, A.F.; Attolico, G. Non-destructive automatic quality evaluation of fresh-cut iceberg lettuce through packaging material. *J. Food Eng.* **2018**, *223*, 46–52. [CrossRef]
66. Cavallo, D.P.; Cefola, M.; Pace, B.; Logrieco, A.F.; Attolico, G. Non-destructive and contactless quality evaluation of table grapes by a computer vision system. *Comput. Electron. Agric.* **2019**, *156*, 558–564. [CrossRef]
67. Amigo, J.M.; Santos, C. Preprocessing of hyperspectral and multispectral images. *Data Handl. Sci. Technol.* **2020**, *32*, 37–53. [CrossRef]
68. Marini, F.; Amigo, J.M. Unsupervised exploration of hyperspectral and multispectral images. *Data Handl. Sci. Technol.* **2020**, *32*, 93–114. [CrossRef]
69. Nicolaï, B.M.; Beullens, K.; Bobelyn, E.; Peirs, A.; Saeys, W.; Theron, K.I.; Lammertyn, J. Nondestructive measurement of fruit and vegetable quality by means of NIR spectroscopy: A review. *Postharvest Biol. Technol.* **2007**, *46*, 99–118. [CrossRef]
70. Hashim, N.; Onwude, D.; Osman, M.S. Evaluation of Chilling Injury in Mangoes Using Multispectral Imaging. *J. Food Sci.* **2018**, *83*, 1271–1279. [CrossRef]
71. Ruiz-Altisent, M.; Ruiz-Garcia, L.; Moreda, G.; Lu, R.; Hernandez-Sanchez, N.; Correa, E.; Diezma, B.; Nicolaï, B.; García-Ramos, J. Sensors for product characterization and quality of specialty crops—A review. *Comput. Electron. Agric.* **2010**, *74*, 176–194. [CrossRef]
72. Lara, M.A.; Lleó, L.; Iglesias, B.D.; Roger, J.; Altisent, M.R. Monitoring spinach shelf-life with hyperspectral image through packaging films. *J. Food Eng.* **2013**, *119*, 353–361. [CrossRef]
73. Taghizadeh, M.; Gowen, A.; Ward, P.; O'Donnell, C.P. Use of hyperspectral imaging for evaluation of the shelf-life of fresh white button mushrooms (Agaricus bisporus) stored in different packaging films. *Innov. Food Sci. Emerg. Technol.* **2010**, *11*, 423–431. [CrossRef]
74. Liu, Y.; Chen, Y.R.; Wang, C.Y.; Chan, D.E.; Kim, M.S. Development of hyperspectral imaging technique for the detection of chilling injury in cucumbers; spectral and image analysis. *Appl. Eng. Agric.* **2006**, *22*, 101–111. [CrossRef]
75. Rivera, N.V.; Gómez-Sanchis, J.; Chanona-Pérez, J.; Carrasco, J.J.; Millán-Giraldo, M.; Lorente, D.; Cubero, S.; Blasco, J. Early detection of mechanical damage in mango using NIR hyperspectral images and machine learning. *Biosyst. Eng.* **2014**, *122*, 91–98. [CrossRef]

76. Panagou, E.Z.; Papadopoulou, O.; Carstensen, J.M.; Stamatiou, A. Potential of multispectral imaging technology for rapid and non-destructive determination of the microbiological quality of beef filets during aerobic storage. *Int. J. Food Microbiol.* **2014**, *174*, 1–11. [CrossRef] [PubMed]
77. Du, D.; Wang, J.; Wang, B.; Zhu, L.; Hong, X. Ripeness Prediction of Postharvest Kiwifruit Using a MOS E-Nose Combined with Chemometrics. *Sensors* **2019**, *19*, 419. [CrossRef]
78. Al-Sanabani, D.G.A.; Solihin, M.I.; Pui, L.P.; Astuti, W.; Ang, C.K.; Hong, L.W. Development of non-destructive mango assessment using Handheld Spectroscopy and Machine Learning Regression. *J. Phys. Conf. Ser.* **2019**, *1367*, 012030. [CrossRef]
79. Van de Looverbosch, T.; Bhuiyan, H.R.; Verboven, P.; Dierick, M.; Van Loo, D.; De Beenbouwer, J.; Sijbers, J.; Nicolaï, B. Nondestructive internal quality inspection of pear fruit by X-ray CT using machine learning. *Food Control* **2020**, *113*, 107170. [CrossRef]
80. Einarsdottir, H. Image Analysis for X-ray Imaging of Food. Ph.D. Thesis, Technical University of Denmark, Kongens Lyngby, Denmark, 2016.
81. Ghasemi-Varnamkhasti, M.; Apetrei, C.; Lozano, J.; Anyogu, A. Potential use of electronic noses, electronic tongues and biosensors as multisensor systems for spoilage examination in foods. *Trends Food Sci. Technol.* **2018**, *80*, 71–92. [CrossRef]
82. Fortuna, L.; Graziani, S.; Rizzo, A.; Xibilia, M.G. *Soft Sensors for Monitoring and Control of Industrial Processes*; Springer Science & Business Media: Berlin, Germany, 2007.
83. Liu, Z.; Jia, X.; Xu, X. Study of shrimp recognition methods using smart networks. *Comput. Electron. Agric.* **2019**, *165*, 104926. [CrossRef]
84. Liu, J.; Zhang, X.; Li, Z.; Zhang, X.; Jemrić, T.; Wang, X. Quality Monitoring and Analysis of Xinjiang 'Korla' Fragrant Pear in Cold Chain Logistics and Home Storage with Multi-Sensor Technology. *Appl. Sci.* **2019**, *9*, 3895. [CrossRef]
85. Feng, C.; Drummond, L.; Zhang, Z.; Sun, D.-W.; Wang, Q. Vacuum Cooling of Meat Products: Current State-of-the-Art Research Advances. *Crit. Rev. Food Sci. Nutr.* **2012**, *52*, 1024–1038. [CrossRef]
86. Lashgari, M.; Maleki, A.; Amiriparian, J. Application of acoustic impulse response in discrimination of apple storage time using neural network. *Int. Food Res. J.* **2017**, *24*, 1075.
87. Zou, Z.; Chen, Q.; Uysal, I.; Zheng, L. Radio frequency identification enabled wireless sensing for intelligent food logistics. *Philos. Trans. R. Soc. A Math. Phys. Eng. Sci.* **2014**, *372*, 20130313. [CrossRef]
88. Badia-Melis, R.; Ruiz-García, L.; Garcia-Hierro, J.; Villalba, J.I.R. Refrigerated Fruit Storage Monitoring Combining Two Different Wireless Sensing Technologies: RFID and WSN. *Sensors* **2015**, *15*, 4781–4795. [CrossRef]
89. Grunow, M.; Piramuthu, S. RFID in highly perishable food supply chains—Remaining shelf life to supplant expiry date? *Int. J. Prod. Econ.* **2013**, *146*, 717–727. [CrossRef]
90. Tian, F. An agri-food supply chain traceability system for China based on RFID & blockchain technology. In Proceedings of the 2016 13th International Conference on Service Systems and Service Management (ICSSSM), Kunming, China, 24–26 June 2016. [CrossRef]
91. Liao, Y.; Zhang, R.; Qian, J. Printed electronics based on inorganic conductive nanomaterials and their applications in intelligent food packaging. *RSC Adv.* **2019**, *9*, 29154–29172. [CrossRef]
92. Lipińska, M.; Tomaszewska, M.; Kołożyn-Krajewska, D. Identifying Factors Associated with Food Losses during Transportation: Potentials for Social Purposes. *Sustainability* **2019**, *11*, 2046. [CrossRef]
93. Badia-Melis, R.; Qian, J.P.; Fan, B.L.; Hoyos-Echevarria, P.; Ruiz-García, L.; Yang, X.T. Artificial Neural Networks and Thermal Image for Temperature Prediction in Apples. *Food Bioprocess Technol.* **2016**, *9*, 1089–1099. [CrossRef]
94. Castell-Palou, A.; Simal, S. Heat pump drying kinetics of a pressed type cheese. *LWT* **2011**, *44*, 489–494. [CrossRef]
95. Hertog, M.L.A.T.M.; Uysal, I.; McCarthy, U.; Verlinden, B.M.; Nicolaï, B.M. Shelf life modelling for first-expired-first-out warehouse management. *Philos. Trans. R. Soc. A Math. Phys. Eng. Sci.* **2014**, *372*, 20130306. [CrossRef]
96. Wu, W.; Cronje, P.J.; Nicolai, B.; Verboven, P.; Opara, U.L.; Defraeye, T. Virtual cold chain method to model the postharvest temperature history and quality evolution of fresh fruit—A case study for citrus fruit packed in a single carton. *Comput. Electron. Agric.* **2018**, *144*, 199–208. [CrossRef]

97. Song, Y.; Hu, Q.; Wu, Y.; Pei, F.; Kimatu, B.M.; Su, A.; Yang, W. Storage time assessment and shelf-life prediction models for postharvest *Agaricus bisporus*. *LWT* **2019**, *101*, 360–365. [CrossRef]
98. Ktenioudaki, A.; O'Donnell, C.P.; Nunes, M.C.D.N. Modelling the biochemical and sensory changes of strawberries during storage under diverse relative humidity conditions. *Postharvest Biol. Technol.* **2019**, *154*, 148–158. [CrossRef]
99. Zou, J.; Li, P. Modelling of litchi shelf life based on the entropy weight method. *Food Packag. Shelf Life* **2020**, *25*, 100509. [CrossRef]
100. Ambaw, A.; Verboven, P.; Defraeye, T.; Tijskens, E.; Schenk, A.; Opara, U.L.; Nicolai, B.M. Porous medium modeling and parameter sensitivity analysis of 1-MCP distribution in boxes with apple fruit. *J. Food Eng.* **2013**, *119*, 13–21. [CrossRef]
101. Delele, M.A.; Ngcobo, M.E.K.; Opara, U.L.; Meyer, C.J. Investigating the Effects of Table Grape Package Components and Stacking on Airflow, Heat and Mass Transfer Using 3-D CFD Modelling. *Food Bioprocess Technol.* **2012**, *6*, 2571–2585. [CrossRef]
102. Kim, W.R.; Aung, M.M.; Chang, Y.S.; Makatsoris, H. Freshness Gauge based cold storage management: A method for adjusting temperature and humidity levels for food quality. *Food Control.* **2015**, *47*, 510–519. [CrossRef]
103. Jedermann, R.; Praeger, U.; Geyer, M.; Lang, W. Remote quality monitoring in the banana chain. *Philos. Trans. R. Soc. A Math. Phys. Eng. Sci.* **2014**, *372*, 20130303. [CrossRef]
104. Hoang, M.H.; Laguerre, O.; Moureh, J.; Flick, D.; Hoang, H. Heat transfer modelling in a ventilated cavity loaded with food product: Application to a refrigerated vehicle. *J. Food Eng.* **2012**, *113*, 389–398. [CrossRef]
105. Defraeye, T.; Lambrecht, R.; Delele, M.; Ambaw, A.; Opara, U.L.; Cronje, P.J.; Verboven, P.; Nicolai, B. Forced-convective cooling of citrus fruit: Cooling conditions and energy consumption in relation to package design. *J. Food Eng.* **2014**, *121*, 118–127. [CrossRef]
106. Laguerre, O.; Düret, S.; Hoang, H.; Guillier, L.; Flick, D. Simplified heat transfer modeling in a cold room filled with food products. *J. Food Eng.* **2015**, *149*, 78–86. [CrossRef]
107. Escuela, G.; Hinze, T.; Dittrich, P.; Schuster, S. Modelling Modified Atmosphere Packaging for Fruits and Vegetables Using Membrane Systems. In Proceedings of the International Conference on Bio-inspired Systems and Signal Processing 2010, Valencia, Spain, 20–23 January 2010; pp. 306–311. [CrossRef]
108. Ahsan, M.; Hussain, A. A comparison of numerical methods used to solve cross flow model for multicomponent membrane gas separation. *World Appl. Sci. J.* **2013**, *22*, 703–711. [CrossRef]
109. Walawender, W.P.; Stern, S.A. Analysis of Membrane Separation Parameters. II. Counter-current and Cocurrent Flow in a Single Permeation Stage. *Sep. Sci.* **1972**, *7*, 553–584. [CrossRef]
110. Szwast, M.; Szwast, Z. A Mathematical Model of Membrane Gas Separation with Energy Transfer by Molecules of Gas Flowing in a Channel to Molecules Penetrating this Channel from the Adjacent Channel. *Chem. Process. Eng.* **2015**, *36*, 151–169. [CrossRef]
111. Xanthopoulos, G.; Koronaki, E.; Boudouvis, A. Mass transport analysis in perforation-mediated modified atmosphere packaging of strawberries. *J. Food Eng.* **2012**, *111*, 326–335. [CrossRef]
112. Dehghannya, J.; Ngadi, M.; Vigneault, C. Mathematical modeling of airflow and heat transfer during forced convection cooling of produce considering various package vent areas. *Food Control* **2011**, *22*, 1393–1399. [CrossRef]
113. Cevoli, C.; Fabbri, A. Heat transfer finite element model of fresh fruit salad insulating packages in non-refrigerated conditions. *Biosyst. Eng.* **2017**, *153*, 89–98. [CrossRef]
114. Choi, S.J.; Burgess, G. Practical mathematical model to predict the performance of insulating packages. *Packag. Technol. Sci.* **2007**, *20*, 369–380. [CrossRef]
115. De Cursi, E.S.; Sampaio, R. Introduction. In *Uncertainty Quantification and Stochastic Modeling with Matlab*; Elsevier Science: Amsterdam, The Netherlands, 2015; pp. 11–13. [CrossRef]
116. Ho, Q.T.; Rogge, S.; Verboven, P.; Verlinden, B.E.; Nicolaï, B.M. Stochastic modelling for virtual engineering of controlled atmosphere storage of fruit. *J. Food Eng.* **2016**, *176*, 77–87. [CrossRef]
117. Valentas, K.; Singh, R.P.; Rotstein, E. *Handbook of Food Engineering Practice*; Informa UK Limited: London, UK, 1997.
118. Zülch, A.; Piringer, O. Measurement and modelling of migration from paper and board into foodstuffs and dry food simulants. *Food Addit. Contam. Part A* **2010**, *27*, 1306–1324. [CrossRef]

119. Montanez, J.C.; Rodríguez, F.A.; Mahajan, P.V.; Frias, J. Modelling the gas exchange rate in perforation-mediated modified atmosphere packaging: Effect of the external air movement and tube dimensions. *J. Food Eng.* **2010**, *97*, 79–86. [CrossRef]
120. Poças, M.F.; Oliveira, J.C.; Pereira, J.R.; Brandsch, R.; Hogg, T. Modelling migration from paper into a food simulant. *Food Control* **2011**, *22*, 303–312. [CrossRef]
121. Xanthopoulos, G.; Mitropoulos, D.; Lambrinos, G. Estimation of Heat and Mass Transfer Coefficients During Air-Freezing of Cucumber. *Int. J. Food Prop.* **2012**, *15*, 221–235. [CrossRef]
122. Pandey, S.K.; Goswami, T.K. Modelling perforated mediated modified atmospheric packaging of capsicum. *Int. J. Food Sci. Technol.* **2012**, *47*, 556–563. [CrossRef]
123. Dehghannya, J.; Ngadi, M.; Vigneault, C. Transport phenomena modelling during produce cooling for optimal package design: Thermal sensitivity analysis. *Biosyst. Eng.* **2012**, *111*, 315–324. [CrossRef]
124. Delele, M.A.; Ngcobo, M.; Getahun, S.; Chen, L.; Mellmann, J.; Opara, U.L. Studying airflow and heat transfer characteristics of a horticultural produce packaging system using a 3-D CFD model. Part I: Model development and validation. *Postharvest Biol. Technol.* **2013**, *86*, 536–545. [CrossRef]
125. Chong, K.L.; Peng, N.; Yin, H.; Lipscomb, G.G.; Chung, T.-S.; Chung, T.-S. Food sustainability by designing and modelling a membrane controlled atmosphere storage system. *J. Food Eng.* **2013**, *114*, 361–374. [CrossRef]
126. Ambaw, A.; Verboven, P.; Defraeye, T.; Tijskens, E.; Schenk, A.; Opara, U.L.; Nicolai, B. Effect of box materials on the distribution of 1-MCP gas during cold storage: A CFD study. *J. Food Eng.* **2013**, *119*, 150–158. [CrossRef]
127. Sousa-Gallagher, M.J.; Mahajan, P.; Mezdad, T. Engineering packaging design accounting for transpiration rate: Model development and validation with strawberries. *J. Food Eng.* **2013**, *119*, 370–376. [CrossRef]
128. Ambaw, A.; Verboven, P.; Delele, M.A.; Defraeye, T.; Tijskens, E.; Schenk, A.; Nicolaï, B.M.; Ambaw, A. CFD Modelling of the 3D Spatial and Temporal Distribution of 1-methylcyclopropene in a Fruit Storage Container. *Food Bioprocess Technol.* **2012**, *6*, 2235–2250. [CrossRef]
129. Chaix, E.; Broyart, B.; Couvert, O.; Guillaume, C.; Gontard, N.; Guillard, V. Mechanistic model coupling gas exchange dynamics and Listeria monocytogenes growth in modified atmosphere packaging of non respiring food. *Food Microbiol.* **2015**, *51*, 192–205. [CrossRef]
130. Castellanos, D.A.; Cerisuelo, J.P.; Hernandez-Muñoz, P.; Herrera, A.; Gavara, R. Modelling the evolution of O2 and CO2 concentrations in MAP of a fresh product: Application to tomato. *J. Food Eng.* **2016**, *168*, 84–95. [CrossRef]
131. Castellanos, D.A.; Herrera, D.R.; Herrera, A.O. Modelling water vapour transport, transpiration and weight loss in a perforated modified atmosphere packaging for feijoa fruits. *Biosyst. Eng.* **2016**, *151*, 218–230. [CrossRef]
132. Fadiji, T.; Coetzee, C.; Opara, U.L. Compression strength of ventilated corrugated paperboard packages: Numerical modelling, experimental validation and effects of vent geometric design. *Biosyst. Eng.* **2016**, *151*, 231–247. [CrossRef]
133. Fadiji, T.S.; Berry, T.M.; Opara, U.L. Investigating the Mechanical Properties of Paperboard Packaging Material for Handling Fresh Produce under Different Environmental Conditions: Experimental Analysis and Finite Element Modelling Investigating the Mechanical Properties of Paperboard Packaging. *J. Appl. Packag. Res.* **2017**, *9*, 1–32.
134. Gruyters, W.; Verboven, P.; Diels, E.; Rogge, S.; Smeets, B.; Ramon, H.; Defraeye, T.; Nicolaï, B.M. Modelling Cooling of Packaged Fruit Using 3D Shape Models. *Food Bioprocess Technol.* **2018**, *11*, 2008–2020. [CrossRef]
135. Han, J.W.; Qian, J.-P.; Zhao, C.-J.; Yang, X.; Fan, B.-L. Mathematical modelling of cooling efficiency of ventilated packaging: Integral performance evaluation. *Int. J. Heat Mass Transf.* **2017**, *111*, 386–397. [CrossRef]
136. Fadiji, T.; Ambaw, A.; Coetzee, C.J.; Berry, T.M.; Opara, U.L. Application of finite element analysis to predict the mechanical strength of ventilated corrugated paperboard packaging for handling fresh produce. *Biosyst. Eng.* **2018**, *174*, 260–281. [CrossRef]
137. Wu, W.; Defraeye, T. Identifying heterogeneities in cooling and quality evolution for a pallet of packed fresh fruit by using virtual cold chains. *Appl. Therm. Eng.* **2018**, *133*, 407–417. [CrossRef]
138. Getahun, S.; Ambaw, A.; Delele, M.; Meyer, C.J.; Opara, U.L. Experimental and Numerical Investigation of Airflow Inside Refrigerated Shipping Containers. *Food Bioprocess Technol.* **2018**, *11*, 1164–1176. [CrossRef]
139. Kolodziejczyk, M.; Smierciew, K.; Gagan, J.; Butrymowicz, D. Numerical Modelling of Heat and Mass Transfer in Vegetables Cold Storage. *Procedia Eng.* **2016**, *157*, 279–284. [CrossRef]

140. Marques, A.C.; Davies, G.; Evans, J.; Maidment, G.; Wood, I. Theoretical modelling and experimental investigation of a thermal energy storage refrigerator. *Energy* **2013**, *55*, 457–465. [CrossRef]
141. Broeze, J.; Guo, X.; Axmann, H.; Vollebregt, M. *A Systemic Approach for Trade-Off Analysis of Food Loss Reduction and Greenhouse Gas Emissions*; CCAFS: Wageningen, The Netherlands, 2019.
142. Pinheiro, J.; Alegria, C.; Abreu, M.; Gonçalves, E.M.; Silva, C.L. Kinetics of changes in the physical quality parameters of fresh tomato fruits (*Solanum lycopersicum*, cv. 'Zinac') during storage. *J. Food Eng.* **2013**, *114*, 338–345. [CrossRef]
143. Aiello, G.; La Scalia, G.; Micale, R. Simulation analysis of cold chain performance based on time–temperature data. *Prod. Plan. Control.* **2011**, *23*, 468–476. [CrossRef]
144. Gonalves, E.M.; Abreu, M.; Brandão, T.R.; Silva, C.L. Degradation kinetics of colour, vitamin C and drip loss in frozen broccoli (*Brassica oleracea* L. ssp. *Italica*) during storage at isothermal and non-isothermal conditions. *Int. J. Refrig.* **2011**, *34*, 2136–2144. [CrossRef]
145. Guo, P.; Han, M.; Cao, N.; Shen, Y. The Research on Innovative Application of E-Commerce in IoT Era. In Proceedings of the 2017 IEEE International Conference on Computational Science and Engineering (CSE) and IEEE International Conference on Embedded and Ubiquitous Computing (EUC), Institute of Electrical and Electronics Engineers (IEEE), Guangzhou, China, 21–24 July 2017; Volume 2, pp. 410–413.
146. Singh, S.; Singh, N. Internet of Things (IoT): Security challenges, business opportunities & reference architecture for E-commerce. In Proceedings of the 2015 International Conference on Green Computing and Internet of Things (ICGCIoT) 2015, Delhi, India, 8–10 October 2015; pp. 1577–1581. [CrossRef]
147. Zhong, R.Y.; Xu, C.; Chen, C.; Huang, G.Q. Big Data Analytics for Physical Internet-based intelligent manufacturing shop floors. *Int. J. Prod. Res.* **2015**, *55*, 2610–2621. [CrossRef]
148. Grgić, K.; Speh, I.; Hedi, I. A web-based IoT solution for monitoring data using MQTT protocol. In Proceedings of the 2016 International Conference on Smart Systems and Technologies (SST), Osijek, Croatia, 12–14 October 2016; pp. 249–253. [CrossRef]
149. Zhamanov, A.; Sakhiyeva, Z.; Suliyev, R.; Kaldykulova, Z. IoT smart campus review and implementation of IoT applications into education process of university. In Proceedings of the 2017 13th International Conference on Electronics, Computer and Computation (ICECCO), Piscataway, NJ, USA, 28–29 November 2017; pp. 1–4. [CrossRef]
150. Bagheri, M.; Movahed, S.H. The Effect of the Internet of Things (IoT) on Education Business Model. In Proceedings of the 2016 12th International Conference on Signal-Image Technology & Internet-Based Systems (SITIS), Naples, Italy, 28 November–1 December 2016; pp. 435–441. [CrossRef]
151. He, J.; Lo, D.C.-T.; Xie, Y.; Lartigue, J. Integrating Internet of Things (IoT) into STEM undergraduate education: Case study of a modern technology infused courseware for embedded system course. In Proceedings of the 2016 IEEE Frontiers in Education Conference (FIE), Eire, PA, USA, 12–15 October 2016; Institute of Electrical and Electronics Engineers (IEEE): New York, NY, USA, 2016; pp. 1–9.
152. Mezghani, E.; Exposito, E.; Drira, K. A Model-Driven Methodology for the Design of Autonomic and Cognitive IoT-Based Systems: Application to Healthcare. *IEEE Trans. Emerg. Top. Comput. Intell.* **2017**, *1*, 224–234. [CrossRef]
153. Amato, A.; Coronato, A. An IoT-Aware architecture for smart healthcare coaching systems. *Proc. Int. Conf. Adv. Inf. Netw. Appl. AINA* **2017**, *2*, 1027–1034. [CrossRef]
154. Doukas, C.; Maglogiannis, I. Bringing IoT and Cloud Computing towards Pervasive Healthcare. In Proceedings of the 2012 Sixth International Conference on Innovative Mobile and Internet Services in Ubiquitous Computing, Palermo, Italy, 4–6 July 2012; pp. 922–926. [CrossRef]
155. Kim, S.; Kim, S. A multi-criteria approach toward discovering killer IoT application in Korea. *Technol. Forecast. Soc. Chang.* **2016**, *102*, 143–155. [CrossRef]
156. Amendola, S.; Lodato, R.; Manzari, S.; Occhiuzzi, C.; Marrocco, G. RFID Technology for IoT-Based Personal Healthcare in Smart Spaces. *IEEE Internet Things J.* **2014**, *1*, 144–152. [CrossRef]
157. Lee, B.M.; Ouyang, J. Intelligent Healthcare Service by using Collaborations between IoT Personal Health Devices. *Int. J. Bio Sci. Bio Technol.* **2014**, *6*, 155–164. [CrossRef]
158. Farahani, B.; Firouzi, F.; Chang, V.; Badaroglu, M.; Constant, N.; Mankodiya, K. Towards fog-driven IoT eHealth: Promises and challenges of IoT in medicine and healthcare. *Future Gener. Comput. Syst.* **2018**, *78*, 659–676. [CrossRef]

159. Shenoy, J.; Pingle, Y. IOT in Agriculture. In Proceedings of the 2016 3rd International Conference on Computing for Sustainable Global Development (INDIACom), New Delhi, India, 16–18 March 2016; pp. 1456–1458.
160. Brewster, C.; Roussaki, I.; Kalatzis, N.; Doolin, K.; Ellis, K. IoT in Agriculture: Designing a Europe-Wide Large-Scale Pilot. *IEEE Commun. Mag.* **2017**, *55*, 26–33. [CrossRef]
161. Pang, Z.; Chen, Q.; Han, W.; Zheng, L. Value-centric design of the internet-of-things solution for food supply chain: Value creation, sensor portfolio and information fusion. *Inf. Syst. Front.* **2012**, *17*, 289–319. [CrossRef]
162. Naeem, H. Detection of Malicious Activities in Internet of Things Environment Based on Binary Visualization and Machine Intelligence. *Wirel. Pers. Commun.* **2019**, *108*, 2609–2629. [CrossRef]
163. European Commission: Information Society and Media. Internet of Things in 2020: A Roadmap for the Future. *Internet Things* **2020**, *22*, 97–114.
164. Dlodlo, N.; Kalezhi, J. The internet of things in agriculture for sustainable rural development. In Proceedings of the 2015 IEEE International Conference on Emerging Trends in Networks and Computer Communications (ETNCC), Windhoek, Namibia, 17–20 May 2015; pp. 13–18.
165. Tsang, Y.P.; Choy, K.; Wu, C.; Ho, G.; Lam, H.; Tang, V. An intelligent model for assuring food quality in managing a multi-temperature food distribution centre. *Food Control* **2018**, *90*, 81–97. [CrossRef]
166. Verdouw, C.; Kruize, J.W. Digital twins in farm management: Illustrations from the FIWARE accelerators SmartAgriFood and Fractals. In Proceedings of the 7th Asian-Australasian Conference on Precision Agriculture, Hamilton, New Zealand, 16 October 2017; pp. 1–5.
167. Ben-Daya, M.; Hassini, E.; Bahroun, Z. Internet of things and supply chain management: A literature review. *Int. J. Prod. Res.* **2017**, *57*, 4719–4742. [CrossRef]
168. Tzounis, A.; Katsoulas, N.; Bartzanas, T.; Kittas, C. Internet of Things in agriculture, recent advances and future challenges. *Biosyst. Eng.* **2017**, *164*, 31–48. [CrossRef]
169. Ramundo, L.; Taisch, M.; Terzi, S. State of the art of technology in the food sector value chain towards the IoT. In Proceedings of the 2016 IEEE 2nd International Forum on Research and Technologies for Society and Industry Leveraging a better tomorrow (RTSI), Piscataway, NJ, USA, 7–9 September 2016; pp. 1–6.
170. Popa, A.; Hnatiuc, M.; Paun, M.; Geman, O.; Hemanth, D.J.; Dorcea, D.; Son, L.H.; Ghita, S. An Intelligent IoT-Based Food Quality Monitoring Approach Using Low-Cost Sensors. *Symmetry* **2019**, *11*, 374. [CrossRef]
171. Zhang, Y.; Zhao, L.; Qian, C. Modeling of an IoT-enabled supply chain for perishable food with two-echelon supply hubs. *Ind. Manag. Data Syst.* **2017**, *117*, 1890–1905. [CrossRef]
172. Ghoshal, G. *Recent Trends in Active, Smart, and Intelligent Packaging for Food Products*; Elsevier Inc.: Amsterdam, The Netherlands, 2018.
173. Porat, R.; Lichter, A.; Terry, L.A.; Harker, R.; Buzby, J. Postharvest losses of fruit and vegetables during retail and in consumers' homes: Quantifications, causes, and means of prevention. *Postharvest Biol. Technol.* **2018**, *139*, 135–149. [CrossRef]
174. Fuertes, G.; Soto, I.; Carrasco, R.R.; Vargas, M.; Sabattin, J.; Lagos, C. Intelligent Packaging Systems: Sensors and Nanosensors to Monitor Food Quality and Safety. *J. Sens.* **2016**, *2016*, 4046061. [CrossRef]
175. Sohail, M.; Sun, D.-W.; Zhu, Z. Recent developments in intelligent packaging for enhancing food quality and safety. *Crit. Rev. Food Sci. Nutr.* **2018**, *58*, 2650–2662. [CrossRef]
176. Borchert, N.B.; Kerry, J.P.; Papkovsky, D.B. A CO_2 sensor based on Pt-porphyrin dye and FRET scheme for food packaging applications. *Sensors Actuators B Chem.* **2013**, *176*, 157–165. [CrossRef]
177. Xiao, X.; Fu, Z.; Zhang, X.; Cheng, J.; Yang, M. Battery-free wireless sensor system with compressed sensing for table grapes cold chain. *Comput. Electron. Agric.* **2019**, *163*, 104869. [CrossRef]
178. Maksimović, M.; Vujović, V.; Omanović-Mikličanin, E. Application of internet of things in food packaging and transportation. *Int. J. Sustain. Agric. Manag. Inform.* **2015**, *1*, 333. [CrossRef]
179. Nukala, R.; Panduru, K.; Shields, A.; Riordan, D.; Doody, P.; Walsh, J. Internet of Things: A review from 'Farm to Fork'. In Proceedings of the 2016 27th IEEE Irish Signals and Systems Conference (ISSC), Derry, Ireland, 21–22 June 2016; pp. 1–6.
180. Bassi, A.; Bauer, M.; Fiedler, M.; Kranenburg, R.V. *Enabling Things to Talk*; Springer-Verlag GmbH: Berlin, Germany, 2013.
181. Verboven, P.; Verboven, P.; Defraeye, T.; Datta, A.K.; Nicolai, B. Digital twins of food process operations: The next step for food process models? *Curr. Opin. Food Sci.* **2020**, *35*, 79–87. [CrossRef]

182. Tao, F.; Cheng, J.; Qi, Q.; Zhang, M.; Zhang, H.; Sui, F. Digital twin-driven product design, manufacturing and service with big data. *Int. J. Adv. Manuf. Technol.* **2018**, *94*, 3563–3576. [CrossRef]
183. Uhlemann, T.H.J.; Schock, C.; Lehmann, C.; Freiberger, S.; Steinhilper, R. The Digital Twin: Demonstrating the Potential of Real Time Data Acquisition in Production Systems. *Procedia Manuf.* **2017**, *9*, 113–120. [CrossRef]
184. Defraeye, T.; Tagliavini, G.; Wu, W.; Prawiranto, K.; Schudel, S.; Kerisima, M.A.; Verboven, P.; Bühlmann, A. Digital twins probe into food cooling and biochemical quality changes for reducing losses in refrigerated supply chains. *Resour. Conserv. Recycl.* **2019**, *149*, 778–794. [CrossRef]
185. Tagliavini, G.; Defraeye, T.; Carmeliet, J. Multiphysics modeling of convective cooling of non-spherical, multi-material fruit to unveil its quality evolution throughout the cold chain. *Food Bioprod. Process.* **2019**, *117*, 310–320. [CrossRef]
186. Furio, A.; Landi, G.; Altavilla, C.; Sofia, D.; Iannace, S.; Sorrentino, A.; Neitzert, H.C. Light irradiation tuning of surface wettability, optical, and electric properties of graphene oxide thin films. *Nanotechnology* **2016**, *28*, 54003. [CrossRef]
187. Cuello, J. Our Fresh-Produce Supply Chain after COVID-19: How vertical farming can play a key role. *Resour. Mag.* **2020**, *27*, 10–12.

Publisher's Note: MDPI stays neutral with regard to jurisdictional claims in published maps and institutional affiliations.

processes

MDPI

Article

The Effectiveness of Different Household Storage Strategies and Plant-Based Preservatives for Dehulled and Sun-Dried Breadfruit Seeds

Macmanus C. Ndukwu [1], Daniel I. Onwude [2,*], James Ehiem [1], Ugochukwu C. Abada [3], Inemesit E. Ekop [4] and Guangnan Chen [5]

1 Department of Agricultural and Bioresources Engineering, Michael Okpara University of Agriculture Umudike, P.M.B. 7267 Umuahia, Nigeria; ndukwumcu@mouau.edu.ng (M.C.N.); ehiem.james@mouau.edu.ng (J.E.)
2 Department of Agricultural and Food Engineering, University of Uyo, Akwa Ibom State, P.M.B. 52021 Uyo, Nigeria
3 Department of Agricultural and Bioresources Engineering, University of Nigeria, 410001 Nsukka, Nigeria; ugochukwu.abada@unn.edu.ng
4 Department of Agricultural Engineering, Akwa Ibom State University, P.M.B. 1167 Uyo, Nigeria; inemesitekop@aksu.edu.ng
5 Faculty of Health, Engineering and Sciences, University of Southern Queensland, Toowoomba, QLD 4350, Australia; Guangnan.Chen@usq.edu.au
* Correspondence: onwudedaniel@uniuyo.edu.ng

Abstract: In a tropical rainforest environment, different storage strategies are often adopted in the preservation of primary processed food crops, such as maize, sorghum, etc., after drying and dehulling to increase shelf-life. For breadfruit seeds (*Treculia Africana*), the current challenge is identifying the most appropriate short-term storage and packaging methods that can retain the quality of stored products and extend shelf-life. In this regard, we compared the performance of a plastic container, a weaved silo bag and a locally developed silo bin for the short-term storage of parboiled, dehulled and dried breadfruit seeds treated with locally sourced and affordable alligator pepper (*Zingiberaceae aframomum melegueta*) and bitter kola (*garcinia*) powder as preservatives. We show that the concentration of CO_2 was lower in the silo bin treated with 150 g alligator pepper and higher in the silo bag-treated with 100 g bitter kola nut. A higher CO_2 concentration resulted in limited oxygen availability, higher water vapor, and a higher heat release rate. Non-treated bag storage had the highest average mold count of 1.093×10^3 CFU/mL, while silo bin-stored breadfruit treated with 150 g of alligator pepper had the lowest mold count of 2.6×10^2 CFU/mL. The storage time and botanical treatments influenced both the crude protein and crude fiber content. Average insect infestations were low (0–4.5) in the silo bin with breadfruits treated with alligator pepper powder, as the seeds seemed to continue to desorb moisture in storage, unlike in other treatments. The obtained results revealed the high potential of alligator pepper (*Zingiberaceae aframomum melegueta*) as a botanical insecticide in preventing insect infestation and mold growth in stored breadfruit instead of using synthetic insecticide. An aluminum silo bin with alligator pepper powder is recommended to store dried and dehulled breadfruit seeds as a baseline for other tropical crops.

Keywords: postharvest storage; food packaging and shelf-life; bitter kola; food preservation; alligator pepper; underutilized seeds

Citation: Ndukwu, M.C.; Onwude, D.I.; Ehiem, J.; Abada, U.C.; Ekop, I.E.; Chen, G. The Effectiveness of Different Household Storage Strategies and Plant-Based Preservatives for Dehulled and Sun-Dried Breadfruit Seeds. *Processes* **2021**, *9*, 380. https://doi.org/10.3390/pr9020380

Academic Editor: Lina Cossignani
Received: 14 December 2020
Accepted: 14 February 2021
Published: 19 February 2021

1. Introduction

Breadfruit belongs to the family of *Moraceae* and is a multipurpose tropical agro-forestry tree crop with about 120 varieties. Prominent among them in Nigeria is African breadfruit (*Treculia africana*) [1,2]. The seeds are rich in oil, carbohydrates, proteins, calcium, phosphorus, minerals, and vitamins [1]. African breadfruit seeds are prepared for consumption by cooking (either fresh or dried), roasting, or breadfruit flour. This highly nutritional but underutilized crop is native to the West Indies, Jamaica, as well as many West African countries, including Nigeria, Ghana and Sierra Leone [3]. In 2011, Nigeria accounted for about 32% of African breadfruits' global production [4]. Food prepared from breadfruit seeds is highly appreciated and sought out because it serves as a common source of calories among over 60 million people in the Central African Republic, southern Nigeria, and southern Cameroun [5]. The processing of this underutilized crop involves the fermentation or dehulling of the fresh fruit to extract the seed, washing, heat treatment, and threshing of the seed to reveal an edible greenish cotyledon [6]. This is then either cooked fresh or dried to a moisture content of about 8–10% wet basis before storage [6]. However, drying unhulled seeds makes dehulling very difficult even if heat treatment is applied. This is because of the adhesive layer that attaches the hull to the kernel [7].

Traditionally, in Nigeria, after drying African breadfruits, various households store dried breadfruits under room conditions in plastic containers, weaved silo bags, bottles, hanging above the cooking attic, spreading on the open floor or on a mat, or keeping them in basins [6,8]. However, storing the dried seeds for a long period is a challenge due to the hygroscopic nature of the seeds even after drying, which makes them prone to quick fungi and insect damage, discoloration, etc. [6,9]. Apart from managing the pest in enclosed storage, temperature increase, high humidity and the reaction of chemical and gaseous constituents from the product can cause considerable deterioration [9]. Investigations were carried out by Adindu and Williams [6] on the nutritional quality of non-dehulled dried breadfruit seed stored in tight plastic buckets, polythene bags, closed and open raffia baskets, and tied jute bags. However, in most homes, breadfruits are parboiled, dehulled and dried before storage, as in rice processing. Therefore, storing pretreated and processed African breadfruit seeds is still a challenge as there is a lack of adequate data for food processors.

Several studies on the different postharvest management practices on stored agro-product exist in the scientific literature [10–12]. However, these studies are limited mainly to grains. Furthermore, many storage approaches have been adopted in the literature for the shelf-life extension of stored products by researchers, ranging from the use of botanical insecticides (oil from Vittallaria paradoxa seed, coconut, mustard seed), to Diatomaceous earth, hermetic bags, metal silos, and charcoal ash [12–19]. However, insight gained from these studies revealed that most of these storage approaches are crop-specific with a particular focus on cereals, including maize, cowpea, wheat, etc. [11–21]. To the best of our knowledge, information on the most suitable storage strategy for primary processed and raw African breadfruit seeds under household conditions is non-existent. Again, the respiratory activities in the interstitial space of the storage ecosystem of biological products influence product storage conditions. Temperature changes in the interstitial space have been adduced to respiratory activities within the storage ecosystem, which is a function of the external conditions. With this knowledge of the interstitial temperature, the gaseous exchange within the space can be accounted for. However, this measurement of the interstitial temperature requires constant measurement and probing of the storage system, which distorts the tightness of the storage system. Therefore, employing a mathematical model that will account for interstitial temperature and all the desired components of gaseous exchange involved in the respiration activities using the external conditions will solve this problem. In this way, the rigorous and cumbersome experimental measurement process is eliminated [22–24]. In addition, using this modeling approach will enhance future optimization and process design [25].

Locally, in Nigeria, dried alligator pepper (*Zingiberaceae aframomum melegueta*) powder has been integrated into the storage bag to drive away insects and pests in maize stored in weaved bags under the market environment [12]. In addition to alligator pepper, bitter kola (*garcinia*) has also been used locally for storage because its extracts have shown antimicrobial, antifungal and anti-pest properties, along with their pungent, peppery and bitter taste [26–31]. Therefore, as food processors reduce the dependency on synthetic insecticides due to the attendant health challenges, evaluating the efficacy of these affordable edible plant seeds as alternatives become indispensable.

This research aims to identify the most effective storage strategies and treatments for an on-the-shelf postharvest preservation approach for breadfruits seeds, as adopted in Nigeria. A plastic container, a weaved bag and a locally developed silo bin were adopted for the short-term storage of parboiled, dehulled, and sun-dried breadfruit seeds treated with alligator pepper (*Zingiberaceae aframomum melegueta*) and bitter kola (*garcinia*) powder as preservatives. The effectiveness of the storage strategies was evaluated in terms of shelf-life duration, nutritional product quality, mold count and insect enumeration. The result of this study will provide valuable information to improve the all-year-round availability and affordability of breadfruits seeds in major breadfruit-producing countries.

2. Materials and Methods

2.1. Description of Storage Strategies

The experiment was carried out for 3 months (14 weeks) (September–December 2019) at the crop processing and storage laboratory of Michael Okpara University of Agriculture Umudike, Nigeria, 5.53° N, 7.49° E. The local storage strategies adopted include a weaved silo bag, a plastic drum and a locally developed aluminum silo bin. Figure 1 shows the schematic view of the aluminum silo and its components. It consists of a storage container equipped with a wooden stirrer force-fitted into the silo through a sealed bearing. Loading and unloading points were created at the top and bottom of the silo and plastic container. The plastic buckets and aluminum silos were airtight with a side opening for material sampling. The loading and unloading opening of the silo was covered with a metal cover sheet, while that of the plastic drum was sealed with masking tape.

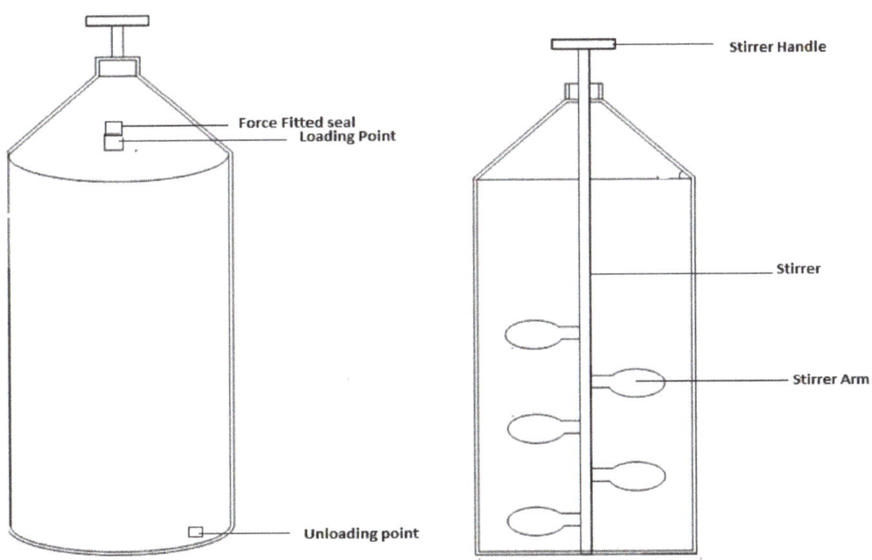

Figure 1. Schematic view of the locally developed aluminum storage silo.

2.2. Sourcing, Heat Treatments and Drying of Breadfruit Seeds

The breadfruit used for this experiment was obtained directly from the farmers in Umudike, southeastern Nigeria. All the seeds were sourced from the same lot as recommended in the International Rules of Seed Testing [32] and there was no requirement for additional homogenization. The seeds were harvested off-season and were not treated with any pesticide. The farmers allow matured fruit to naturally fall from the tree in the wild, after which it is allowed to ferment from two to three weeks. Fermented fruits were washed with clean water to remove the pulp from the seeds. The seeds were manually selected to remove any bad seed or dirt, after which they were spread on a stainless tray and kept under the sun for three hours. The seeds were parboiled in hot water for 5 min with clean water, after which the water was filtered out. The parboiled seeds were spread under the sun and allowed to cool before manual threshing using a local wooden thresher to reveal the cotyledon. Parboiling the seed unstuck the chaff from the cotyledon for easy separation, and it also kills any larvae of pests from the farm or the fermentation process. Parboiling of the seed for a short time of less than 10 min causes no significant difference in the seed's nutritional composition [33]. The threshed mixture of cotyledon and chaff was separated manually by winnowing using a stainless tray. After winnowing, broken cotyledon was manually removed while the rest was dried under the sun until the moisture content reached about 8 $\pm$ 2.2%. Initial moisture content was checked with a grain moisture meter (Kongskilde, series 20849) initially calibrated with a laboratory oven (DHG-9053A Ocean med+ England).

2.3. *Preparation of Botanical Treatments*

Dried pulverized alligator pepper seed (*Zingiberaceae aframomum melegueta*) at 8% moisture content and dried pulverized bitter kola seed (*Garcinia*) at 10% moisture content were introduced as botanical preservatives. These two plant materials were harvested from the trees in southeastern Nigeria, and care was taken to select seeds of premium quality. Before milling, the two seeds were dried in the sun for 7 days and further in the oven at 40 °C. They were milled separately using a hammer mill (Animal ration shredder hammer mill foliage, Model: TRF 400; 1.5 kW; 10 swinging hammers). The milled seeds were sieved with 90 μm sieve openings and kept in an airtight vessel before application.

2.4. *Experimental Test*

Each of the storage strategies received two botanical treatments (pest management) consisting of mixing 2 kg of dried breadfruit with 100 and 150 g of dried pulverized alligator pepper (*Zingiberaceae aframomum melegueta*), and 100 and 150 g dried pulverized bitter kola (*Garcinia*), respectively. In addition, a similar experiment without treatment was set up to serve as a control. Therefore, a total of 15 experiments were set up. The treatments were categorized as follows: breadfruits stored in the aluminum silo and treated with various mass of pulverized alligator pepper, labeled as $SLAP_{100}$ and $SLAP_{150}$; breadfruits stored in the aluminum silo and treated with pulverized bitter kola were labeled as $SLBK_{100}$ and $SLBK_{150}$, while SL_0 was the control. In addition, breadfruits stored in the plastic drum and treated with pulverized alligator pepper were labeled as $RBAP_{100}$ and $RBAP_{150}$, and breadfruits stored in the plastic drum and treated with pulverized bitter kola were labeled as $RBBK_{100}$ and $RBBK_{150}$. At the same time, RB_0 serves as the control. In addition, breadfruit stored in the woven bag and treated with pulverized alligator pepper were labeled as $BGAP_{100}$ and $BGAP_{150}$, and breadfruit stored in the woven bag and treated with pulverized bitter kola were labeled as $BGBK_{100}$ and $BGBK_{150}$, while BG_0 served as the control. All containers received 2 kg of heat-treated, dehulled and dried breadfruit seed with the botanical treatment applied once at the beginning of storage. Figure 2 shows the storage strategies.

Figure 2. The different storage strategies for household preservation of breadfruit seeds.

2.5. Weight Loss and Moisture Content

The moisture content (MC) of the seeds was determined using a grain moisture meter (Kongskilde, series 20849) previously calibrated with a laboratory oven for accuracy. MC was respectively measured at 0, 6, and 12 weeks of storage. Weight loss in storage was determined by direct weighing (Camry weighing balance, ACS, 56; China) of the entire sample at the beginning of storage and after 4, 8, and 12 weeks of sampling.

2.6. Enumeration of Insects

The number of adult insects present in the stored product was counted at 6 weeks and 12 weeks of storage by sifting the 100 g sample through a sieve pan. The insects were retrieved and counted manually.

2.7. Total Mold Count

Sabouraud dextrose agar and Yeast extract agar (BDH, London, UK) for fungi and yeast count were used as the media for isolating the microorganisms. The media were prepared according to the manufacturer's instructions by sterilization at a temperature of 121 °C for 15 min in an autoclave. According to the dilution-plating method described by Cowan and Steel [34], the total mold count was determined. Tenfold serial dilution of the sample was prepared with distilled water, and 9 mL measures of diluent were placed into 9 sterile test-tubes. About 1 g of each sample was uniformly mixed and 1.0 mL was transferred into the first tube of diluent. This was done for the remaining dilutions using a fresh pipette for each to obtain different concentrations. Using the third and fifth (10^{-3} and 10^{-5} concentration) dilutions, 0.1 mL amounts of each dilution were pipetted into each of three Petri dishes. The Sabouraud dextrose agar and Yeast extract agar were added using the spread plate method. The plates were incubated at 37 °C for 24 h aerobically. Counts were taken of the colonies in the three plates that were inoculated with the dilutions of 50 and 500 colonies per plate. The discrete colonies were purified by repeated subculturing onto Sabouraud dextrose agar and Yeast extract agar until pure cultures were obtained. Pure mold isolates were streaked onto new culture plates and incubated for 48 h. The pure isolates were stored on Sabouraud dextrose agar. The average number per plate was multiplied by the dilution factor to obtain the viable count per gram of the sample [34]. The cut-off point for contamination was taken as >10^2 CFU/mL [35]. The samples were subjected to microbiological analysis to monitor the dynamic changes in the fungal populations. Mold isolates were identified and characterized by microscopic morphological observations. The morphological characteristics of mold include the shape of colonies, colonial outline, colonial evaluation, color, consistency and size.

2.8. Product Quality

Fresh and stored dried breadfruit seeds were analyzed for ash, carbohydrate, fat, protein, and crude fiber content. Prior to nutritional analysis, the seeds were washed with clean water to remove any traces of the botanicals used for the treatment. The ash content was deduced by the furnace incineration gravimetric method, while the fat content was inferred via the continuous solvent extraction gravimetric method using Soxhlet apparatus as described in the Association of Official Analytical Chemists (AOAC 2007) manual [36]. The crude fiber content was deduced using the Wende method described by James [37], while the crude protein was determined by the Kjedahl method [38]. The carbohydrate content was calculated via the nitrogen-free extraction method as described by James [37].

2.9. Determination of Minerals

The dried breadfruit's mineral content was determined using the dry ash acid extraction method, as described by Carpenter and Hendricks [39]. A 5g measure of each breadfruit sample was burned to ashes in a muffle furnace at 500 °C. The ash produced was dissolved in 10 mL of 2M HCl solution and diluted to 100 mL with distilled water in a volumetric flask and filtered. The filtrate was used for mineral analysis. The mineral analysis was carried out according to Ref. [40] by employing atomic absorption spectroscopy for magnesium (Mg), calcium (Ca), phosphorus (P), iron (Fe) and zinc (Zn), and flame emission photometry for sodium (Na) and potassium (K).

2.10. Data Analysis

All the accumulated data were subjected to two-way ANOVA to test the variation in the mean between different storage approaches, periods and treatments. Turkey's HSD ($p < 0.05$) test was used to detect the mean differences between the examined traits.

3. Mathematical Modeling

3.1. Modeling of Interstitial Gaseous Exchange

According to Abelon et al. [41], the concentration of gas in a storage system is dependent on the degree of consumption or entrance of oxygen, which leads to the total combustion of carbohydrates, so the production or loss of carbon dioxide, water and heat within the storage system is as follows:

$$C_6H_{12}O_6 + 6O_2 \rightarrow 6CO_2 + 6H_2O + 2835\text{kJ/mol} \tag{1}$$

$$180\text{g} + 192\text{g} \rightarrow 264\text{g} + 108\text{g} + 2835\text{kJ/mol} \tag{2}$$

The gas exchange between the system and the environment is a function of the partial pressure difference and the permeability of the systems. Assuming uniformly distributed temperature and moisture, and ignoring respiration from insects and carbon dioxide sorption [42], the rate of heat produced during respiration, the water vapor produced and the oxygen consumed were calculated from Equations (1) and (2). These equations were used to deduce Equations (3)–(5) considering the molecular weight of carbohydrate, oxygen, carbon dioxide and water, and the heat produced according to Gaston et al. [42], as follows:

$$Y_{res} = q_h Y_{CO_2} \tag{3}$$

$$Y_{H_2O} = q_w Y_{CO_2} \tag{4}$$

$$Y_{O_2} = q_o Y_{CO_2} \tag{5}$$

where q_h, q_w, and q_o are the parameters deduced from Equation (1), while Y_{CO2} is the CO_2 release rate in mg (CO_2) kg^{-1} (dry matter) in 24 h. The Y_{CO2} was given from a generic equation presented in Ref. [43] for the oxidation of hexose, as follows:

$$\log Y_{CO_2} = q_1 + q_2 T_C - q_3\theta + q_4\theta^2 + q_5 M \tag{6}$$

where q_1, q_2, q_3, q_4 and q_5 are the regression constants presented in Table 1. Equation (6) can be adapted to suit any stored product by adding a non-unite parameter ζ, as follows:

$$\log Y_{CO_2} = \zeta \left(q_1 + q_2 T_C - q_3\theta + q_4\theta^2 + q_5 M\right) \tag{7}$$

The total amount of CO_2 produced or oxygen consumed over the storage period (% v/v) was calculated by the integration of Equation (8) overtime in days, and the result is presented in Equation (9), according to Gaston et al. [42] and Abelon et al. [41].

$$m_{CO_2}(t) = \int_0^t \overline{Y}_{CO_2(t')} dt' \tag{8}$$

$$X_{CO_2} = \frac{Y_{CO_2}(t)}{1000 M_{CO_2}} \rho_b \times \frac{RT\ (t)}{\varepsilon\ P_{at}} \tag{9}$$

where M_{CO2} is molecular weight (44 g/mol), T is the intergranular temperature (K), P_{at} is the atmospheric pressure (1 atm = 101 325 Pa), R is the gas constant (8.314J/mol K), ε is the porosity and ρ_b is the bulk density of the stored product (kg/m^3), given as follows:

$$\rho_b = (1 - \varepsilon)\rho_{sp} + \varepsilon\rho_a \tag{10}$$

where ϱ_{sp} is the density of the parboiled breadfruit, ε is the porosity of parboiled breadfruit, and ϱ_a is the density of intergranular air (kg/m^3) adapted from the density of a heated space given by Simo-Tagne et al. [44], as follows:

$$\rho_a = \frac{b}{T} \tag{11}$$

where T is the temperature (K) and b is given as 353 (kg K/m^3).

Table 1. Input parameters for the determination of gaseous exchange.

Input Parameters	Values	Source
q_1	−4.0540	[43]
q_2	0.0406	[43]
q_3	−0.0165	[43]
q_4	0–0001	[43]
q_5	0.2389	[43]
ε	0.49 for 5 min blanching	[45]
ϱ_b	632.87 kg/m^3 for 5 min blanching	[45]
ϱ_{sp}	1262.38 kg/m^3 for 5 min blanching	[45]
q_h	10.738 J/mg (CO_2)	[42]
q_w	4.09×10^{-5} kg (H_2O)/mg (CO_2)	[42]
q_o	0.7272 mg (O_2)/kg (dry matter)	[42]

3.2. Intergranular Temperature and Relative Humidity

The building (ambient) temperature and relative humidity were measured with a temperature and humidity clock (DTH-82; TLX, Guandong China). The intergranular dry bulb temperature was determined three times daily with a temperature probe (Extech, Taiwan China) inserted through the small hole made at the top of the silo and plastic container. In contrast, for the bag storage, it was inserted carefully through the body. After each measurement, the hole was sealed correctly. However, the interstitial relative humidity was determined from the vapor pressure equations, according to Abe and Basunia [46] for a stored product, as follows:

$$RH = \frac{h_{vp}}{h_{ds}} \tag{12}$$

where h_{vp} and h_{ds} are the vapor pressure and saturated vapor pressure at the dry bulb temperatures given in Equations (13) and (15), respectively, by Abe and Basunia [46] as follows:

$$h_{vp} = h_{ws} - 0.5(T_{db} - T_{wb}) \times \left(\frac{760}{755}\right) \quad (13)$$

where T_{db} and T_{wb} are the intergranular dry and wet bulb temperature, and h_{ws} is the saturated vapor pressure at the wet bulb temperature, given as:

$$h_{ws} = 4.58 \times 10^{(7.5T_{wb})/(237+T_{wb})} \quad (14)$$

$$h_{ds} = 4.58 \times 10^{(7.5T_{db})/(237+T_{db})} \quad (15)$$

The wet bulb temperature (°C) of the intergranular air was calculated from the empirical relationship presented by Fouda and Melikyan [47] for moist air, as follows,

$$T_{wb} = 2.65(1.97 + 4.3T_{db} + 1000d)^{1/2} - 14.85 \quad (16)$$

where d is the pressure parameter given in Equation (17), as follows,

$$d = \frac{0.622p_{st}}{p_{atm} - p_{st}} \quad (17)$$

where p_{atm} is the atmospheric pressure (Pa) and p_{st} is the saturated vapor pressure at the surface of the storage containers, given as Equation (18) as follows,

$$p_{st} = \exp\left(23.196 - \frac{3816.44}{T - 46.13}\right) \quad (18)$$

where T is the temperature (K).

4. Results and Discussions

4.1. Temperature, Relative Humidity and Moisture Distribution

The initial breadfruit seeds moisture was 10.2% w.b. (wet basis), while the intergranular temperature of bulk seed was recorded as 28.3 °C at the beginning of storage. The intergranular temperature variations among the storage strategies and botanical treatments are shown in Figure 3A–F. The building temperature (ambient temperature) was marginally higher than the intergranular temperatures for the botanically treated seeds, but lower than the temperature of the untreated seeds stored by different methods. This showed the stability of seeds treated botanically. Monitoring the storage temperature in bin silos has been reported as an effective way of measuring stored products' storage conditions [41]. A sharp rise in core temperature is an indication of localized heating as a result of spoilage, as shown in non-treated bin silos (Figure 3A,B), while in the case of bags, it is as a result of gaseous exchange between the core and the environment, which is influenced by the climatic conditions [48]. The mean value for the building temperature was determined as 28.78 °C, while SL_0, $SLAP_{100}$, $SLAP_{150}$, $SLBK_{100}$, $SLBK_{150}$, RB_0, $RBAP_{100}$, $RBAP_{150}$, $RBBK_{100}$, $RBBK_{150}$, BG_0, $BGAP_{100}$, $BGAP_{150}$, $BGBK_{100}$ and $BGBK_{150}$ were recorded as 29.49, 28.59, 28.66, 28.61, 28.69, 29.42, 28.64,28.73, 28.68, 28.62, 29.38, 28.68,28.72, 28.67 and 28.77 °C, respectively, with a mean building relative humidity of 73%. The observed temperature is within the temperature range of seeds stored in bins and silo bags, which is less than 30 °C [48]. It was also observed with the inclusion of non-treated samples that there was a mean temperature increase from 28.25 to 29.46 °C in the first week to a range of 28.62–29.63 °C in the fourth week. However, this average value decreased to a range of 28.59–29.49 °C in the 12th week. Nevertheless, for the storage strategies adopted, the maximum temperature difference between the building (storage room) and the intergranular temperature is less than 1.0 °C. This could be due to the small size of storage [46].

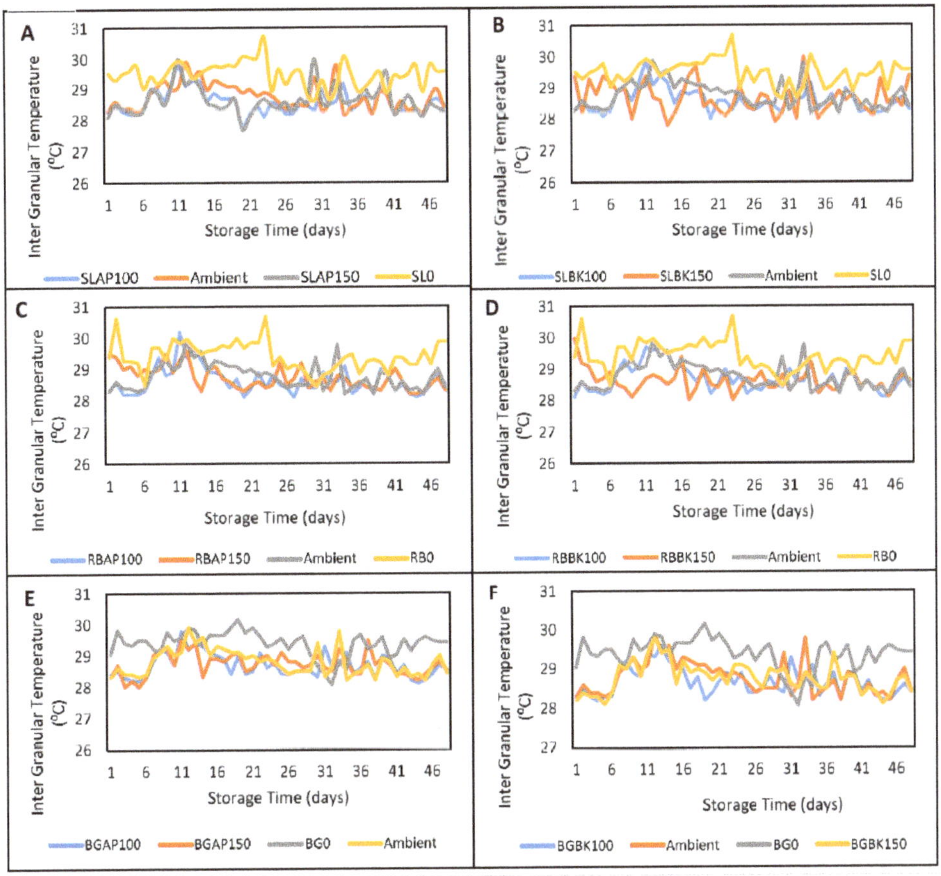

Figure 3. Intergranular temperature of the stored breadfruits for different storage methods and treatments. (**A**) Breadfruit stored in the silo and treated with aligator pepper. (**B**) Breadfruit stored in the silo and treated with bitter kola. (**C**) Breadfruit stored in the rubber and treated with aligator pepper. (**D**) Breadfruit stored in the rubber and treated with bitter kola. (**E**) Breadfruit stored in the bag and treated with bitter kola. (**F**) Breadfruit stored in the bag and treated with aligator pepper.

Stored food's shelf-life can be influenced by the initial product moisture, temperature, and weather variation during storage, which is modified by the building conditions [49–51]. Moisture uptick was recorded for all non-botanical treated stored breadfruits, as shown in Figure 4. The increase ranged from 0.09 to 5.04%, with the product stored in the silo recording the lowest moisture increase, while the silo bag recorded the highest value after 12 weeks of storage. This might result from the marginally increased temperature observed in non-botanical-treated storage compared to the ambient values (Figure 3). However, considering the moisture content of the botanically treated breadfruits, the moisture contents of those treated with alligator pepper decreased consistently. In contrast, those treated with bitter kola increased within the first six weeks but decreased later when checked at 12 weeks. The variation in the seed moisture content of the different storage methods and pest management strategies could be due to the possible condensation of air occasioned by the temperature drop in the intergranular spaces, or between the storage spaces [52]. Besides this, the vapor emitted by biomaterials due to biological processes might have contributed to the uptick in seed moisture [53,54]. On the other hand, seed moisture with alligator pepper tends to drop due to the lower biological activities. Therefore, it can be stated that the alligator pepper can serve as a drying agent in stored products. A significant ($p < 0.05$) moisture decrease was observed between those treated with alligator pepper when the amount of alligator pepper increased from 100 to 150 g, while it was not significant ($p < 0.05$) when compared, under the same storage method, with those treated with bitter kola. However, a significant difference ($p < 0.05$) exists between the moisture levels measured for those treated with the same amounts of alligator pepper and bitter kola. Again, for different silo storage strategies, there were significant differences ($p > 0.05$) among the same amounts of treated material.

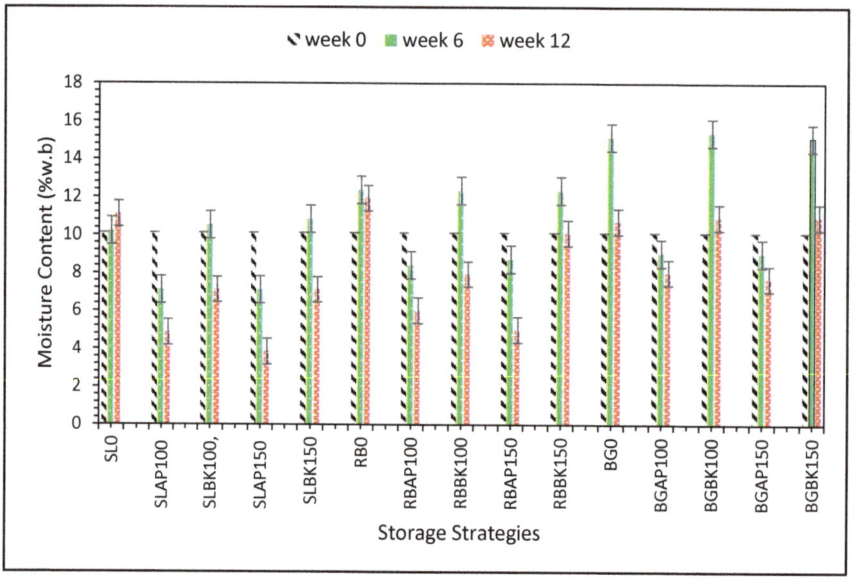

Figure 4. Variation in moisture content for different storage strategies within selected weeks.

The ambient relative humidity is shown in Figure 5. The weather pattern highly influences relative humidity in the storage room. The room relative humidity will increase during the ambient peak humidity, while the temperature drops, and vice versa. The ambient relative humidity ranged from 71 to 76%, with an average value of 73%. This is relatively high, which is common in most tropical rainforest zones, making it difficult to achieve stability in natural storage because of the influence of relative humidity on microbial developments. The value of the ambient relative humidity will also influence the interstitial humidity due to environmental interactions between the storage space and the external environment. In addition, the internal heat of the storage product can also affect the relative humidity of the intergranular space.

The interstitial relative humidity was predicted using Equations (12)–(18), and is presented in Figure 6A–F for different storage strategies. The simulated relative humidity and wet bulb temperature were deduced from the measured interstitial temperature, and therefore varied as the interstitial temperature varied. In addition, from Figure 4, we can see that the products desorb or adsorb moisture from the interacting environment due to the storage strategies and management approach adopted, which might have affected the intergranular temperature vis-a-vis the interstitial relative humidity profile. For all the storage strategies, the interstitial relative humidity ranged from 29.86 to 34.25%. The lower relative humidity obtained is a result of the dried nature of the product. However, due to the high ambient relative humidity and temperature, it is possible to have water accumulate on the walls of the silo bin and the plastic containers. The introduction of alligator pepper, in particular, seemed to function as a drying agent for the product.

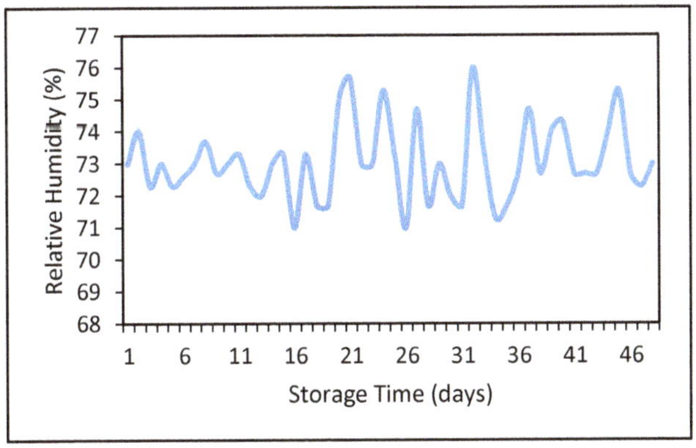

Figure 5. Ambient relative humidity during the storage of breadfruit seeds.

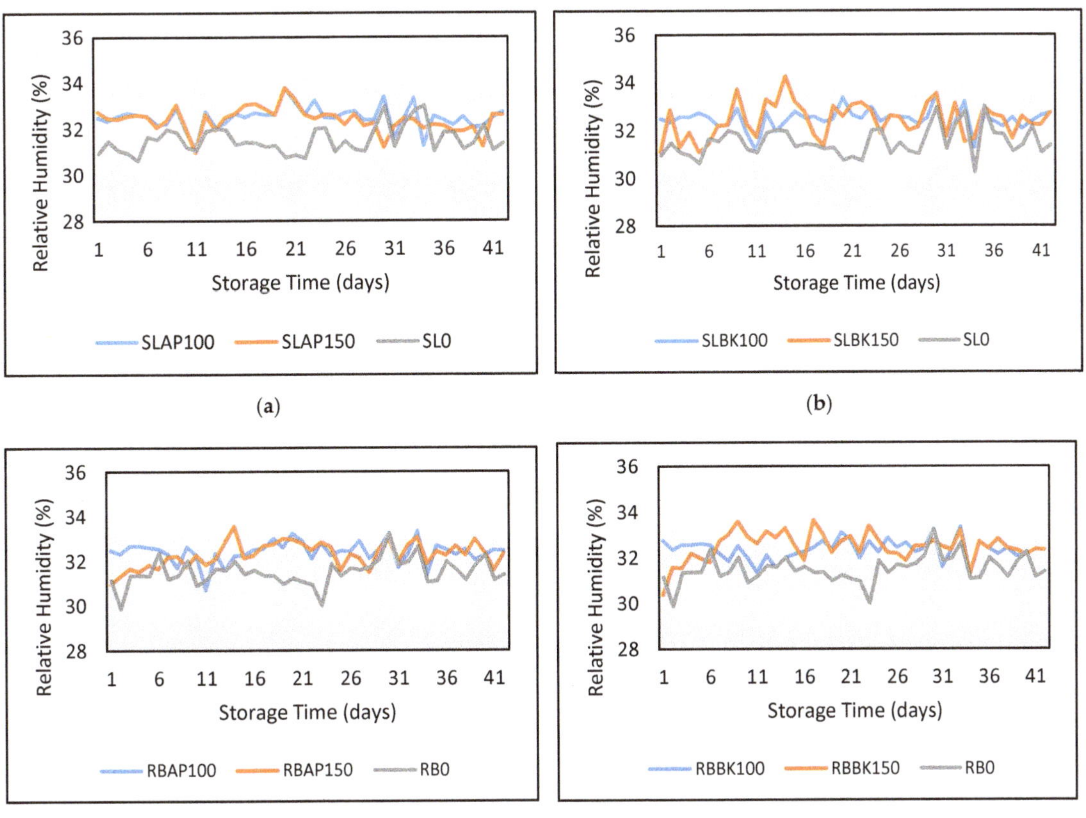

Figure 6. *Cont.*

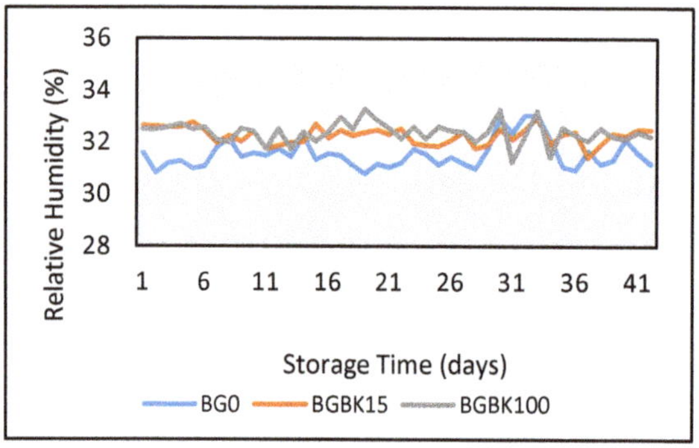

(**e**)

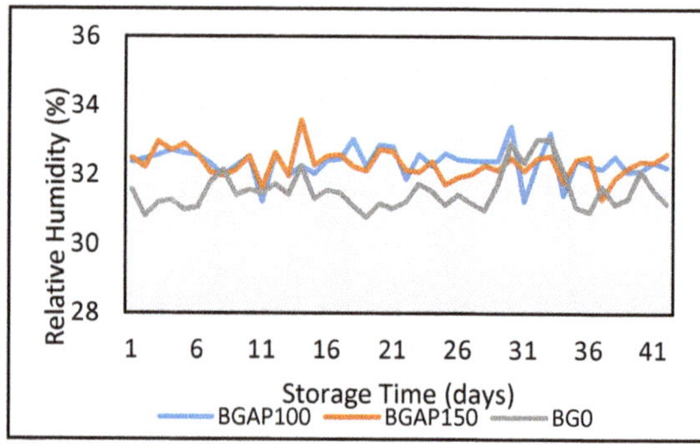

(**f**)

Figure 6. Relative humidity of different storage methods and treatments. (**a**) Breadfruit stored in the silo and treated with aligator pepper. (**b**) Breadfruit stored in the silo and treated with bitter kola. (**c**) Breadfruit stored in the rubber and treated with aligator pepper. (**d**) Breadfruit stored in the rubber and treated with bitter kola. (**e**) Breadfruit stored in the bag and treated with aligator pepper. (**f**) Breadfruit stored in the bag and treated with bitter kola.

4.2. Insect Enumeration

At the beginning of the storage period, no insect infestations were observed, but after four weeks, *Tribolium casteneum* (Figure 7) was observed in all storage strategies except for the untreated and alligator pepper-treated breadfruits stored in silo, as shown in Table 2. These values significantly ($p < 0.05$) increased by 12 weeks of storage, with infestations now noticed in all storage strategies. However, during the counting of the insects, it was observed that not all the insects were alive, even though the ratio of dead and living insects was not separated; the total insects in the infestations were enumerated together. Quantifying the ratio of dead and living insects for future insect mortality prediction in the treated samples could be a future gap to be filled.

Nonetheless, silo storage treatment with alligator pepper demonstrated lower total insect infestation, followed by rubber containers, while infestations were very high in bag storage. It has been reported that metal silos can eliminate insects from the stored product due to a shortage of oxygen [19]. However, since the silo is not fully filled and there is the possibility of space between the bearing and the stirring shaft, air could have been introduced into the silo. This is because as little as 15% oxygen is enough for insects to thrive [55]. Therefore, from this study, we deduce that none of the storage strategies could stop the proliferation of insects beyond four weeks.

Figure 7. *Tribolium casteneum* on the stored breadfruit seeds after four weeks.

Table 2. Average live and dead insects (*Tribolium casteneum*) for different storage strategies for selected weeks.

Sample	4 Weeks	12 Weeks
SL_0	0.0 ± 0.0	3.3 ± 0.8
$SLAP_{100}$	0.0 ± 0.0	4.5 ± 0.5
$SLAP_{150}$	0.0 ± 0.0	3.0 ± 0.4
$SLBK_{100}$	3.2 ± 1.5	8.2 ± 2.5
$SLBK_{150}$	3.4 ± 1.2	7.4 ± 0.9
RB_0	2.1 ± 0.5	4.0 ± 1.5
$RBAP_{100}$	2.4 ± 1.1	5.0 ± 1.5
$RBAP_{150}$	5.2 ± 1.5	9.4 ± 0.5
$RBBK_{100}$	4.4 ± 1.1	10.3 ± 3.6
$RBBK_{150}$	2.0 ± 0.3	6.2 ± 0.4
BG_0	3.0 ± 0.7	6.0 ± 0.2
$BGAP_{100}$	4.1 ± 1.5	6.0 ± 1.5
$BGAP_{150}$	2.8 ± 0.6	6.4 ± 0.7
$BGBK_{100}$	2.0 ± 0.2	7.3 ± 1.6
$BGBK_{150}$	3.2 ± 0.4	9.4 ± 1.8

4.3. Microbial Analysis

The mold counts after 4, 8 and 12 weeks of storage are presented in Table 3. The mold counts increased as the weeks went by. While the increase in carbon dioxide or the lower oxygen levels slowed the insect infestation on a modified atmosphere-stored product, this was not efficient in stopping mold proliferation. However, the mold's ability to attack the product tissues can be delayed [48,56]. From the results presented in Table 3, we see that treating the product with 100 g of alligator pepper and bitter kola increased the mold count. However, when the treatments were increased to 150 g, the fungi count decreased under the same storage condition. However, non-treated bag storage had the highest average mold count of 1.093×10^3 CFU/mL, while silo-stored breadfruit treated with 150 g alligator pepper had the lowet mold count of 0.26×10^3 CFU/mL. The lower mold count observed in alligator pepper-treated samples can be associated with its lower moisture content compared to others, as shown in Figure 4. When products treated with 100 g botanicals were compared to non-treated products for silo and plastic container storage, we saw that non-treated products had lower mold count. This implies that the botanicals, being biological material, are subject to decay over time, and might have contributed to the increased mold count. Three mold species were predominantly isolated from the mold analysis for each storage strategy, i.e., *Aspergillus niger*, *Aspergillus sp* and *Rhodotorula sp.* However, *Aspergillus sp,* was predominantly isolated for most of the storage methods and treatments. The predominance of *Rhodotorula sp* in the plastic container ($RBAP_{100}$), unlike other storage methods, might be due to its opportunistic infection nature and affinity to plastic, as reported in the literature [57]. However, none of the storage strategies could stop mold growth after four weeks of storage completely.

Table 3. Average fungi counts for different storage strategies and dominant mold species detected.

Sample	Mean Mold Count ($\times 10^3$ CFU/mL)	Dominant Mold Specie Detected
SL_0	0.443 ± 0.00312	*Aspergillus niger*
$SLAP_{100}$	0.577 ± 0.00249	*Aspergillus niger*
$SLAP_{150}$	0.26 ± 0.00196	*Aspergillus sp*
$SLBK_{100}$	0.49 ± 0.00330	*Aspergillus niger*
$SLBK_{150}$	0.33 ± 0.00312	*Aspergillus sp*
RB_0	0.347 ± 0.00220	*Aspergillus niger*
$RBAP_{100}$	0.703 ± 0.00401	*Rhodotorula sp*
$RBAP_{150}$	0.677 ± 0.00596	*Aspergillus sp*

Table 3. *Cont.*

Sample	Mean Mold Count ($\times 10^3$ CFU/mL)	Dominant Mold Specie Detected
$RBBK_{100}$	0.833 ± 0.00605	*Aspergillus sp*
$RBBK_{150}$	0.697 ± 0.00420	*Aspergillus sp*
BG_0	1.093 ± 0.00681	*Aspergillus niger*
$BGAP_{100}$	0.93 ± 0.0040	*Aspergillus sp*
$BGAP_{150}$	0.36 ± 0.00157	*Aspergillus sp*
$BGBK_{100}$	0.283 ± 0.00159	*Aspergillus sp*
$BGBK_{150}$	0.59 ± 0.00467	*Aspergillus sp*

4.4. Proximate Analysis and Mineral Composition

Proximate composition assessment was carried out at the beginning, and after 6 weeks and 12 weeks of storage, and the concentrations of crude protein, fat, ash, and carbohydrates are presented in Table 4. The maximum decrease in crude protein concentration from the initial value was about 30.22% in the silo storage-treated samples with and without alligator pepper, and the reduction was about 75.27% for those treated with bitter kola within the first six weeks of storage. A similar magnitude of depreciation in crude protein was observed for other treatments with the same storage methods. Storage time influenced the nutritional composition. The chemical process occurs during the storage of crops due to either respiration or chemical decomposition, reducing the nutritional values [58]. Among storage methods with the same types of botanical treatments, crude protein values were not significant ($p < 0.05$), but they were significant ($p > 0.05$) when compared among the different botanical treatments of the same storage method (Table 4). The crude fiber increased for all storage strategies, while crude fat decreased. The loss in crude fat for botanical treatment with and without alligator pepper was steeper, reaching a 61.42 to 65.64% loss in crude fat for all storage strategies. Samples treated with bitter kola lost between 25.4 and 34.84%, which shows that the bioconversion effect of bitter kola on crude fat is high, and might have contributed to the smaller reduction observed than in the non-treated samples. The ash content increased for those treated with or without alligator pepper, but decreased for those treated with bitter kola. However, the carbohydrate and crude fiber contents did not significantly ($p > 0.05$) change for all treatments. The mineral compositions of sodium (Na), magnesium (Mg), calcium (Ca), phosphorus (P), potassium (K), iron (Fe) and zinc (Zn) of the stored products remained inferior to the initial values, except for phosphorous, which increased as shown in Table 5. The values of P, K, and Fe were higher in the alligator pepper-treated products stored in the silo than in other storage methods. This might be due to more negligible microbial and fungi attacks on products treated with alligator pepper. Alligator pepper contains alkaloids, tannins, saponin, steroids, cardiacglycoside, flavonoid and terpenoids, among other antimicrobial and antifungal agents [27]. Therefore, due to the antimicrobial and antifungal potency, alligator pepper-treated products stored in the silo will contain higher mineral contents than products stored using other methods in this study. Similar observations have been made by Ubani et al. [59] in the storage of oilseeds. An adequately stored oil seed with 6–8% moisture content always shows chemical and mineral stability [59].

Table 4. Proximate composition of stored breadfruits with different storage strategies.

Storage Method	Crude Protein (%)	Crude Fiber (%)	Crude Fat (%)	Ash (%)	Carbohydrates (%)	KcalEV
Initial values (0 weeks)	10.72 ± 1.5	4.69 ± 1.15	6.17 ± 1.02	2.21 ± 0.08	78.11 ± 1.5	342.79 ± 23.3
6 weeks						
SL_0	7.48 ± 0.15	5.20 ± 0.35	2.38 ± 0.00	2.42 ± 0.10	75.40 ± 3.3	352.94 ± 18.9
$SLAP_{100}$	7.49 ± 0.02	5.21 ± 0.45	2.39 ± 0.51	2.43 ± 0.00	75.39 ± 2.5	352.95 ± 34.2
$SLAP_{150}$	7.49 ± 0.11	5.22 ± 0.05	2.39 ± 0.45	2.43 ± 0.05	75.37 ± 4.1	352.95 ± 40.3
$SLBK_{100}$	2.66 ± 0.05	5.15 ± 0.85	4.61 ± 0.15	1.18 ± 0.01	76.19 ± 6.8	356.87 ± 42.2
$SLBK_{150}$	2.65 ± 0.05	5.36 ± 0.63	4.6 ± 0.72	1.17 ± 0.01	76.21 ± 4.5	356.84 ± 52.0
RB_0	7.26 ± 0.24	5.14 ± 0.44	2.32 ± 0.43	2.36 ± 0.04	74.51 ± 2.5	347.96 ± 11.20
$RBAP_{100}$	7.24 ± 0.35	5.13 ± 0.41	2.33 ± 0.11	2.35 ± 0.05	74.54 ± 3.7	348.03 ± 1.80
$RBAP_{150}$	7.22 ± 0.05	5.12 ± 0.51	2.33 ± 0.86	2.34 ± 0.06	74.56 ± 4.3	348.09 ± 11.11
$RBBK_{100}$	2.48 ± 0.15	5.06 ± 0.53	4.55 ± 0.54	1.15 ± 0.01	74.42 ± 8.5	348.49 ± 18.9
$RBBK_{150}$	2.47 ± 0.01	5.09 ± 0.51	4.56 ± 0.51	1.16 ± 0.02	74.40 ± 6.5	348.52 ± 36.7
BG_0	7.20 ± 0.16	5.1 ± 0.57	2.24 ± 0.36	2.3 ± 0.00	74.10 ± 5.7	345.36 ± 44.4
$BGAP_{100}$	7.22 ± 0.22	5.11 ± 0.41	2.23 ± 0.29	2.31 ± 0.10	74.10 ± 7.5	345.31 ± 51.9
$BGAP_{150}$	7.23 ± 0.42	5.11 ± 0.81	2.22 ± 0.33	2.31 ± 0.08	74.09 ± 5.5	345.26 ± 45.8
$BGBK_{100}$	2.42 ± 0.02	4.92 ± 0.64	4.41 ± 0.22	1.13 ± 0.02	72.00 ± 3.6	337.29 ± 61.8
$BGBK_{150}$	2.43 ± 0.02	4.92 ± 0.53	4.4 ± 0.51	1.14 ± 0.02	71.99 ± 1.8	337.28 ± 21.1
12 weeks						
SL_0	9.14 ± 0.11	5.49 ± 0.27	2.27 ± 0.08	2.53 ± 0.12	76.68 ± 8.5	363.71 ± 5.59
$SLAP_{100}$	9.15 ± 0.08	5.50 ± 0.47	2.28 ± 0.04	2.53 ± 0.05	76.69 ± 4.5	363.86 ± 2.51
$SLAP_{150}$	9.16 ± 1.01	5.5 ± 0.52	2.29 ± 0.27	2.52 ± 0.03	76.69 ± 6.3	364.01 ± 6.53
$SLBK_{100}$	3.21 ± 0.07	5.41 ± 0.35	4.27 ± 0.22	1.7 ± 0.01	78.31 ± 8.8	364.45 ± 10.1
$SLBK_{150}$	3.2 ± 0.030	5.41 ± 0.05	4.25 ± 0.31	1.71 ± 0.05	78.31 ± 6.1	364.29 ± 18.24
RB_0	9.03 ± 0.080	5.37 ± 0.75	2.21 ± 0.44	2.3 ± 0.06	76.06 ± 7.4	360.25 ± 27.7
$RBAP_{100}$	9.04 ± 1.12	5.38 ± 0.11	2.21 ± 0.18	2.32 ± 0.01	76.05 ± 4.6	360.19 ± 8.8
$RBAP_{150}$	9.04 ± 1.09	5.39 ± 0.64	2.2 ± 0.59	2.33 ± 0.01	76.04 ± 9.2	360.12 ± 38.2
$RBBK_{100}$	3.15 ± 0.02	5.32 ± 0.62	4.17 ± 0.15	1.05 ± 0.01	78.39 ± 4.0	363.69 ± 41.2
$RBBK_{150}$	3.16 ± 0.25	5.31 ± 0.29	4.18 ± 0.52	1.06 ± 0.03	78.38 ± 3.5	363.78 ± 19.34
BG_0	7.05 ± 0.15	5.3 ± 0.880	2.14 ± 0.43	2.14 ± 0.10	77.62 ± 6.1	357.94 ± 10.8
$BGAP_{100}$	7.06 ± 0.15	5.32 ± 0.12	2.13 ± 0.50	2.15 ± 0.02	77.62 ± 7.3	357.90 ± 15.5
$BGAP_{150}$	7.06 ± 1.20	5.33 ± 0.32	2.12 ± 0.59	2.16 ± 0.02	77.62 ± 5.8	357.8 ± 12.35
$BGBK_{100}$	2.21 ± 0.00	5.23 ± 0.09	4.03 ± 0.06	1.01 ± 0.03	78.60 ± 6.2	359.45 ± 28.3
$BGBK_{150}$	2.21 ± 0.03	5.21 ± 0.86	4.02 ± 0.17	1.00 ± 0.02	78.61 ± 1.5	359.46 ± 22.97

Table 5. Variation in the mineral compositions of the stored breadfruits using different strategies.

Storage Method	Na	Mg	Ca	P	K	Fe	Zn
Initial values	9.32 ± 0.85	1.71 ± 0.11	1.40 ± 0.40	1.13 ± 0.01	11.34 ± 2.5	1.88 ± 0.22	2.03 ± 0.42
6 weeks							
SL_0	1.01 ± 0.00	1.14 ± 0.05	1.25 ± 0.13	1.38 ± 0.21	2.06 ± 0.85	0.42 ± 0.10	0.83 ± 0.00
$SLAP_{100}$	1.02 ± 0.00	1.14 ± 0.22	1.25 ± 0.10	1.38 ± 0.20	2.05 ± 0.25	0.41 ± 0.00	0.82 ± 0.04
$SLAP_{150}$	1.02 ± 0.01	1.13 ± 0.12	1.24 ± 0.13	1.37 ± 0.10	2.04 ± 0.56	0.04 ± 0.01	0.81 ± 0.01
$SLBK_{100}$	1.93 ± 0.01	1.31 ± 0.08	1.86 ± 0.08	4.17 ± 0.73	10.33 ± 3.6	2.38 ± 0.09	0.53 ± 0.01
$SLBK_{150}$	1.92 ± 0.01	1.30 ± 0.22	1.85 ± 0.20	4.15 ± 0.87	10.31 ± 2.5	2.37 ± 0.05	0.52 ± 0.00
RB_0	0.92 ± 0.02	1.10 ± 0.05	1.20 ± 0.00	1.29 ± 0.02	2.01 ± 0.08	0.40 ± 0.04	0.80 ± 0.01
$RBAP_{100}$	0.92 ± 0.03	1.13 ± 0.03	1.21 ± 0.84	1.28 ± 0.02	2.02 ± 0.50	0.41 ± 0.05	0.80 ± 0.00
$RBAP_{150}$	0.91 ± 0.01	1.13 ± 0.15	1.22 ± 0.25	1.27 ± 0.05	2.02 ± 0.15	0.42 ± 0.05	0.80 ± 0.00
$RBBK_{100}$	1.92 ± 0.05	1.27 ± 0.06	1.82 ± 0.09	4.01 ± 1.50	10.12 ± 3.33	2.31 ± 0.25	0.52 ± 0.10
$RBBK_{150}$	1.93 ± 0.03	1.27 ± 0.18	1.81 ± 0.16	4.0 ± 1.55	10.13 ± 2.50	2.32 ± 0.21	0.52 ± 0.00
BG_0	0.84 ± 0.10	1.10 ± 0.11	1.12 ± 0.08	1.20 ± 0.11	1.93 ± 0.25	0.32 ± 0.00	0.72 ± 0.08
$BGAP_{100}$	0.83 ± 0.03	1.11 ± 0.01	1.12 ± 0.33	1.22 ± 0.00	1.92 ± 0.53	0.32 ± 0.00	0.72 ± 0.11

Table 5. *Cont.*

Storage Method	Na	Mg	Ca	P	K	Fe	Zn
$BGAP_{150}$	0.82 ± 0.02	1.12 ± 0.07	1.12 ± 0.06	1.23 ± 0.02	1.91 ± 0.51	0.31 ± 0.00	0.71 ± 0.13
$BGBK_{100}$	1.81 ± 0.04	1.27 ± 0.08	1.81 ± 0.06	3.92 ± 0.84	10.02 ± 1.5	2.24 ± 0.80	0.46 ± 0.01
$BGBK_{150}$	1.80 ± 0.02	1.26 ± 0.11	1.82 ± 0.12	3.92 ± 0.11	10.01 ± 2.5	2.23 ± 0.53	0.46 ± 0.00
12 weeks							
SL_0	1.02 ± 0.00	1.13 ± 0.05	1.25 ± 0.13	1.34 ± 0.21	2.03 ± 0.85	0.43 ± 0.10	0.81 ± 0.00
$SLAP_{100}$	1.07 ± 0.00	1.21 ± 0.17	1.35 ± 0.34	1.61 ± 0.04	2.17 ± 0.35	0.44 ± 0.00	0.88 ± 0..23
$SLAP_{150}$	1.06 ± 0.01	1.20 ± 0.16	1.34 ± 0.18	1.60 ± 0.06	2.15 ± 0.80	0.43 ± 0.00	0.88 ± 0.01
$SLBK_{100}$	2.08 ± 0.00	1.36 ± 0.31	1.95 ± 0.19	4.31 ± 1.80	10.82 ± 2.11	2.45 ± 0.06	0.61 ± 0.02
$SLBK_{150}$	2.07 ± 0.01	1.34 ± 0.05	1.94 ± 0.28	4.32 ± 0.04	10.82 ± 2.6	2.44 ± 0.04	0.61 ± 0.06
RB_0	0.92 ± 0.02	1.12 ± 0.05	1.20 ± 0.00	1.29 ± 0.02	2.01 ± 0.08	0.40 ± 0.04	0.80 ± 0.01
$RBAP_{100}$	0.92 ± 0.08	1.12 ± 0.00	1.18 ± 0.03	1.24 ± 0.05	1.94 ± 0.03	0.35 ± 0.00	0.83 ± 0.01
$RBAP_{150}$	0.91 ± 0.2	1.10 ± 0.08	1.17 ± 0.03	1.22 ± 0.00	1.93 ± 0.11	0.33 ± 0.01	0.82 ± 0.01
$RBBK_{100}$	2.01 ± 0.6	1.22 ± 0.16	1.82 ± 0.12	3.92 ± 0.94	9.87 ± 1.34	2.20 ± 0.12	0.46 ± 0.00
$RBBK_{150}$	2.01 ± 0.08	1.21 ± 0.08	1.83 ± 0.17	3.10 ± 0.11	9.85 ± 0.98	2.20 ± 0.43	0.46 ± 0.01
BG_0	0.84 ± 0.10	1.10 ± 0.11	1.12 ± 0.08	1.20 ± 0.11	1.93 ± 0.25	0.32 ± 0.00	0.72 ± 0.08
$BGAP_{100}$	0.75 ± 0.23	0.84 ± 0.13	1.01 ± 0.03	1.11 ± 0.02	1.72 ± 0.44	0.25 ± 0.00	0.64 ± 0.09
$BGAP_{150}$	0.73 ± 0.11	0.83 ± 0.20	1.00 ± 0.00	1.10 ± 0.07	0.72 ± 0.35	0.24 ± 0.02	0.64 ± 0.11
$BGBK_{100}$	1.63 ± 0.21	1.04 ± 0.00	1.63 ± 0.01	3.62 ± 1.1	8.68 ± 2.87	2.02 ± 0.50	0.40 ± 0.05
$BGBK_{150}$	1.64 ± 0.10	1.02 ± 0.02	1.62 ± 0.02	3.61 ± 0.03	8.66 ± 2.08	2.00 ± 0.42	0.40 ± 0.02

4.5. Analysis of Gaseous Exchange

The gaseous concentrations in the stored product showed changes as the days progressed. The CO_2 concentration at the beginning of storage was 0, but increased to 0.13–0.14 (% v/v) in the first week of storage, as shown in Figure 8A. The CO_2 concentration ranged from 0.16 to 0.47 (% v/v) at the sixth weeks, and 0.57 to 3.98 (% v/v) in the twelfth week for all treatments in the silo bin storage. Other storage strategies ranged from 0.12–1.09 (% v/v) to 1.09–9.6 (% v/v) for plastic container, and 0.19–6.55 (% v/v) to 5.12–10.81 (% v/v) for silo bag, respectively. The concentration of CO_2 was lower in the silo bin-treated samples with 150 g of alligator pepper, and higher in the silo bag-treated samples with 100 g bitter kola. A higher CO_2 concentration results in limited oxygen availability as more oxygen is consumed, higher water vapor production, and a higher heat release rate, as shown in Figure 8B–D for blanched and dehulled breadfruits storage. The concentration of CO_2 and O_2 availability depends on the respiratory capacity of the stored seed, the infiltration of external O_2 and the loss of CO_2 to the storage building [41].

Storage strategies with higher moisture content showed higher CO_2 concentration, higher O_2 consumption with more water vapor, and higher energy release rate. As such, a greater tendency to spoilage from fungi infestation could have resulted in the higher fungi count observed in Table 2. This is in tandem with other similar studies that reported that higher moisture content increases the CO_2 concentration of stored bio-products [48,60]. In the first week of storage, the moisture content was almost the same. The observed CO_2 concentration showed no significant difference ($p < 0.05$), but a significant difference was observed as the storage time increased to 6 weeks and 12 weeks. Higher respiration occurs due to the increased moisture content and depletion of oxygen, as observed in Figure 8B,C. For the weeks analyzed, respiration reached its peak at 6 weeks and decreased in the 12th week, which was reflected in the high oxygen depletion peaking in the same week (6) before going down in week 12 for all storage strategies. Due to the lower moisture content of the silo bin with alligator pepper, they also had lower respiration heat and CO_2 concentration values. At the same time, the bag storage method had a higher CO_2 concentration due to the higher moisture content, which resulted probably from environmental interference.

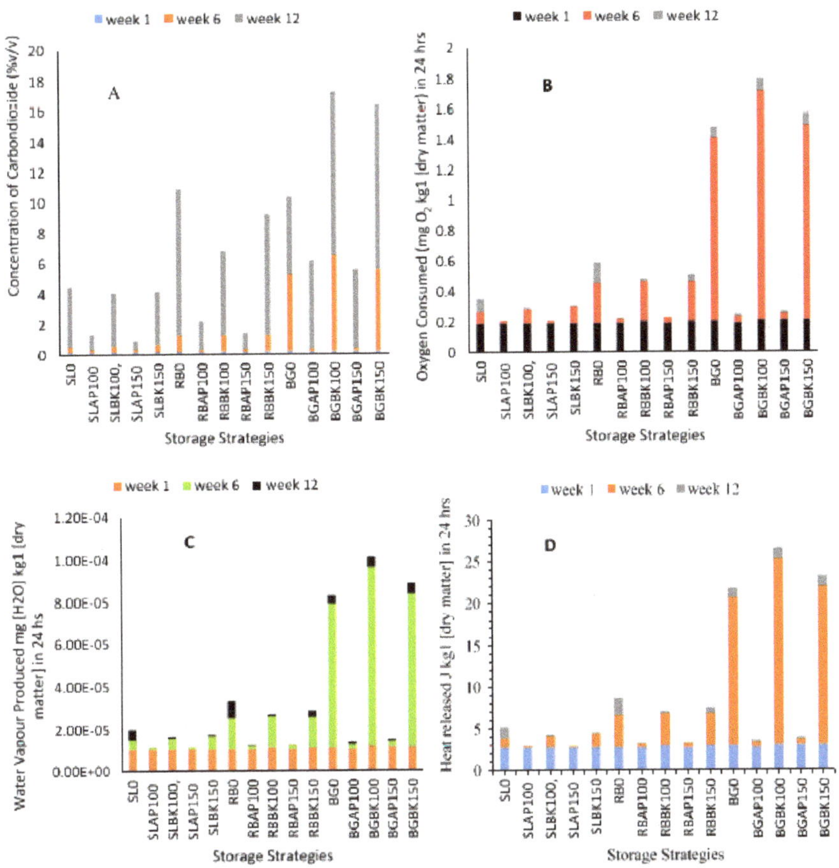

Figure 8. Gaseous exchange of breadfruit seeds for different storage strategies. (**A**) CO_2 concentration. (**B**) O_2 consumption. (**C**) Water vaper production. (**D**) Heat release.

5. Conclusions

The present study evaluates the effectiveness of different local storage methods and botanical treatments to preserve blanched, dehulled, and dried breadfruit seeds under small-scale household conditions in a tropical rainforest environment with very high ambient humidity. Non-treated bag storage had the highest average mold count of 1.093×10^2 CFU/mL, while silo bin-stored breadfruit treated with 150 g of alligator pepper had the lowest mold count of 0.26×10^2 CFU/mL. *Aspergillus niger*, *Aspergillus sp* and *Rhodotorula sp* were the predominant mold species isolated; however, *Aspergillus sp* was dominant. The storage time influenced both the nutritional and mineral quality; regardless, products stored in a plastic container and weaved silo bags were inferior compared to those stored in the aluminum silo bin. Insect infestations were low in alligator pepper-treated seeds, as the seed continued to desorb moisture in storage, unlike other treatments. Silo bins had a comparatively lower *Tribolium casteneum* attack, and this did not happen until after 2 months. The concentration of CO_2 was lower in the silo bin samples treated with 150 g alligator pepper, and higher in the silo bag-treated samples with 100 g bitter kola nut. A higher CO_2 concentration resulted in limited oxygen availability, higher water vapor production, and a higher heat release rate. The analysis of the various results showed that adopting the mini silo bin storage with treated alligator pepper can help in the short-term storage of blanched breadfruit, achieving better quality and an extended shelf-life, which most households aim to achieve.

Author Contributions: M.C.N.: conceptualization, methodology, supervision, project administration, data analysis, writing—original draft, review and editing. D.I.O.: methodology, review and editing. J.E.: review and editing: U.C.A.: review and editing. I.E.E.: review and editing. G.C.: review and editing. All authors have read and agreed to the published version of the manuscript.

Funding: This research received no external funding.

Institutional Review Board Statement: Not applicable.

Informed Consent Statement: Not applicable.

Data Availability Statement: Data is contained within the article.

Acknowledgments: The authors acknowledge Ndkuwu's 2018/2019 undergraduate project students and Postgraduate Diploma students for helping in data collection for this project.

Conflicts of Interest: The authors declare no conflict of interest.

References

1. Onyekwelu, J.C.; Fayose, O.J. Effect of storage methods on germination and proximate composition of Treculia africana seeds. In Proceedings of the Conference on International Agricultural Research for Development, University of Kassel-I-Witzen-hausen and University of Gottigen, Tropentag, October 2007.
2. Osuji, J.O.; Owei, S.D. Mitotic index studies on Treculia africana Decne in Nigeria. *Aust. J. Agric. Eng.* **2010**, *1*, 25–28.
3. Osukoya, O.; Nwoye-Ossy, M.; Olayide, I.; Ojo, O.; Adewale, O.; Kuku, A. Antioxidant activities of peptide hydrolysates obtained from the seeds of Treculia africana Decne (African breadfruit). *Prep. Biochem. Biotechnol.* **2020**, *50*, 504–510. [CrossRef]
4. Adewole, E.; Ogunmodede, O.T.; Adeniran, O.A.; Talabi, J.; Lajide, L. Rheological properties and fatty acid of seedless bread fruit (Artocarpus altilis). *Arch. Appl. Sci. Res.* **2011**, *3*, 80–86.
5. Okafor, V.C.; Chinweuba, D.C.; Ejesu, P.K.; Nwakuba, R.N. Some Selected Physical Properties of Parboiled Breadfruit Seeds (Treculia africana). *Greener J. Sci. Eng. Technol. Res.* **2016**, *6*, 19–26. [CrossRef]
6. Adindu, M.N.; Williams, J.O. Effect of storage on dehydrated African breadfruit seeds (Treculia africana Decne). *Plant Foods Hum. Nutr.* **2004**, *58*, 1–8. [CrossRef]
7. Nwabueze, T. Kernel extraction and machine efficiency in dehulling parboiled african breadfruit (Treculia africanadecne) whole seeds. *J. Food Qual.* **2009**, *32*, 669–683. [CrossRef]
8. Ugwu, C.S.; Iwuchukwu, J.C. Processing and preservation of African bread fruit (Treculia africana) by women in Enugu North agricultural zone, Enugu State, Nigeria. *Afr. J. Agric. Res.* **2013**, *8*, 984–994. [CrossRef]
9. Ijeh, I.I.; Ejike, E.C.; Nkwonta, O.M.; Njoku, B.C. Effect of Traditional Processing Techniques on the Nutri-tional and Phytochemical Composition of African Bread-Fruit (Treculia africana) Seeds. *J. Appl. Sci. Environ. Manag.* **2010**, *14*, 169–173.
10. Mugo, N.C.; Nderi, M.J.; Kimondo, M. Effect of Diatomaceous Earths on Mortality, Progeny and Weight Loss Caused by Three Primary Pests of Maize and Wheat in Kenya. *Int. J. Sci. Technol. Soc.* **2015**, *3*, 76. [CrossRef]
11. Amadou, L.; Baoua, I.B.; Baributsa, D.; Williams, S.B.; Murdock, L.L. Triple bag hermetic technology for control-ling a bruchid (Spermophagus sp.) (Coleoptera: Chrysomelidae) in stored Hibiscus sabdariffa grain. *J. Stored Prod. Res.* **2016**, *69*, 22–25. [CrossRef]
12. Nwaubani, S.I.; Otitodun, G.O.; Ajao, S.K.; Opit, G.P.; Ala, A.A.; Omobowale, M.O.; Ogwumike, J.C.; Abel, G.I.; Ogundare, M.O.; Braimah, J.A.; et al. Assessing efficacies of in-sect pest management methods for stored bagged maize preservation in storehouses located in Nigerian markets. *J. Stored Prod. Res.* **2020**, *86*, 101566. [CrossRef]
13. Ramzan, M. Efficacy of edible oils against pulse beetle, callosobruchus maculatus. *J. Insect Sci.* **1994**, *7*, 37–39.
14. Uddin, R.O.; Abdulazeez, R.W. Comparative efficacy of neem (Azadirachta indica), False sesame (Ceratotheca sesamoides) Endl. And the Physic nut (Jatropha curcas) in the protection of stored cowpea (Vigna unguiculata) L. Walp against the seed beetle Callosobruchus maculatus (F.). *Ethiop. J. Environ. Stud. Manag.* **2013**, *6*, 827–834. [CrossRef]
15. Kalsa, K.K.; Subramanyam, B.; Demissie, G.; Mahroof, R.; Gabbiye, N. Efficacy of filter cake against Sitophilus granarius L. and Rhyzopertha Dominica F. In stored wheat. In Proceedings of the 1st All Africa Post-harvest Congress and Exhibition, Nairobi, Kenya, 29 March 2017; pp. 118–120.
16. Adarkwah, C.; Obeng-Ofori, D.; Ulrichs, C.; Schöller, M. Insecticidal efficacy of botanical food by-products against selected stored-grain beetles by the combined action with modified diatomaceous earth. *J. Plant Dis. Prot.* **2016**, *124*, 255–267. [CrossRef]
17. Tadesse, T.M.; Subramanyam, B. Efficacy of filter cake and Triplex powders from Ethiopia applied to concrete arenas against Sitophilus zeamais. *J. Stored Prod. Res.* **2018**, *76*, 140–150. [CrossRef]
18. Kalsa, K.K.; Subramanyam, B.; Demissie, G.; Mahroof, R.; Worku, A.; Gabbiye, N. Evaluation of postharvest preservation strategies for stored wheat seed in Ethiopia. *J. Stored Prod. Res.* **2019**, *81*, 53–61. [CrossRef]

19. Korunić, Z.; Liška, A.; Lucić, P.; Hamel, D.; Rozman, V. Evaluation of diatomaceous earth formulations enhanced with natural products against stored product insects. *J. Stored Prod. Res.* **2020**, *86*, 101565. [CrossRef]
20. Tefera, T.; Mugo, S.; Likhayo, P. Effects of insect population density and storage time on grain damage and weight loss in the due to the maize weevil Sitophilus zeamais and larger grain borer Prostephanus truncatus. *Afr. J. Agric. Res.* **2011**, *6*, 2249–2254.
21. Crop Production and Global Environmental Issues. In *Crop Production and Global Environmental Issues*; Springer International Publishing: Cham, Switzerland, 2015; pp. 153–168.
22. Ndukwu, M.C.; Manuwa, S.I.; Olukunle, O.J.; Oluwalana, I.B. Mathematical model for direct evaporative space cooling systems. *Niger. J. Technol.* **2013**, *32*, 403–409.
23. Ndukwu, C.; Manuwa, S.; Olukunle, O.; Oluwalana, B. A simple model for evaporative cooling of a storage space in a tropical climate. *Sci. J. Agric. Eng.* **2013**, *3*, 27–39.
24. Igboayaka, E.C.; Ndukwu, M.C.; Ernest, I.C. A Modelling approach for determining the throughput capacity and en-ergy consumption of a cassava tuber shredder. *J. Chin. Adv. Mater. Soc.* **2018**, *6*, 801–816. [CrossRef]
25. Onwude, D.I.; Chen, G.; Eke-Emezie, N.; Kabutey, A.; Khaled, A.Y.; Sturm, B. Recent Advances in Reducing Food Losses in the Supply Chain of Fresh Agricultural Produce. *Processes* **2020**, *8*, 1431. [CrossRef]
26. Onunkwo, G.C.; Egeonu, H.C.; Adikwu, M.U.; Ojile, J.E.; Olowosulu, A.K. Some Physical Properties of Tablet-ted Seed of Garcinia kola (HECKEL). *Chem. Pharm. Bull.* **2004**, *52*, 649–653. [CrossRef] [PubMed]
27. Doherty, F.V.; Olaniran, O.O.; Kanife, U.C. Antimicrobial Activities of Aframomum Melegueta (Alligator Pepper). *Int. J. Biol.* **2010**, *2*, 126. [CrossRef]
28. Gbadamosi, I.T.; Moody, J.O.; Lawal, A.M. Phytochemical screening and proximate analysis of eight ethnobotani-cals used as antimalaria remedies in Ibadan, Nigeria. *J. Appl. Biosci.* **2011**, *44*, 2967–2971.
29. Arekemase, M.O.; Aliyu, M.B.; Kayode, R.M.; Ajiboye, A.E.; Ajijolakewu, A.K. Antimicrobial Effects of Garcinia Kola (Bitter Kola) on Some Selected Pathogens from University of Ilorin Teaching Hospital Ilorin, Nigeria. *J. Asian Sci. Res.* **2012**, *2*, 159–169.
30. Ea, M.; Ka, O.; Uk, I. Physico-Chemical and Nutritive Properties of Bitter Kola (Garcinia Kola). *J. Nutr. Food Sci.* **2013**, *3*, 1–4. [CrossRef]
31. Badmos, A.A.; Ahmad-el, I.A.; Annongu, A.A.; Yusuff, A.T.; Kayode, R.M.O.; Salami, K.O.; Ahutu, I.A.; Lawal, A.O. Preservative effects of aqueous and ether extracts of Aframomum melegueta on West African soft cheese. *Bang. J. Anim. Sci.* **2017**, *46*, 51–55. [CrossRef]
32. ISTA. International Rules for seed testing. In *International Seed Testing Association (ISTA)*; Zürichstr. 50; ISTA: Bassersdorf, Switzerland, 2014.
33. Badifu, G.I.O.; Akuboil, P.I.; Akpapunam, M.A. Chemical, Functional and Organoleptic Evaluation of African-Breadfruit (Treculia a.fricanil Ikne) Kernel Flour for Making Cooki. *Tanzan. J. Agric. Sci.* **2000**, *3*, 31–38.
34. Cowan, S.T. Steel, Microscopic Examination of Micro-organisms. In *Microbiological Methods*; Collins: Cambridge, UK, 1990.
35. Chessbrough, M. District Laboratory Practice. In *Tropical Countries*; Cambridge University Press: Cambridge, UK, 2002; Part 2.
36. Association of Official Analytical Chemists (AOAC). *Official Methods of Analysis*, 18th ed.; AOAC: Rockville, MD, USA, 2007.
37. James, C.S. Analytical Chemistry of Foods. *Anal. Chem. Foods* **1995**, 140–144. [CrossRef]
38. Association of Official Analytical Chemists (AOAC). Copper catalyst kjeldahl method (984.13). In *Official Methods of Analysis*, 15th ed.; AOAC: Rockville, MD, USA, 1990.
39. Carpenter, C.E.; Hendriide, D.G. Mineral Analysis. In *Food Analysis*, 3rd ed.; Kluwer Academy, Plenum Publisher: New York, NY, USA, 2003; pp. 191–198.
40. Ensminger, L.G. The Association of Official Analytical Chemists. *Clin. Toxicol.* **1976**, *9*, 471. [CrossRef]
41. Abalone, R.; Gastón, A.; Bartosik, R.E.; Rodríguez, J.C. Mathematical Modelling of Oxygen and Carbon Diox-ide Composition in the Interstitial Atmosphere of Silo-Bags". In Proceedings of the International Commission of Agricultural and Biological Engineers, Section V, Conference "Technology and Management to Increase the Efficiency in Sustainable Agricultural Systems", Rosario, Argentina, 1–4 September 2009.
42. Gastón, A.; Abalone, R.; Bartosik, R.E.; Rodríguez, J.C. Mathematical modelling of heat and moisture transfer of wheat stored in plastic bags (silo-bags). *Biosyst. Eng.* **2009**, *104*, 72–85. [CrossRef]
43. White, N.D.G.; Sinha, R.N.; Muir, W.E. Intergranular carbon dioxide as an indicator of biological activity associ-ated with the spoilage of stored wheat. *Can. Agric. Eng.* **1982**, *24*, 35–42.
44. Simo-Tagne, M.; Ndukwu, M.C.; Zoulalian, A.; Bennamoun, L.; Kifani-Sahban, F.; Rogaume, Y. Numerical analysis and validation of a natural convection mix-mode solar dryer for drying red chilli under variable conditions. *Renew. Energy* **2020**, *151*, 659–673. [CrossRef]
45. Obi, O.F.; Okechukwu, M.E. Parboiling duration effects on physical properties of African breadfruit seed. Agricultural Engineering International. *CIGR J.* **2020**, *22*, 186–193.
46. Abe, T.; Basunia, M.A. Simulation of temperature and moisture content changes during storage of rough rice in cy-lindrical bins owing to weather variability. *J. Agric. Eng. Res.* **1996**, *65*, 223–233. [CrossRef]
47. Fouda, A.; Melikyan, Z. A simplified model for analysis of heat and mass transfer in a direct evaporative cooler. *Appl. Therm. Eng.* **2011**, *31*, 932–936. [CrossRef]
48. Bartosik, R.; Rodríguez, J.; Cardoso, L. Storage of corn, wheat soybean and sunflower in hermetic plastic bags. In Proceedings of the 2008 International Grain Quality & Technology Congress Proceedings, Chicago, IL, USA, 15–18 July 2008.

49. Nithya, U.; Chelladurai, V.; Jayas, D.; White, N. Safe storage guidelines for durum wheat. *J. Stored Prod. Res.* **2011**, *47*, 328–333. [CrossRef]
50. Martin, D.T.T.; Baributsa, D.; Huesing, J.E.E.; Williams, S.B.B.; Murdock, L.L.L. PICS bags protect wheat grain, Trit-icum aestivum (L.), against rice weevil, Sitophilus oryzae (L.) (Coleoptera: Curculionidae). *J. Stored Prod. Res.* **2015**, *63*, 22–30. [CrossRef]
51. Angelovič, M.; Krištof, K.; Jobbágy, J.; Findura, P.; Križan, M. The effect of conditions and storage time on course of moisture and temperature of maize grains. *BIO Web Conf.* **2018**, *10*, 02001. [CrossRef]
52. Sun, D.-W.; Woods, J. The Moisture Content/Relative Humidity Equilibrium Relationship of Wheat—A Review. *Dry. Technol.* **1993**, *11*, 1523–1551. [CrossRef]
53. Ben, D.C.; Lieu, B.; Liem, R.; Van, P. Effect of hermetic storage in the super bag on seed quality and milled rice quality of different varieties in Bac Lieu. *Vietnam Agric. Eng.* **2006**, *31*, 55–56. [CrossRef]
54. Njoroge, A.; Affognon, H.; Mutungi, C.; Manono, J.; Lamuka, P.; Murdock, L. Triple bag hermetic storage delivers a lethal punch to Prostephanus truncatus (Horn) (Coleoptera: Bostrichidae) in stored maize. *J. Stored Prod. Res.* **2014**, *58*, 12–19. [CrossRef]
55. Dowell, F.; Dowell, C. Reducing grain storage losses in developing countries. *Qual. Assur. Saf. Crop. Foods* **2017**, *9*, 93–100. [CrossRef]
56. Navarro, S.; Donahaye, J. Innovative environmentally friendly technologies to maintain quality of durable agricultural produce. In *Environmentally Friendly Technologies for Agricultural Produce Quality*; Yeoshua, S.B., Ed.; CRC Press: Boca Ratón, FL, USA, 2005; pp. 203–226.
57. Bakker, N.D. All You Need to Know about Rhodotorula. Available online: https://www.yeastinfection.org/all-you-need-to-know-about-rhodotorula/ (accessed on 20 August 2020).
58. Paraginski, R.T.; Rockenbach, B.A.; Dos Santos, R.F.; Elias, M.C.; De Oliveira, M. Quality of second-crop corn according to the period between harvest and drying. *Rev. Bras. Eng. Agric. Ambient.* **2015**, *19*, 358–363. [CrossRef]
59. Ubani, O.N.; Williams, J.O.; Okpadokun, J.S.; Akano, D.A.; Ikeorah, J.N. *Storage of Melon Seeds in Various Containers. Nigerian Stored Products Research Institute 22nd Annual Report*; Technical Report No. 7; NSPRI: Ilorin, Nigeria, 1991; pp. 55–58.
60. Weinberg, Z.G.; Yan, Y.; Chen, Y.; Finkelman, S.; Ashbell, G.; Navarro, S. The effect of moisture level on high moisture maize (Zea mays L.) under hermetic storage conditions—In vitro studies. *J. Stored Prod. Res.* **2008**, *44*, 136–144. [CrossRef]

processes

MDPI

Article

Enhancing the Shelf-Life of Fresh Cassava Roots: A Field Evaluation of Simple Storage Bags

Keith Tomlins [1], Aditya Parmar [1,*], Celestina Ibitayo Omohimi [2], Lateef Oladimeji Sanni [2], Adekola Felix Adegoke [2], Abdul-Rasaq Adesola Adebowale [2] and Ben Bennett [1]

1 Food and Markets Department, Natural Resources Institute, University of Greenwich, Central Avenue, Chatham Maritime, Kent ME4 4TB, UK; k.i.tomlins@greenwich.ac.uk (K.T.); ben.bennett@greenwich.ac.uk (B.B.)
2 Department of Food Science and Technology, Federal University of Agriculture, PMB 2240, Abeokuta 110001, Nigeria; get2tina2@yahoo.com (C.I.O.); sannilateef5@gmail.com (L.O.S.); saintadekolaadegoke@yahoo.com (A.F.A.); rasaq.debo@gmail.com (A.-R.A.A.)
* Correspondence: a.parmar@greenwich.ac.uk; Tel.: +44-0-1634-883070

Abstract: Postharvest physiological deterioration (PPD) of fresh cassava roots limits their shelf-life to about 48 h. There is a demand for simple, cheap, and logistically feasible solutions for extending the shelf life of fresh cassava roots in industrial processes. In this study, three different types of bag materials were tested, namely woven polypropylene, tarpaulin, and jute as a potential storage solution for cassava roots with different levels of mechanical damage. Microclimate related to temperature, humidity, and carbon di-oxide (CO_2) was monitored in order to understand the storage conditions for up to 12 days. The results showed that fresh cassava roots could be stored for 8 days, with minimal PPD and starch loss (2.4%). However, roots with significant mechanical damage in the form of cuts and breakages had a considerably shorter shelf life in the storage bag, compared to whole roots and roots with retained stalk (peduncle) where roots are connected to the main plant. Wetting of the roots and bag material were not significant factors in determining the shelf life and starch loss. Carbon dioxide concentration in the stores was significantly correlated with the starch loss in fresh cassava roots and is proposed as a possible method for continuously and remotely monitoring starch loss in large-scale commercial operations and reducing postharvest losses.

Keywords: cassava; storage; PPD; starch; shelf-life; postharvest losses

Citation: Tomlins, K.; Parmar, A.; Omohimi, C.I.; Sanni, L.O.; Adegoke, A.F.; Adebowale, A.-R.A.; Bennett, B. Enhancing the Shelf-Life of Fresh Cassava Roots: A Field Evaluation of Simple Storage Bags. *Processes* **2021**, *9*, 577. https://doi.org/doi:10.3390/pr9040577

Academic Editors: Daniel I. Onwude and Guangnan Chen

Received: 9 March 2021
Accepted: 23 March 2021
Published: 26 March 2021

1. Introduction

The short shelf-life of cassava (*Manihot esculenta* C.) roots is primarily attributed to postharvest physiological deterioration (PPD), which is triggered as a wound response shortly after harvest [1–3]. PPD reduces the quality and quantity of starch and renders the cassava roots unmarketable and inedible. PPD is a complex process, and its exact mechanism is still not fully understood [2,4,5]; however, it is known that it involves enzymatic stress responses to wounds and changes in gene expression. PPD can be accompanied by moisture and starch loss [4,6,7]. It results in the formation of blue-black internal root discolouration (vascular streaking) because of the combination of insoluble precipitates formed from scopoletin reacting with hydrogen peroxide. Cassava roots can also suffer from fungal rots. Cassava varieties have been reported to differ in storability [2,8] concerning PPD.

Various attempts have been made in the past to store fresh cassava in various conditions to control PPD and enhance the overall shelf life of fresh cassava roots. Several factors appear to be important for storing roots under ambient tropical conditions; these include curing, cassava variety, chemical treatments, the container, or bag that the roots are stored in, and physical damage to the roots due to harvesting and handling. Curing of the fresh cassava roots is a critical factor; it is the process of wound healing and has been shown to reduce storage losses of slightly damaged roots [4,7]. The optimum conditions for curing are a humidity of 80–85% and a temperature of 25 to 35 °C [9–11]. Previous studies on the storage of fresh cassava roots have reported durations of 11 days [12–14], 11 to 21 days [15], 14 days [16], one month, and two months under cooler conditions. However, these experiments were undertaken in laboratory conditions with small quantities and lacked practical applications in a commercial setting where tons of roots arrive every day at the processing plant. The limited research at a larger scale, such as the storage of 300 kg of roots in Colombia [12,13] and 500 kg in Tanzania, indicates that the roots can store well in larger quantities, but there was no replication and hence the information is limited.

Since cassava is a low-value crop, the challenge is to find a reliable solution at a low cost. The hypothesis for this study was that PPD in fresh cassava can be restricted at scale by the application of a low-cost simple bag storage method along with control measures that might include bag design and the control of moisture, root damage, and variety. To test this, we addressed the following questions for storage at scale:

1. Does the type of storage bag material used to store fresh cassava roots influence storability?
2. Is wetting the roots important?
3. What is the maximum level of fresh root damage (including the peduncle) while still offering acceptable fresh root quality after storage?
4. Does the cassava root variety affect PPD (in industry roots are often mixed, so sorting would be difficult)?
5. What is the impact on starch loss (a critical parameter for commercial-scale operation), and can methods to continuously monitor starch loss during the storage of fresh cassava roots be suggested?

2. Materials and Methods

The research was undertaken during the main harvest period in Nigeria in 2018 and 2019. All experiments were conducted at the Federal University of Agriculture, Abeokuta, Ogun State, Nigeria in an experimental barn for root and tuber crops and at ambient temperature (23–32 °C) and humidity (56–100%). Due to the need to harvest, sort, weigh, transport, and handle large quantities of fresh cassava roots, a stepwise approach to minimize logistical challenges was adopted (Figure 1).

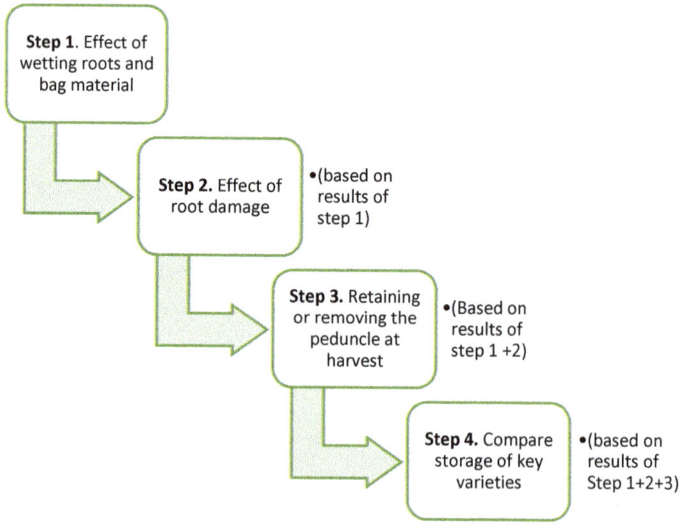

Figure 1. Stepwise progression in experiments to investigate shelf-life at scale.

2.1. Cassava Roots

Fresh roots of variety TME (Tropical Manihot Esculenta) 419 were used in the experiments unless stated otherwise. They were manually harvested and sorted. In the field, exposure of the roots to direct sunlight was avoided and roots were transported from the field to the research station within 2–3 h of harvest. The variety TME 419 was selected to develop the storage method because it is mostly preferred by farmers and commercial industries in Nigeria. It has high yields, higher starch content, and matures early in the growing season [3]. Two additional varieties, namely, TMS (Tropical Manihot Species) 0581 and TMS 1632, were used for the varietal comparison.

2.2. Bag Material and the Effect of Wetting

Three types of bag material were compared: woven polypropylene (1390 mm long by 789 mm wide by 0.7 mm thick), tarpaulin (1180 mm long by 853 mm wide by 0.6 mm thick), and jute (1083 mm long by 823 mm wide by 2.1 mm thick). These were selected based on previous research [13] and availability in Nigeria. Bags were locally purchased. Additionally, fresh roots were either wetted with water (to provide high humidity) or not wetted. Controls were roots heaped on the ground without using any bag materials. Table 1 illustrates the experimental design for bag material and the effect of wetting.

Table 1. Experiment design for bag material and wetting of roots.

Bag Material	Wetted (W)/ Non-Wetted (NW)	Replicates	Weight of Root for Each Replicate (kg)
Woven polypropylene (WP)	Wetted	5	50
	Not wetted	5	50
Tarpaulin (T)	Wetted	5	50
	Not wetted	5	50
Jute (J)	Wetted	5	50
	Not wetted	5	50
Control (no bag) (C)	Wetted	5	50
	Not wetted	5	50

Fresh cassava root samples were either wetted (dipping in clean water) or not wetted. Roots (50 kg) were placed in three bag types, constructed of either polypropylene, jute, or tarpaulin, and control roots (wetted and unwetted) were heaped on the floor. All treatments were replicated five times. Roots from each treatment (5 kg) were randomly sampled at 48 h intervals for weight loss, PPD, fungal rot, and starch content measurements for a duration of up to 12 days.

The carbon dioxide (CO_2), temperature (°C), and relative humidity (RH %) levels were measured using a Rotronics CP11 datalogger (Rotronics Instruments (UK) Ltd., Crawley, West Sussex, UK). They were set to record continuously during the experiments. The data loggers were fitted inside a ventilated plastic tube (68 mm × 300 mm) which was situated in the centre of each sack or a pile of roots.

2.3. Effect of Mechanical Root Damage

Woven polypropylene sacks were filled with 50 kg wetted fresh cassava roots (TME 419) with the following treatments being either one break, two breaks, no breaks but slight bruising, no breaks with extensive bruising, and roots with no breaks and no bruising. The breaks were obtained by cutting roots with a knife. All treatments were replicated five times. Cassava roots (5 kg) were sampled for a period of 12 days, every two days. Table 2 illustrates the categories of mechanical damage.

Table 2. Categories of mechanical damage to cassava roots.

Damage Class	Definition	Pictorial Representation *
One break	One break only and a significant proportion of the root remaining	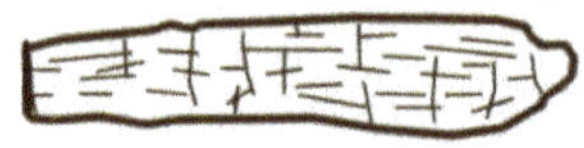
Two breaks	Two breaks in the root but with a significant proportion of the root remaining	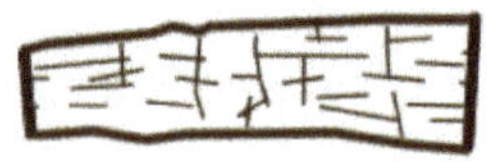
No breaks	The root is intact and has no visible bruising	
Slight bruising	The surface of the root is only slightly bruised and estimated to be 10% or less of the surface area	
Severe bruising	The root is severely bruised and affects most of the visible surface	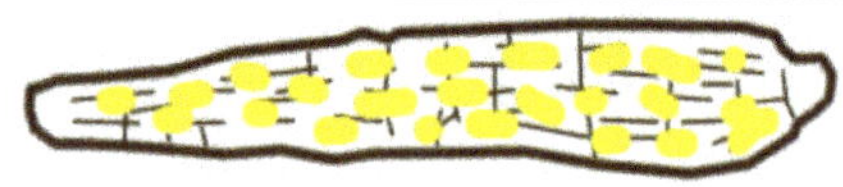

* Where yellow area represents bruising of the surface of the root.

2.4. Harvesting with and without the Peduncle

Woven polypropylene sacks were filled with 50 kg wetted fresh cassava roots (TME 419) with the following treatments being either root with the peduncle still attached or root without (Table 3). Control roots were not stored in sacks. All treatments were replicated five times. Cassava roots (5 kg) were sampled at zero, two, four, six and eight, ten, and twelve days.

Table 3. Cassava roots harvested with and without the peduncle.

Damage Class	Definition	Pictorial Representation
Without peduncle	Stalk bearing cassava roots (attached to the main stem) is removed during harvest	
With peduncle	Stalk bearing cassava roots (attached to the main stem) is retained during harvest	

2.5. Varietal Comparison

Woven polypropylene sacks were filled with 50 kg wetted fresh cassava roots (TMS 0581, TMS 1632, and TME 419) with the following treatments being either root with the peduncle still attached or root without peduncle. Control sample cassava roots were not stored in sacks. All treatments were replicated five times. Cassava roots (5 kg) were sampled at zero, two, four, six, eight, ten, and twelve days.

2.6. Weight Loss, PPD, and Fungal Rot Measurement

The weight of the fresh cassava roots was recorded every 24 h for a period of 12 days to measure the weight loss during storage in different bag materials and without any bags (control). The PPD for each treatment was measured based on seven transversal slices of 2 cm thickness which were cut along the root, starting from the proximal end [7]. A score of 0–1 was assigned to each slice by an expert panel of three people (cassava crop physiologists and postharvest scientists), corresponding to the percentage of the transversal cut surface showing vascular discolouration (which represents PPD) (0.1 = 10%, 0.2 = 20%, etc.). The mean score of PPD for each root was calculated as the final measurement. For fungal rotting incidence, roots were visually scored from 0 to 5 (by the same expert panel), where 0 means no rotting and 5 severe rotting.

2.7. Starch Content Measurement

Starch was measured using the same method as used by the industry in Nigeria, that is, the Reimann Balance (Specific Gravity) Method (1986) (Figure 2). This is recommended by the International Starch Institute, Science Park, Aarhus, Denmark. The calculated data is in per cent, and deviates less than 0.05 from the values readout of the EU table enforced by the European Commission covering potatoes with 13% to 23% starch dry matter. A calibrated Kern CH 15K20 weighing balance was used (RS Components Ltd., Northants, UK).

Figure 2. Reimann Balance (Specific Gravity) Method as constructed at the Federal University of Agriculture, Abeokuta, Nigeria.

2.8. Statistical Analysis

Data were analysed by analysis of covariance (ANCOVA) procedures using R, (Version 3.6.3, R Core team, Vienna, Austria, 2020) and IBM SPSS Statistics, Version 25.0. (Armonk, NY, USA: IBM Corp) statistical packages with the least significant difference (LSD) at $p = 0.05$ and linear regression.

3. Results

3.1. Comparison of the Storage Bag Materials and Wetting

Loading bags with either dry cassava roots or roots that had been wetted with water did not result in any significant differences in root quality concerning PPD, fungal rot, starch content, or root moisture content ($p < 0.05$). Therefore, wet and non-wet scores were combined for the subsequent analyses. This lack of a significant effect of wetting is probably because the humidity within the sacks of cassava roots remained above 85% RH [1] regardless of whether the roots were wetted or not, as illustrated in Figure 3. It is speculated that wetting the roots is probably important to maintain relative humidity above 85% in a small-scale laboratory context due to the small number of roots involved; where larger qualities of roots are involved (in this case 50 kg), the natural respiration is probably sufficient to keep the relative humidity above 85%, as demonstrated in this study.

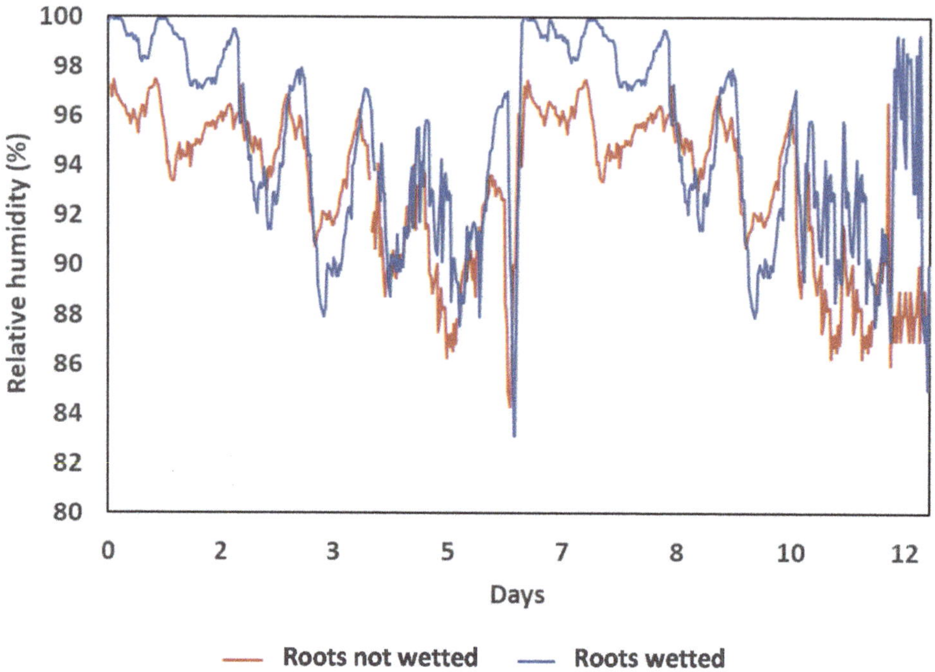

Figure 3. Relative humidity (RH %) within the sacks of cassava roots during storage, either wetted by adding water or not wetted. On average, wetting increased the weight of roots by 0.7%.

Considering the bag type, polypropylene bags resulted in the least PPD compared to the control, but this difference was only significant for the 12 day duration (Figure 4a). Considering the fungal rotting, the bags made from tarpaulin resulted in a significant improvement in shelf-life compared to the control, such that for just perceptible rots (5%) the storage time was 8 days (Figure 4b). After eight days, the rotting was extensive in the roots, and the results for 10 and 12 days storage already exceeded the threshold rotting of 40% of roots and above. The other bag types (jute and tarpaulin) did not differ from the control. Regarding the roots' moisture content, bags made from tarpaulin resulted in the least reduction in moisture content during storage, followed by jute and lastly polypropylene ($p < 0.05$) (Figure 4c). Most of the decline in moisture content was seen after 6 to 8 days. Regarding the starch content, storage in bags led to a lesser loss in starch content than in the control. Although there were significant differences between the bags with respect to starch content change, in practice the differences were small and hence the polypropylene bag was used for other experiments, as this is already used by the industry to transport cassava roots. In general, the decline in starch content started to occur after 2 days, but the loss did not accelerate until eight days of storage (Figure 4d); the starch loss after two days when stored in bags was about 0.3%, and at eight days it was 2.4%. If the roots were not stored in a bag, the starch loss was higher, at 4.3%.

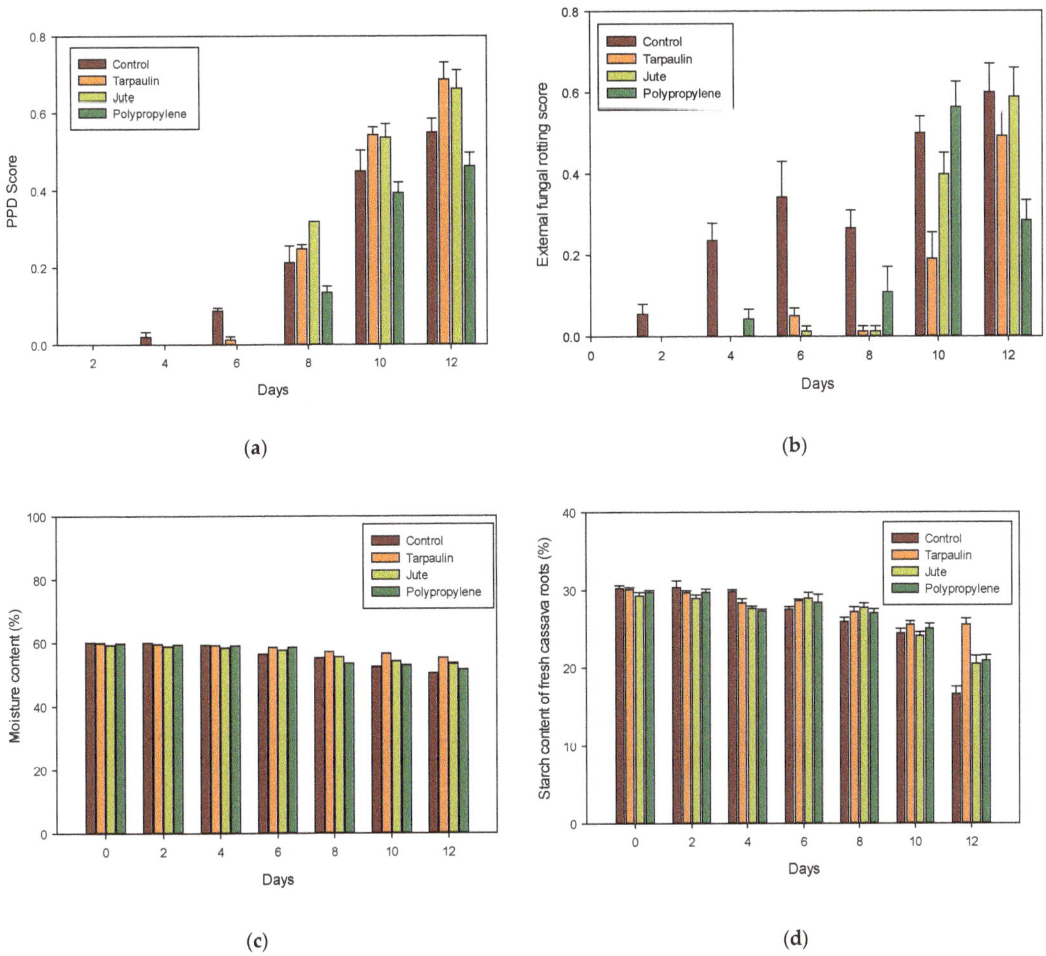

Figure 4. Variation in PPD (Postharvest physiological deterioration) score (**a**), fungal rot (**b**), moisture (**c**), and starch content (**d**) of fresh cassava roots with storage time in different types of bags.

3.2. Effect of Root Damage on Shelf-Life

Considering PPD measurements of the fresh cassava roots, after eight days (Figure 5) only roots with slight bruising or no bruising and no breaks had minimal discolouration. It should be noted that there was a slight increase in discolouration, but this was considered acceptable. However, roots that that were broken (either 1 or 2 breaks) or had extensive bruising resulted in significant discolouration. Regarding the extent of fungal rot, after eight days of storage roots with bruising (slight or extensive) and no breaks did not differ from the control sample (no breaks or bruising). However, the inclusion of extensive bruising was probably due to the wide variation between the replicates. However, any breaks (either 1 or 2) resulted in significant fungal rot. Considering the extent of starch loss after eight days of storage, where the roots were free of breaks or bruising (or had only slight bruising), the starch loss was 2.1% to 2.3%. However, if the roots had any breaks or extensive bruising, the starch loss was higher, at 4.4% to 5.8% on average.

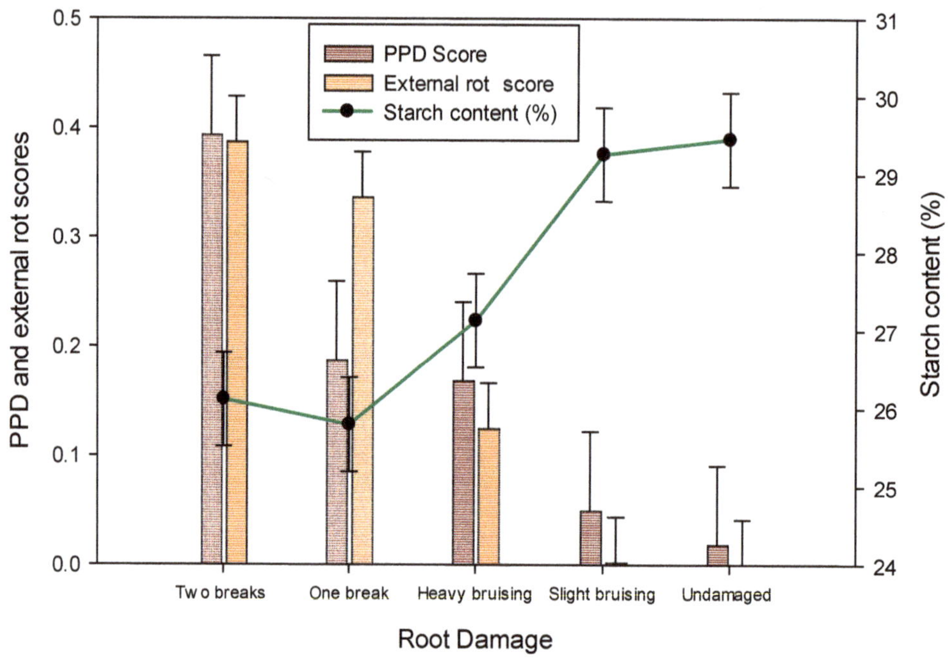

Figure 5. Changes in postharvest physiological deterioration (PPD), fungal rot, and starch content of fresh cassava roots with varying degrees of damage after 8 days of storage.

How the root is disconnected from the stem during harvesting might have an impact on storage. Processors of starch prefer to harvest without the peduncle because this part of the root is more fibrous, but keeping the peduncle attached is thought to increase the shelf-life (unpublished). Harvesting the roots with the peduncle did not have a significant impact on the starch content. It is speculated that there may have been no significant loss in starch content due to cutting without the peduncle because the additional damage to the fresh root was relatively minor, and as shown in the previous section, minor root damage did not adversely affect the shelf-life when storing for up to eight days.

3.3. Effect of Cassava Variety

The cassava variety (TME 419, TMS 0581 and TMS 1632) did not have a significant effect on root quality (PPD, fungal rot, starch content, or weight loss). Previously, weight loss occurring during storage was reported to be about 10% after two weeks of storage [1], but this is the first time that comparisons between these varieties have been reported. The starch content of the three varieties did change during storage, but the initial starch content of the different varieties were also different (Figure 6). Previously, during the storage of fresh roots, the starch loss was observed to occur at a rate of about 1% per day [1]. This research agrees with the loss in the starch of about 1% per day, but also shows that the curve is not linear, with little loss during the first 48 h and then increasing losses after this point. Zainuddin et al. [2] reported that starch losses might vary with variety; in this study, variety did not have an effect. A gap in this research was that possible differences might exist in starch gel clarity and swelling power and gel viscosity between varieties [1]. This would require further investigation.

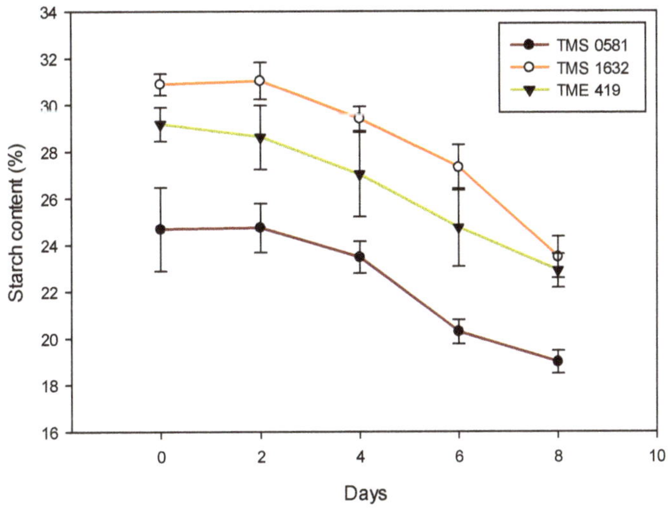

Figure 6. Variation in starch content of three varieties of cassava during storage of 50 kg in sacks.

3.4. CO_2 Concentration and Starch Losses

Root damage is known to increase the rate of respiration in fresh cassava roots due to PPD and wound response along with a corresponding loss in starch [1,9]. This increase in respiration is associated with an increase in CO_2 concentration, temperature, and RH % measured during storage, and could potentially be used to monitor changes in starch content during storage. While root damage did result in an increase in temperature and humidity in the stored fresh cassava roots, this did not significantly correlate with the extent of the starch loss. A significant correlation ($p < 0.05$) was found between starch loss (%) and maximum CO_2 concentration measured at eight days (Figure 7). The maximum CO_2 concentration was measured to account for the diurnal effect. While CO_2 is known to be produced by the respiration of cassava roots, this is the first time that a relationship between CO_2 concentration and starch loss in fresh cassava roots during storage has been reported. This research suggests that CO_2 concentration in a cassava root store might be a suitable means to measure losses in starch, but would require further development.

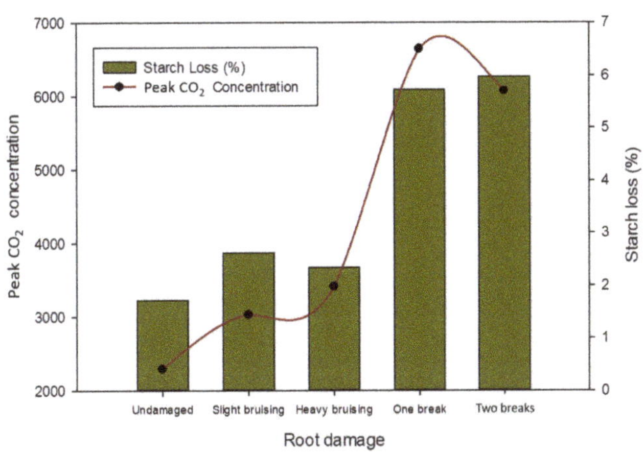

Figure 7. Relationship between starch loss (%) and peak carbon dioxide concentration after eight days of storage.

4. Discussion

Previously, more costly and logistically challenging chemical treatments (oxidising agents such as calcium hypochlorite, disinfectants such as ethanol, and fungicides such as benomyl, dicloran, and thiabendazole) have been used to increase the shelf-life of cassava roots by reducing PPD. Most of these chemicals probably only provide relief from secondary deterioration, which is microbial (mostly fungal rotting). Wax coatings [17,18] have been reported to be successful in storing fresh cassava roots, which has been tested in Uganda. Edible surface coatings of fresh cassava roots are effective in preserving the quality of various perishable food products [19,20]. By using 1.5% xanthan guar/gum as an edible coating, cassava shelf life could be extended by up to 20 days at 25 °C. However, these approaches using chemical treatments were not included in this research because on a commercial scale, these methods would only be practical for high-value markets. Previous research has demonstrated that during the storage of fresh roots, starch loss occurs at a rate of about 1% per day [6] and might vary by variety and storage conditions [21,22]. Reported changes in the starch qualities during storage include a reduction in gel clarity and swelling power and an increase in gel viscosity; it is not known how these changes reflect the quality of the starch or its commercial value [4,23]. In this research, the effect of gel clarity and swelling power was not measured, and this would require further research.

The temperature, RH %, and CO_2 in the sacks were similar regardless of the bag material, and this supports the findings that the bag material and addition of water had little effect. The average temperature varied between 27.1 and 29.1 °C and was within the range of 25 to 35 °C recommended by [12,24], while the average relative humidity varied between 92% and 98%, which is above the value of 85% for curing [25,26]. However, as the ambient conditions for humidity were also high at 74%, the higher humidities were probably acceptable. Previously, different bag materials have been compared for storing fresh cassava roots, such as open weave (manmade fibres), polyethylene (0.13 mm thick), and recycled rice/flour sacks made of tightly woven polyethylene. Of these, the polyethylene and recycled rice sacks were reported to retard deterioration the most [13]. In this research, 50 kg of cassava roots could be kept for up to eight days, irrespective of whether the roots were wet or dry and irrespective of the bag material. The loss in starch content, a key commercial product requirement, was reduced if the fresh roots were stored in bags, and was about 2.4% on average over eight days. This is less than the 8% over eight days (i.e., 1% per day of storage) reported by Sánchez et al. [6]. The difference in starch losses might be because we stored only a few roots and each was stored individually rather than in bags.

Root damage had an impact on the quality of the roots with respect to PPD, fungal rot, and starch content. Previous research has demonstrated that physical damage to the fresh roots (cuts, breaks, and bruising) increased PPD and reduced shelf-life [1,2,4,27]. After 11 days of storage, roots with slight damage lost 9.6% in weight compared to 18.2% for roots with severe damage [6]. In response to wounding, such as cuts and abrasions, cassava storage roots show early physiological changes, including increased respiratory rate and water loss. Increased respiration induces the conversion of starch to sugar [6,28]. During storage, the starch content of unwounded cassava storage roots gradually decreases. However, starch loss and sugar production during storage are significantly higher if the roots are wounded. This research is new in that the type of root damage (bruising or breaks) was considered. As the fresh cassava root remains biologically active, PPD is accompanied by an increase in respiration and the root can initiate a wound response, and discolouration of the roots (vascular streaking) occurs due to oxidation of secondary metabolites [1]. However, previous research was undertaken on only a few cassava roots and hence, little was known about the response with larger sample sizes or how much root damage was possible before PPD became unacceptable. This research, using larger sample sizes of 50 kg, shows that while root damage does reduce the shelf-life, slight bruising of the root is acceptable provided the roots are not cut or broken.

5. Conclusions

Currently, for the processing of cassava roots in the production of high-quality cassava flour or starch, a limitation is the rapid deterioration of the roots within 48 h after harvesting. It can be concluded that fresh cassava roots (50 kg) can be stored in a bag at ambient temperature in West Africa for up to eight days with minimal deterioration or loss in starch. The advantage would be reduced product handling and potentially lower costs when transporting the fresh cassava roots from the farm to the processing factory.

Several control measures would be required to support the management of the process to ensure that roots can be consistently stored. This includes minimizing root damage. Our results suggest that there is no need to wet the roots, harvest with the peduncle attached, or separate different varieties (TME 419, TMS 0581 and TMS 1632). Other varieties may have different storage characteristics, and this would need to be tested.

A new approach to monitoring starch losses in fresh cassava roots during storage using a simple carbon dioxide sensory is suggested, but would require further testing. In industrial applications, this could provide a potential way to continuously and remotely monitor large quantities of cassava roots, avoid postharvest losses, and maintain profitability.

Author Contributions: K.T., C.I.O., L.O.S., A.P., and B.B. conceived and planned the experiments. C.I.O., A.F.A., and A.-R.A.A. carried out the experiments. K.T. and A.P. took the lead in writing the manuscript. All authors provided critical feedback and helped shape the research, analysis, and manuscript. All authors have read and agreed to the published version of the manuscript.

Funding: This research was funded by Rockefeller Foundation, Cassava Challenge Fund 2018.

Institutional Review Board Statement: Not Applicable.

Informed Consent Statement: Not applicable.

Acknowledgments: The authors would like to thank the Rockefeller Foundation for providing the financial support for these experiments. These laboratory studies would not have been possible without the support of numerous people in Nigeria. We particularly want to mention the staff of Federal University of Agriculture, Abeokuta (FUNAAB). Notwithstanding, the content of this report represents the views of the authors only.

Conflicts of Interest: The authors declare no conflict of interest.

References

1. Reilly, K.; Gómez-Vásquez, R.; Buschmann, H.; Tohme, J.; Beeching, J.R. Oxidative stress responses during cassava post-harvest physiological deterioration. *Plant Mol. Biol.* **2004**, *56*, 621–637. [CrossRef] [PubMed]
2. Zainuddin, I.M.; Fathoni, A.; Sudarmonowati, E.; Beeching, J.R.; Gruissem, W.; Vanderschuren, H. Cassava post-harvest physiological deterioration: From triggers to symptoms. *Postharvest Biol. Technol.* **2017**, *142*, 115–123. [CrossRef]
3. Parmar, A.; Sturm, B.; Hensel, O. Crops that feed the world: Production and improvement of cassava for food, feed, and industrial uses. *Food Secur.* **2017**, *9*, 5. [CrossRef]
4. Blagbrough, I.S.; Bayoumi, S.A.L.; Rowan, M.G.; Beeching, J.R. Cassava: An appraisal of its phytochemistry and its biotechnological prospects. *Phytochemistry* **2010**, *71*, 1940–1951. [CrossRef]
5. Bayoumi, S.A.L.; Rowan, M.G.; Beeching, J.R.; Blagbrough, I.S. Constituents and secondary metabolite natural products in fresh and deteriorated cassava roots. *Phytochemistry* **2010**, *71*, 598–604. [CrossRef]
6. Sánchez, T.; Dufour, D.; Moreno, J.L.; Pizarro, M.; Aragón, I.J.; Domínguez, M.; Ceballos, H. Changes in extended shelf life of cassava roots during storage in ambient conditions. *Postharvest Biol. Technol.* **2013**. [CrossRef]
7. Mahmod, N.; Beeching, J. Correlation of scopoletin-induced fluorescence with visible PPD symptoms in greenhouse cassava. *Postharvest Biol. Technol.* **2018**, *145*, 10–14. [CrossRef]
8. Carter, S.E.; Fresco, L.O.; Jones, P.G.; Fairbairn, J.N. Introduction and Diffusion of Cassava in Africa. IITA (International Institute of Tropical Agriculture), Ibadan, Nigeria. 1995. Available online: https://www.researchgate.net/publication/40207427_Introduction_and_diffusion_of_cassava_in_Africa (accessed on 20 February 2020).
9. Parmar, A.; Fikre, A.; Sturm, B.; Hensel, O. Post-harvest management and associated food losses and by-products of cassava in southern Ethiopia. *Food Secur.* **2018**, *10*, 419–435. [CrossRef]
10. Osunde, Z.D.; Fadeyibi, A. Storage Methods and Some Uses of Cassava in Nigeria. *Cont. J. Agric. Sci.* **2012**, *5*, 12–18.
11. Reynolds, T.W.; Waddington, S.R.; Anderson, C.L.; Chew, A.; True, Z.; Cullen, A. Environmental impacts and constraints associated with the production of major food crops in Sub-Saharan Africa and South Asia. *Food Secur.* **2015**, *7*, 795–822. [CrossRef]

12. Booth, R.H. Storage of fresh cassava (*Manihot esculenta*). I. Post-harvest deterioration and its control. *Exp. Agric.* **1976**, *12*, 103–111. [CrossRef]
13. Bancroft, R.D.; Crentsil, R.D. Application of a Low-Cost Storage Technique for Fresh Cassava (Manihot esculenta) Roots in Ghana. In *Transformation Alimentaire du Manioc*; AgborEgbe, T., Brauman, A., Griffon, D., Treche, S., Eds.; ORSTOM Ed.: Paris, France, 1995; pp. 547–555.
14. Rees, D.; van Oirschot, Q.E.A.; Aked, J. The role of carbohydrates in wound-healing of sweetpotato roots at low humidity. *Postharvest Biol. Technol.* **2008**, *50*, 79–86. [CrossRef]
15. Atieno, L.; Owino, W.; Ateka, E.M.; Ambuko, J. Effect of surface coatings on the shelf life and quality of cassava. *J. Food Res.* **2018**, *7*, 46–60.
16. Zidenga, T. Delaying Post-Harvest Physiological Deterioration in Cassava. Virginia Tech Blacksburg, VA. 2012. Available online: http://www.isb.vt.edu/news/2012/Aug/Zidenga.pdf (accessed on 20 February 2020).
17. Uchechukwu-Agua, A.D.; Caleb, O.J.; Opara, U.L. Postharvest Handling and Storage of Fresh Cassava Root and Products: A Review. *Food Bioprocess Technol.* **2015**, *8*, 729–748. [CrossRef]
18. Naziri, D.; Quaye, W.; Siwoku, B.; Wanlapatit, S.; Phu, T.V.; Bennett, B. The diversity of postharvest losses in cassava value chains in selected developing countries. *J. Agric. Rural Dev. Trop. Subtrop.* **2014**, *115*, 111–123.
19. El-Mogy, M.M.; Parmar, A.; Ali, M.R.; Abdel-Aziz, M.E.; Abdeldaym, E.A. Improving postharvest storage of fresh artichoke bottoms by an edible coating of Cordia myxa gum. *Postharvest Biol. Technol.* **2020**, *163*, 111143. [CrossRef]
20. Haq, M.A.; Azam, M.; Hasnain, A. Gum cordia as carrier of antioxidants: Effects on lipid oxidation of peanuts. *J. Food Sci. Technol.* **2015**, *52*, 2366–2372. [CrossRef] [PubMed]
21. Campo, B.V.H.; Hyman, G.; Bellotti, A. Threats to cassava production: Known and potential geographic distribution of four key biotic constraints. *Food Secur.* **2011**, *3*, 329–345. [CrossRef]
22. Rees, D.; Wesby, A.; Tomlins, K.; van Oirschot, Q.; Cheema, U.; Cornelius, E.W.; Amjad, M. Tropical Root Crops. In *Crop Post-Harvest: Science and Technology: Perishables*, 1st ed.; Rees, D., Farrel, G., Orchard, J., Eds.; Wiley Blackwell Publishing Ltd.: Sussex, UK, 2012; pp. 392–396.
23. Brimer, L. Chapter 10—Cassava Production and Processing and Impact on Biological Compounds. In *Processing and Impact on Active Components in Food*; Preedy, V., Ed.; Academic Press: San Diego, CA, USA, 2015; pp. 81–87.
24. Noon, R.A.; Booth, R.H. Nature of post-harvest deterioration of cassava roots. *Trans. Br. Mycol. Soc.* **1977**, *69*, 287–290. [CrossRef]
25. Ray, R.C.; Ravi, V. Post harvest spoilage of sweetpotato in tropics and control measures. *Crit. Rev. Food Sci. Nutr.* **2005**, *45*, 623–644. [CrossRef]
26. Kader, A.A. Postharvest Technology of Horticultural Crops—An Overview from Farm to Fork. *J. Appl. Sci. Technol.* **2013**, *1*, 1–8.
27. Uarrota, V.G.; Moresco, R.; Coelho, B.; da Costa Nunes, E.; Peruch, L.A.M.; de Oliverra Neubert, E.; Rocha, M.; Maraschin, M. Metabolomics combined with chemometric tools (PCA, HCA, PLS-DA and SVM) for screening cassava (Manihot esculenta Crantz) roots during postharvest physiological deterioration. *Food Chem.* **2014**, *161*, 67–78. [CrossRef] [PubMed]
28. El-Sharkawy, M.A. Physiological characteristics of cassava tolerance to prolonged drought in the tropics: Implications for breeding cultivars adapted to seasonally dry and semiarid environments. *Braz. J. Plant Physiol.* **2007**, *19*, 257–286. [CrossRef]

processes MDPI

Review

A Review on Individual and Combination Technologies of UV-C Radiation and Ultrasound in Postharvest Handling of Fruits and Vegetables

Okon Johnson Esua [1,2], Nyuk Ling Chin [1,*], Yus Aniza Yusof [1] and Rashidah Sukor [3]

1 Department of Process and Food Engineering, Faculty of Engineering, Universiti Putra Malaysia, UPM Serdang, Selangor 43400, Malaysia; johnsonesua@uniuyo.edu.ng (O.J.E.); yus.aniza@upm.edu.my (Y.A.Y.)

2 Department of Agricultural and Food Engineering, Faculty of Engineering, University of Uyo, Uyo 520101, Akwa Ibom, Nigeria

3 Department of Food Science, Faculty of Food Science and Technology, Universiti Putra Malaysia, UPM Serdang, Selangor 43400, Malaysia; rashidah@upm.edu.my

* Correspondence: chinnl@upm.edu.my; Tel.: +60-39-7696353

Received: 24 September 2020; Accepted: 29 October 2020; Published: 10 November 2020

Abstract: Ultraviolet-C radiation and ultrasound technology are widely accepted and continuously being appraised as alternatives to conventional thermal techniques for decontamination of fruits and vegetables. However, studies in these areas have presented challenges related to quality, safety, limited capability, and cost of energy. This review paper presents an up-to-date summary of applications of ultraviolet-C radiation and ultrasound technology for postharvest handling of fruits and vegetables from relevant literature. The limitations associated with applications of ultraviolet-C radiation and ultrasound technology individually has prompted their combination alongside other antimicrobial strategies for enhanced bactericidal effect. The combination of ultraviolet-C radiation and ultrasound technology as a hurdle approach also provides enhanced efficiency, cost effectiveness, and reduced processing time without compromising quality. The review includes further scope of industrial-led collaboration and commercialization of ultraviolet-C radiation and ultrasound technology such as scale-up studies and process optimization.

Keywords: deterioration; cavitation; dosage; hurdle technology; microorganisms; nonthermal; decontamination

1. Introduction

Fruits and vegetables are an essential component of our diet and their consumption increases yearly due to perceived interest of consumers in healthy and functional foods [1,2]. However, they are susceptible to postharvest moisture loss, improper handling, mechanical damage, and microbial contamination. Postharvest moisture loss affects maturation, which can be inferred from evaluation of important quality parameters such as weight, texture, acidity, sugars, carotenoids, vitamins, and phenolic compounds. Microbial contamination is accompanied by postharvest losses and prevalence of foodborne disease outbreaks. Data suggest that about one-third of food (1.3 billion tons) fit for human consumption is wasted yearly, with 44% attributed to fruits and vegetables, while numerous foodborne illnesses and deaths linked to fresh produce contamination have been reported in the last two decades [3–6]. Food losses and wastes pose serious concern at a global level; hence, the United Nations agenda has initiatives with development investments towards preventing and reducing postharvest losses [6,7].

The common practices to reduce spoilage and enhance shelf life of fruits and vegetables include washing with sanitizers, refrigeration, controlled atmosphere storage using natural and synthetic preservative agents, and drying [8–14]. The majority of these approaches are marginally effective and may alter nutritional properties while the corrosive nature of some sanitizers has been linked with environmental and health concerns [15,16]. As consumer demands for minimally processed products with fresh-like characteristics has increased in recent times, this preoccupation has encouraged continuous evaluation of non-thermal technologies for postharvest handling of fruits and vegetables.

The non-thermal technologies of ultraviolet-C (UV-C) radiation and ultrasound (US) technology have been extensively studied for surface decontamination of fruits and vegetables, and are considered highly efficient, non-toxic, and environmentally friendly [17]. UV-C radiation is generated at wavelengths of 250–280 nm and is reported to disrupt functionality and integrity of microorganisms' DNA in addition to being linked with generation of reactive oxygen species that regulates physiological processes to induce secondary metabolite production [18]. On the other hand, US technology is a form of pressure waves beyond human hearing (>20 kHz), and high intensity US at low frequencies of 20–100 kHz can induce acoustic cavitation that generates reactive hydroxyl radicals associated with microbial inactivation and stimulation of secondary metabolite accumulation in fruits and vegetables [17,19,20].

Nevertheless, UV-C radiation can only penetrate up to several millimeters and US technology consumes high energy coupled with long treatment time, which limits their applications. This has encouraged the use of UV-C radiation and US technology in combination with other techniques for postharvest handling of fruits and vegetables with encouraging results [1,19,21–23]. The current review, drawn from selected past and current literature, presents research progress of the individual and combination approaches of UV-C radiation and US technologies in postharvest handling of fruits and vegetables. The types of disinfecting systems typically utilized for postharvest UV-C radiation and US technology decontamination of fruits and vegetables is also summarized. This review also included discussion on limitations associated with individual applications of UV-C radiation and US technology for postharvest decontamination of fruits and vegetables. The potential of a hurdle technology utilizing UV-C radiation and US technology in combination or simultaneously for maximizing and harnessing their advantages is also presented.

2. Applications of UV-C Radiation in Postharvest Handling of Fruits and Vegetables

2.1. Disinfecting Systems and Factors Affecting Efficiency

The use of UV-C radiation for food preservation was discovered in the 1930s, and the first studies on fruits and vegetables was recorded in 1977 when resveratrol and viniferins were induced in grape vine from UV-C exposure [24,25]. The efficacy of UV-C radiation for decontamination of fruits and vegetables is heavily dependent upon the type of disinfection systems, mode of applications, pulse intensity, morphology and location of microorganisms on fruits and vegetables, distance from irradiation source, and type of microorganism.

Typical systems for exposing fruits and vegetables to UV-C radiation consist of self-contained chambers with fluorescent germicidal lamps and have been widely reported in the literature. However, modifications to the conventional self-contained chambers have been reported to induce improved microbial reductions and maintain quality properties even at low UV-C radiation doses. For instance, Collazo et al. [23] showed that low doses of 0.1–0.3 kJ/m^2 from water-assisted UV-C radiation equipment with pressurized water sprinklers were enough to reduce respiration, maintain quality, and provoke 0.9–2.0 log CFU/g (log colony forming unit/g) reductions of *Listeria monocytogenes* (*L. monocytogenes*) and *Salmonella enterica* (*S. enterica*) in lettuce and spinach (Figure 1a). The study revealed that dual action of irradiation by immersion and simultaneous actions of process water from modification of the conventional treatment chambers with pressurized water sprinklers accounted for higher efficiency. Yan et al. [26] demonstrated that it is possible to reduce large variations and maximize UV-C radiation dose on fruit surfaces in disinfecting systems. In this study, conveyor belts were placed at a distance of 14 cm from UV-C lamps to convey and rotate fruits between two treatment chambers connected by an inclined belt (Figure 1b). This inclusion to the conventional UV-C lamp self-contained chambers ensured dose uniformity such that fruits received an average dose of 1 kJ/m^2, compared with 0.2 kJ/m^2 received by fruits without rotation. The modified rotation device ensured 1.3–1.8 and 1.0–1.2 log CFU/fruit reductions in *Escherichia coli* O157:H7 (*E. coli* O157:H7) and *E. coli* ATCC 25922, respectively, compared with 0.5–0.7 log CFU/fruit reduction obtained without the rotation device, revealing the importance of dose uniformity for maximum effect of UV-C radiation in disinfection systems. Likewise, a benchtop UV-C radiation system that allowed upward and downward movement of fruits from a rotating roller was employed by Taze and Unluturk [27] to ensure uniformity of intensity (Figure 1c). They observed 3.0 and 2.38 log CFU/g reductions in total aerobic mesophilic bacteria and yeast and mold, respectively, at doses of 31.01 and 7.75 kJ/m^2, and reported that light intensity was almost constant at the lamp center and lower at both ends. Ensuring the uniformity of exposure on food surfaces due to the low penetrative ability of UV-C radiation is seen as the most important challenge to commercialization, and these studies have provided an avenue to ensure dose uniformity for maximum impact.

The applied dosage of UV-C radiation, the type of microorganism, and their location on fruits and vegetables have been reported to have significant effect on decontamination efficiency of UV-C radiation. Mukhopadhyay et al. [28] showed that increasing UV-C radiation doses from 0.6 to 6.0 kJ/m^2 resulted in 2.2–3.1 and 2.3–3.5 log CFU/fruit reductions, respectively, for cocktail mixtures of *S. enterica* and *E. coli* O157:H7 on the surfaces of tomato. Lower reductions of 1.9–2.8 and 1.7–3.2 log CFU/fruit, respectively, were recorded on the stem scar under the same conditions, indicating that specific location of microorganisms on produce may influence efficacy of treatment and should be considered during the design of UV-C radiation disinfection systems. Further analysis showed that color and texture of tomato were not affected by the increasing dosages. In contrast, Lim and Harrison [29] observed a reduction range of 3.22–4.39 log CFU/g in *S. enterica* for a dosage range of 0.223–1.785 kJ/m^2, irrespective of the contamination location on tomato placed in a close end reflector lined with aluminum foil. The high reduction from such low doses was attributed to the bacterial strain and low initial inoculum level utilized in the study, which corroborates the assertion that the sensitivity of bacteria to UV-C radiation may vary among species and strains of the same species.

The mode of application of UV-C radiation, pulse intensity, and distance of samples from irradiation source have also been reported to influence the actions of UV-C radiation during decontamination of fruits and vegetables. For example, short pulses of incident UV-C irradiance was not effective in reducing microbial population, which could be linked to ability of resistant strains of microorganism to produce protein protection mechanism against UV-C radiation treatment, especially at short treatment duration [30]. The study showed that there was no significant difference in log reductions (1.0 log CFU/g) of resistant strains of *Enterococcus faecalis* on zucchini squashes when a UV-C radiation system was operated in continuous incidence radiation of 0.011 kJ/m^2 or pulse irradiation of 0.067 kJ/m^2 in a discontinuous fluence. In addition, a continuous UV-C incidence radiation of 0.011 kJ/m^2 for 90 s did not have any significant effect on the reduction of *Deinococcus radiodurans* on zucchini squashes. However, higher inactivation of *E. coli* was reported on tomatoes and lettuce when the intensity of UV-C radiation was increased from 6.5 to 16 W/m^2 in a treatment cabinet [31]. Enhanced reductions of 2.8 and 1.7 log CFU/g were also observed in tomatoes and lettuce, respectively, for samples closer to the radiation source at 31 cm compared with samples placed at a distance of 70 cm. In the same vein, Cote et al. [32] observed that increasing intensity for a given dose could maximize the benefits of UV-C radiation. For an applied dosage of 4 kJ/m^2, only 5% of tomatoes irradiated with higher intensity of 33 W/m^2 showed incidences of postharvest rot caused by *Botrytis*, compared with 8% from lower intensity of 3 W/m^2, after 9 days of storage. On the other hand, 12% of strawberries irradiated with higher intensity showed decay symptoms after 5 days, compared with 46% of strawberries irradiated with lower intensity. The treatment also delayed ripening and maintained better quality of the fruits, while titratable acidity, soluble solids, and antioxidants properties were unaffected.

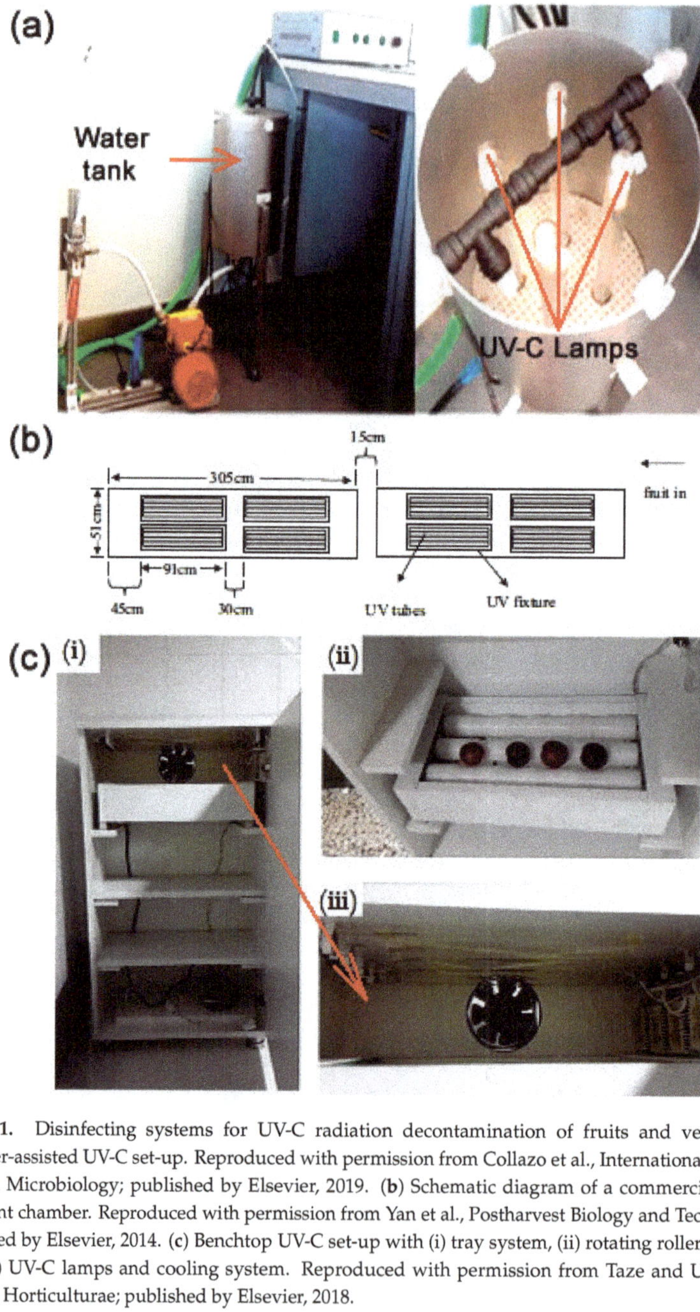

Figure 1. Disinfecting systems for UV-C radiation decontamination of fruits and vegetables. (**a**) Water-assisted UV-C set-up. Reproduced with permission from Collazo et al., International Journal of Food Microbiology; published by Elsevier, 2019. (**b**) Schematic diagram of a commercial UV-C treatment chamber. Reproduced with permission from Yan et al., Postharvest Biology and Technology; published by Elsevier, 2014. (**c**) Benchtop UV-C set-up with (i) tray system, (ii) rotating roller bearing, and (iii) UV-C lamps and cooling system. Reproduced with permission from Taze and Unluturk, Scientia Horticulturae; published by Elsevier, 2018.

The influence of fruit surface characteristics on the inactivation efficiency of UV-C radiation has also been studied, with results demonstrating that microorganisms on fruits' surfaces responded differently to UV-C radiation [33]. It was observed that microbial inactivation was lower for fruits with rougher surfaces (strawberry, cantaloupe, and raspberry) compared to less hydrophobic fruits with smoother surfaces (pear and apple). Reductions of 2.9 and 2.1 log CFU/g were recorded for

E. coli O157:H7 on apple and pear surfaces, respectively, at 0.9 kJ/m^2 for 60 s compared with 2.0 and 1.1 log CFU/g recorded for strawberry and raspberry at 7.2 and 10.5 kJ/m^2 for 480 and 720 s, respectively. On the other hand, *L. monocytogenes* showed more resistance than *E. coli* O157:H7, revealing 1.6 and 1.7 log CFU/g reductions, respectively, on apple and pear at 3.75 kJ/m^2 and 11.9 kJ/m^2. The log reduction of *L. monocytogenes* was 1.0 log CFU/g on both cantaloupe and strawberry at 11.9 kJ/m^2. The findings from this study were corroborated by Syamaladevi et al. [34] by using a similar device when they observed that trichomes on peach surfaces and wounds on pears protected and shielded microorganisms from irradiation, while smaller surface roughness and spreading coefficient of pear surfaces compared with peach likely caused a more uniform distribution of microbial cells on pear surfaces. Maximum reductions of 3.70 log CFU/g were recorded for *E. coli* on intact pear surfaces, with lower reductions of 3.10 and 2.91 log CFU/g achieved on wounded pear and peach surfaces, respectively, after 4 min at 7.56 kJ/m^2.

2.2. Effects of UV-C Radiation on Microorganism Inactivation and Quality of Fruits and Vegetables

Table 1 lists studies that have used UV-C radiation to inactivate microorganisms and subsequent effects on quality parameters for a wide variety of fruits and vegetables. Collectively, these studies demonstrate the potential of UV-C radiation at low doses of 0.00176–1 kJ/m^2 for enhancing bioactive compound production, reducing enzymatic activities, extending shelf life, and having degradative effects on biological process-regulating proteins responsible for deteriorative changes of postharvest produce [33,35–39]. The range of doses were also responsible for 1.2–2.9 log CFU/g reduction in *E. coli* O157:H7, *Salmonella* spp., and *Listeria* spp., as well as reduced rot and decay development. In contrast, Wang et al. [40] observed that a low dose of 0.3 kJ/m^2 had no effect on total bacteria number and browning-related enzymes of fresh cut lotus during storage, and resulted in a high degree of browning, but the authors maintained that a moderate dose of 1.5–3 kJ/m^2 significantly retarded microbial growth and suppressed the activities of browning-related enzymes. Pinheiro et al. [41] also reported that low doses of 0.97 kJ/m^2 was not adequate for metabolic degradation delay in tomatoes but presented 2.1 log CFU/g reduction in mesophilic load, delayed red color development, and caused increases in firmness and phenolic content. The variations in results could be attributed to the factors discussed earlier, in addition to the type of disinfection systems used.

It is apparent that increasing dosage applications could result in increased microbial reduction. For instance, Sommers et al. [42] observed a progressive increase from 2.59 to 3.79 log CFU/g in the inactivation of *L. monocytogenes*, *Salmonella* spp., and *Staphylococcus aureus* (*S. aureus*) on Roma tomatoes and jalapeño peppers when the dosage was increased from 5 to 40 kJ/m^2. Similarly, successive increases in UV-C radiation doses evoked greater reductions of *E. coli*, *S. aureus*, *S.* Enteritidis, and *Listeria innocua* (*L. innocua*) on tomato, lettuce, and strawberry, with no modifications in texture and appearance [19,43]. Doubling UV-C radiation dose from 32 to 72 kJ/m^2 also doubled the reduction of human adenoviruses from 0.92 to 2.22 and 1.26 to 3.98 logs for tomato and strawberry, respectively.

Table 1. List of fruits and vegetables treated with UV-C radiation

Fruit/ Vegetable	UV-C Parameters	Main Findings		Sources
		Quality Improvement Attributes	Microbial Log Reduction and Disease Incidence	
Apple	7.5 kJ/m^2 + blanching 1.0 kJ/m^2	Retained color	*L. innocua*, *E. coli*, *Saccharomyces cerevisiae*: 1–1.19	[44]
			E. coli: 1.89, *L. innocua*, *S. enterica*: 1.5	[38]
	0.92–5 kJ/m^2, 60–300 s	Retained color and physicochemical properties	*E. coli*: 2.9, *L. monocytogenes*: 1.6	[33]
Apricot	0.74 kJ/m^2, 10 s	-	*E. coli*: 1.2, *Salmonella* spp.: 1.5	[37]
	1.0–2.54 kJ/m^2	-	*E. coli* O157:H7: 0.5–1.8	[26]
	7.75 kJ/m^2	-	Yeast and mold: 2.38, TAPC: 3.0	[27]
Blueberry	2 kJ/m^2, 600 s	Increased anthocyanin, phenols, and antioxidants	*Colletotrichum acutatum*: 10% reduction	[45]
	9.54 kJ/km^2, 120 s, + O_3, 60 s		*E. coli*: 3.08–3.8	[46]
Broccoli	2.5–7.5 kJ/m^2, + 100 mg/L NEW + MAP + PA	Retained total phenols and antioxidants; increased enzyme activity, reduced steric and α-linolenic acid	*L. monocytogenes*: 0.72, *E. coli*, *S.* Enteritidis: 1.22–3.0, psychrophilic bacteria: 1.3	[11,47]
Carrot	0.78 kJ/m^2	Increased bioactive compounds and whiteness index; reduced enzyme activity and O_2/CO_2 rate	Aerobic mesophile: 1.7	[36]
	57.6 kJ/m^2, 3600 s	Increased enzyme reaction and whiteness index	-	[31]
Cauliflower	5 kJ/m^2, 30 s + AMF		*L. monocytogenes*: 1.5, *E. coli*: 2.0, yeast and mold: 2.0	[13]
Date palm	6.22 kJ/m^2, + O_3	Increased antioxidants and TA; reduced sugar and water activity	Yeast and mold: 1.63, coliform: 0.82, mesophilic counts: 1.05	[48]
Grape	6 kJ/m^2, + 0.5% chitosan	Increased resveratrol, reduced respiration, weight loss	*Botrytis cinerea*: growth inhibition	[49]
Lettuce	57.6 kJ/m^2, 3600 s	Overall quality parameters maintained	*E. coli*: 1.7	[31]
	24–72 kJ/m^2, 1200–3600 s		*E coli*: 1.75, *L. innocua*: 1.27, *S. aureus*: 1.21, *S.* Enteritidis: 1.39, HAdV: 2.13	[19,43]
	0.1–0.5 kJ/m^2, 60–300 s, PA + MAP	Oxidative discoloration, enhanced respiration	*L. monocytogenes*: 1.2–2.1, MAM: 1.3, *S. enterica*: 1.8–2.5	[23]

Table 1. *Cont.*

Fruit/ Vegetable	UV-C Parameters	Main Findings		Sources
		Quality Improvement Attributes	Microbial Log Reduction and Disease Incidence	
Lotus root (fresh cut)	0.3–12 kJ/m^2, 60–2400 s	Low browning degree and soluble quinone content; inactivation of PAL, POD, TSS; hardness maintained	Total bacteria: 1.0–1.4	[40]
Mango	0.00176–0.00706 kJ/m^2, 900–1800 s	Increased flavonoid and antioxidant activity	Microbial load under acceptable limit	[39]
	5 kJ/m^2	-	*E. coli*: 1.98, *Cronobacter sakazakii*: 2.60	[50]
Melons	11.9 kJ/m^2, 840 s	-	*L. monocytogenes*: 1.0	[33]
	1.2 kJ/m^2, 60 s	Retained physicochemical properties	TVC: 2.16, *Enterobacteriaceae*: 2.62	[51]
	2.8 kJ/m^2	Color, lycopene, vitamin C preserved	*Enterobacteria*, AMB, APB: 2.0	[52]
Peach	4 kJm^2, 120 s		*E. coli*: 2.5	[34]
Pear	1.7–4 kJ/m^2, 90–180 s	Maintained weight, TSS, and texture	*Penicillium expansum*: 2.7–2.8, *E. coli*: 2.6–3.59	[34,53]
	0.92–11.9 kJ/m^2, 60–840 s	-	*E. coli*: 2.1, *L. monocytogenes*: 1.7	[33]
	87 kJ/m^2, 1200 s	-	*L. monocytogenes*, *E. coli*: 2.6–3.4	[54]
	7.5 kJm^2	Increased firmness; color, TA, and TSS maintained	*E. coli*: 3.2, *L. innocua*: 2.9, *S. enterica*: 2.8	[55]
Pepper	5–40 kJ/m^2	-	*S. aureus*: 3.09–3.73, *L. monocytogenes*: 3.11–3.72	[42]
	1.483 kJ/m^2, 120 s	Improved firmness; color preserved	*L. innocua*: 1.05	[56]
Pineapple	0.2–4.8 kJ/m^2 + PET pouches	Maintained quality and color	LAB, yeast and mold: 0.20–2.0, *Enterobacteriaceae*: undetected	[57]
Plum	10 kJ/m^2 + ClO_2 + fumaric acid	-	*E. coli*: 2.07–5.48, *L. monocytogenes*: 1.62–6.26	[58]
Pomegranate	1.13 kJ/m^2, + MAP	Anthocyanin; antioxidant activity maintained	LAB, AMB, *Enterobacteriaceae*: reduced growth	[59]
Raspberry	7.2 kJ/m^2, 480 s	-	*E. coli*: 1.0	[33]
Spinach	4.54 kJ/m^2	Chlorophyll maintained; low ethylene production	Mesophilic, psychrophilic, coliform, enterobacteria: 1.1	[60]
	0.212–0.848 kJ/m^2, + ClO_2	Color and texture maintained	*E. coli*: 2.01–5.17, *Salmonella* Typhimurium: 2.10–5.47, *L. monocytogenes*: 1.70–4.32	[16]
	0.1–0.5 kJ/m^2, 60–300 s, PA + MAP	Reduced respiration; maintained quality	*L. monocytogenes*: 0.5–2.2, MAM: 1.3, *S. enterica*: 0.8–2.0	[23]

Table 1. *Cont.*

Fruit/ Vegetable	UV-C Parameters	Main Findings		Sources
		Quality Improvement Attributes	Microbial Log Reduction and Disease Incidence	
Strawberry	0.5 kJ/m^2	Increased PAL and ethylene production	*B. cinera, Rhizopus stolonifier, Mucor* sp.: reduced incidence	[35]
	0.92–11.9 kJ/m^2, 60–840 s	-	*E. coli*: 2.0, *L. monocytogenes*: 1.0	[33]
	24–72 kJ/m^2, 1200–3600 s	-	*E coli*, *S. aureus*, *S.* Enteritidis, *L. innocua*: 1–1.4, HAdV: 2.85–3.98	[19,43]
	3, 33 W/m^2, 4 kJ/m^2	Low weight loss and respiration rate; increased anthocyanin	*Botrytis cinera*: reduced incidence	[32]
	1.483 kJ/m^2, 120 s	Improved firmness; anthocyanin retained	Total mesophiles: 0.26	[56]
Tomato	5–40 kJ/m^2	-	*S. aureus*: 3.13–3.63, *Salmonella* spp.: 3.02–3.79, *L. monocytogenes*: 2.59–3.60	[42]
	0.97 kJ/m^2, 180 s	Low weight loss; improved firmness and phenolic content	Total microbial count: 0.6–2.1	[41]
	0.6–6.0 kJ/m^2, 10–100 s	Firmness and color maintained	*E. coli*: 2.3–3.5, *S. enterica*: 2.2–3.1	[28]
	57.6 kJ/m^2, 3600 s	Increase in tomato color index	*E. coli*: 2.8	[31]
	2 kJ/m^2, 500 s + MAP	Increased lycopene; color maintained	*S.* Typhimurium: 1.3	[61]
	3, 33 W/m^2, 4 kJ/m^2, 120 s	Low weight loss and respiration rate; bioactive and physicochemical content retained	*B. cinerea*: reduced incidence	[32]
	3 kJ/m^2	Higher total phenol and firmness; reduced ethylene production and enzyme activity	Decrease in decay severity	[62]
	0.223–1.785 kJ/m^2	Color and texture maintained	*S. enterica*: 3.22–4.39	[29]
	0.848 kJ/m^2, + ClO_2	-	*E. coli*: 2.27, *S.* Typhimurium: 2.20, *L. monocytogens*: 1.66	[16]
Watercress	1.483 kJ/m^2, 120 s	Improved firmness; color maintained	Total coliforms: 0.53	[56]

TAPC: total aerobic plate count, NEW: neutral electrolyzed water, MAP: modified atmosphere packaging, PA: peroxyacetic acid, AMF: antimicrobial formulations, TA: titratable acidity, HAdV: human adenovirus, MAM: mesophilic aerobic microbes, PAL: phenylalanine, POD: peroxidase, TSS: total soluble solids, LAB: lactic acid bacteria, AMB: aerobic mesophilic bacteria, APB: aerobic psychrotrophic bacteria.

The synergistic or additive effects of combining UV-C radiation with other techniques for improved efficiency have also been studied. Enhanced reductions of 3 log CFU/g were recorded in the populations of *E. coli* and *S.* Enteritidis when UV-C radiation was combined with either neutral electrolyzed water or peroxyacetic acid for broccoli decontamination, compared with 1.3–2.2 log CFU/g obtained for individual application of UV-C radiation at 7.5 kJ/m^2 [11,47]. Combination treatment also maintained quality, retained bioactive and fatty acid contents, and was seen as a potential eco-friendly alternative to conventional NaClO application. Similar enhanced reductions in *E. coli* on blueberry, and *L. monocytogenes*, *E. coli*, and yeast and mold on cauliflower were also reported when low UV-C doses were combined with ozone and antimicrobial formulations [13,46]. The effects of combining UV-C radiation and ClO_2 gas was also observed to be significantly greater than the sum of individual inactivation for the decontamination of spinach and tomato [16]. Individual UV-C radiation application at 0.848 kJ/m^2 reduced *E. coli* O157:H7 and *Salmonella* Typhimurium by 1.85 and 2.02 log CFU/g, respectively, while combined treatment reduced the population by 5.17 and 5.41 log CFU/g on spinach and 5.62 and 5.46 log CFU/g on tomato. Enhanced microbial reduction was attributed to cell membrane damage and changes to membrane permeability as a result of the synergistic actions of UV-C radiation and ClO_2 gas. This position was also maintained by Kim and Song [58] during the sequential application of ClO_2 gas and UV-C radiation on plum. The synergistic effect produced 4 and 3.27 log CFU/g reductions in *L. monocytogenes* and *E. coli* O157:H7, respectively, compared with 1.62 and 2.07 log CFU/g reductions obtained for individual UV-C radiation application. The authors however upheld that sequential applications of ClO_2 gas, UV-C radiation, and fumaric acid, which produced 5.48 and 6.26 log CFU/g reductions, respectively, in *E. coli* O157:H7 and *L. monocytogenes*, was the most effective combination treatment.

3. Applications of US Technology in Postharvest Handling of Fruits and Vegetables

3.1. Disinfecting Systems and Factors Affecting Efficiency

The antimicrobial effect of US technology was first documented 80 years ago, but its prospective practical application for microbial inactivation began in the 1960s [63]. The chemical reactivity of a system is affected by US technology through actions of cavitation and sonolysis from mechanical vibrations, typically produced from transformation of high-frequency electrical energy by piezoelectric transducers in probe or tank systems [15,19]. The threshold of cavitation bubble in the tank or probe systems is usually influenced by US frequency, power, intensity, and exposure time. The sonochemist typically uses frequencies of 20–50 kHz, as it is easier to obtain cavitation at these frequencies [64]. Together with the type of microorganism, mode of application in terms of pulse or continuous modes, and dual-mode or mono-mode frequency, these factors affect the efficiency of US technology during decontamination of fruits and vegetables.

For the US probe systems, a pulse mode application of 10 s on and 5 s off from a 15 mm diameter, 20 kHz, 400 W, and 226 W/cm^2 probe system submerged in water (Figure 2a) was capable of reducing initial total number of colonies and yeast and mold on cucumber by 0.49–1.02 and 0.41–0.84 log CFU/g, respectively, with increasing treatment time from 5 to 15 min under modified atmosphere packaging storage [65]. Treatment up to 10 min effectively reduced losses of firmness, ascorbic acid, color, soluble solids, and weight, and maintained cell wall integrity. Further increase in treatment time led to deterioration of quality properties. Similarly, Millan-Sango et al. [66] reported 2.56 log CFU/cm^2 reduction in *E. coli* O157:H7 on lettuce leaves at pulse mode application of 10 s on and 6 s off from a 14 mm probe, 26 kHz, 90 μm, and 200 W US system operating in both continuous and pulsed mode configurations. Results showed no significant differences between continuous and pulsed mode applications for up to 25 min of application, suggesting that decontamination efficacy was dependent upon treatment time rather than the configuration of the probe system utilized.

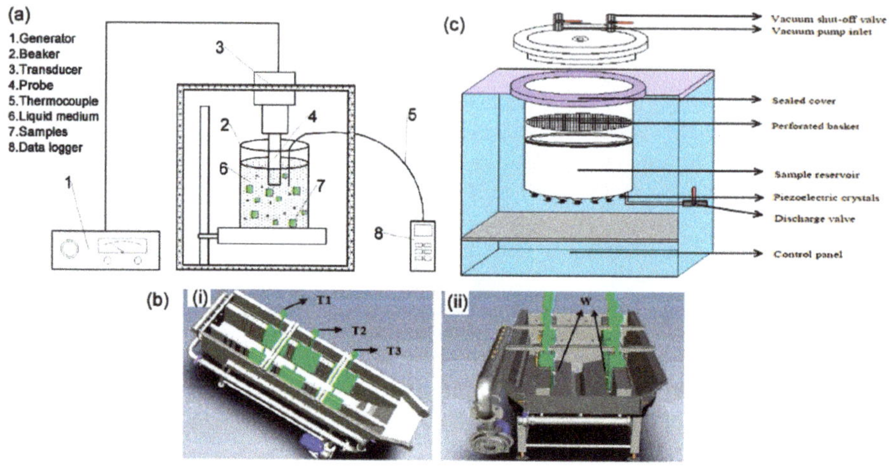

Figure 2. Typical disinfecting systems for ultrasound (US) technology decontamination of fruits and vegetables. (**a**) Schematic representation of US probe system. Reproduced with permission from Fan et al., Ultrasonics Sonochemistry; published by Elsevier, 2019. (**b**) Schematic diagram of US vacuum impregnation system. Reproduced with permission from Yilmaz and Bilek, Ultrasonics Sonochemistry; published by Elsevier, 2018. (**c**) Continuous-flow US bath equipment depicting (i) different transducers T1, T2, and T3, and (ii) washing area, W. Reproduced with permission from Zhou et al., Innovative Food Science and Emerging Technologies; published by Elsevier, 2012.

Increasing US intensity in a probe system has also been reportedly associated with enhanced antimicrobial activities while higher intensities had negative effects on product quality during storage. For example, reduced decay incidence of gray mold on strawberry with increasing intensity from 10.6–31.8 W/cm^2 was reported by Aday et al. [67] from a 1.9 cm diameter probe system operating at 20 kHz. Results also show that pH, color, soluble solids, and sugar content of strawberry treated with 10.6–21.2 W/cm^2 were better when compared with those treated with 31.8 W/cm^2, indicating that higher US intensity was detrimental to the quality of strawberry. In the same vein, Wang et al. [68] showed that the population of total bacteria and yeast and mold on tomato reduced to a range of 0.42–1.04 and 0.41–0.93 log CFU/g, respectively, with increasing US intensity from 66.64 to 145.74 W/L. Likewise, lower intensity levels inhibited ethylene production and respiration rates, and maintained firmness and antioxidant properties, while higher intensity had negative effects on tomato quality, demonstrating that appropriate US intensity can inhibit decay and maintain nutritional properties and flavor of tomato during storage.

For the US bath systems, research has demonstrated that it is more challenging to obtain cavitation at higher frequencies, with US penetration inversely proportional to frequency, while dual-mode frequency can be associated with improved cavitation when compared with mono-mode frequency applications. In this manner, Zhou et al. [63] fabricated a closed-tank continuous-flow system that allowed for frequency and nominal power variation and observed that 25, 40, and 75 kHz transducers dissipated 95, 85, and 45% rated power, corresponding to intensities of 79.41, 68.95, and 42.36 W/L, respectively (Figure 2b). The use of agitation ensured spatial uniformity of US treatment, and the populations of *E. coli* O157:H7 on spinach leaves were reduced by 4.45, 4.21, and 2.42 log CFU/g, respectively, from 25, 40, and 75 kHz transducers, suggesting that non-uniform cavitation from ultrasonic field variations may contribute to variations in microbial inactivation and cross-contamination in the US bath system. In a recent study, Mustapha et al. [20] showed that the application of dual-mode frequency of 20/40 kHz during US treatment of tomato presented higher microbial reductions of 1.3–2.6 log CFU/g for mesophilic bacteria and molds and yeasts, compared with <1 log CFU/g obtained when mono-mode frequency of 20 or 40 kHz was applied. There were no detrimental effects on the bioactive compounds and physicochemical properties of tomato during storage, substantiating the enhanced cavitation characterized by dual-mode frequency applications.

It has also been demonstrated that substances can be incorporated into samples from the US bath system for better effects. Yilmaz and Bilek [69] combined vacuum impregnation and US (72–840 W, 35 kHz) in custom-made equipment, observing that calcium and black carrot phenolics can be incorporated into apple tissues to produce natural colorant-fortified products without disturbing cellular integrity (Figure 2c). Treatment reduced psychrophilic and mesophilic bacteria on apples by 1.0–2.6 log CFU/g during storage and led to increases in firmness and bioactive compounds, while higher power levels resulted in apple cell rupture, which was evident from increased ion leakage in apple tissues. Wiktor et al. [70] showed that immersion treatment from a 40 kHz, 180 W bath system presented apples with lower mass loss of 0.5–1.2%, dry matter of 0.115 kg dm/kg, higher bioactive compound contents, and less color change when compared with contact treatment from a 24 kHz, 100% amplitude ring sonotrode probe system. The surrounding water during immersion treatment played a significant role in water gain and cooling of apples, hence lower weight loss, while localized oxygen on apple tissue pores and enhanced enzymatic activities during contact treatment could degrade bioactive compounds. Although this may appear to suggest that treatment using the US bath system has shown a better quality of sample properties when compared with the US probe system, this does however need more detailed and careful comparison in terms of same usage of US energy frequency and intensity.

3.2. Effects of US Technology on Microorganism Inactivation and Quality of Fruits and Vegetables

Table 2 presents a summary of some studies that have applied US technology to inactivate microorganisms and subsequent effects on quality parameters for a number of fruits and vegetables. The majority of the applications were carried out in combination with sanitizers so as to avoid possible cross-contamination, since microorganisms are usually released into the wash water during US decontamination. However, among the individual US applications, Gani et al. [71] reported 2.0 and 1.2 log reductions in bacterial counts and yeast and mold, respectively, in addition to retention in color, firmness, vitamin C, and antioxidant activity. Prolonged exposure time after 40 min, 33 kHz, and 60 W did not further extend shelf life but may have caused injuries to strawberry samples. In another study, the action of US at 25 kHz and 26 W/L was reported as similar to the actions of abiotic stress such as wounding in triggering and eliciting defense systems in Romaine lettuce [72]. A response mechanism was suggested for such actions (Figure 3a), and treated lettuce displayed enhanced antioxidant activity, higher firmness, delayed enzymatic browning, and absence of deterioration during storage. Preservative treatments are usually known to be associated with the degradation of important freshness-related attributes that define the quality of fresh produce. This perception needs to be critically considered, especially the effects on health-stimulating phytonutrients and other related compounds during the design and selection of processing parameters.

For combination applications, the synergistic effects of US and chemical sanitizers, surfactants, organic acids, and electrolyzed water for improved decontamination efficiency and enhanced quality attributes have been reported [10,73–84]. Increased efficiency was observed without synergistic effect, and also with both synergistic and antagonistic effects in some cases. Park et al. [79] reported antagonistic effects against *Cronobacter sakazakii* when US at 37 kHz and 380 W was combined with 150 ppm of NaOCl. Although the plausible reason for antagonistic effect was not clear, synergistic combination mostly eliminated bacteria cells from the surface of lettuce and in the stomata, as observed in the scanning emission spectroscopy (Figure 3b), yielding 4.44 log CFU/g reduction of *C. sakazakii* compared with 0.01–2.71 log CFU/g reduction for individual applications of US and NaOCl. These studies demonstrated that US alone was not greatly effective, even with long treatment time, while the cavitation effect of US detached microorganisms from fresh produce, especially at inaccessible regions, increasing their susceptibility to sanitizers. Thus, the chemical composition of the liquid in the US system can influence decontamination actions. The US system also promoted the rapid dispersion of organic matter, which increased their contact time with sanitizer constituents, leading to minimal residues in post-treatment solutions. The presence of microorganisms in treatment solutions were also very low and, in some cases, nonexistent, indicating that US and adequate sanitizer concentrations could prevent and/or reduce cross-contamination.

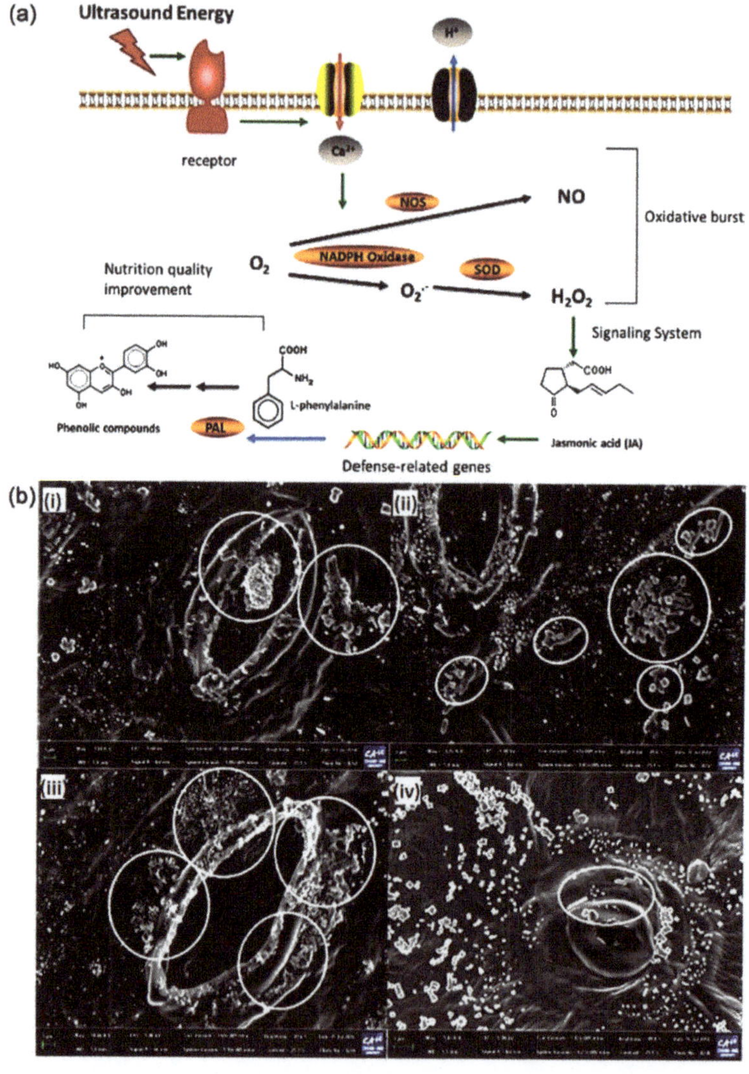

Figure 3. (**a**) Proposed mechanism of secondary metabolism response in Romaine lettuce during US. Reproduced with permission from Yu et al., Food and Bioprocess Technology; published by Springer Nature, 2016. (**b**) Field emission scanning electron microscopy of lettuce head showing aggregation and elimination of *Cronobacter sakazakii* during treatment: (i) control, (ii) US treatment, (iii) NaOCl treatment, and (iv) combined US and NaOCl treatment. Reproduced with permission from Park et al., Food Control; published by Elsevier, 2016.

Table 2. List of fruits and vegetables treated with US technology.

Fruit/ Vegetable	Ultrasound Parameters	Main Findings		Sources
		Quality Attributes	Microbial Log Reduction and Disease Incidence	
Apple	170 kHz, + 20 ppm ClO_2, 360 s	-	*E. coli*: 3.3, *S. enterica*: 4.0.	[73]
	40 kHz, 180 W	Higher antioxidant and phenolic content	-	[70]
	35 kHz, 72–840 W, + VI	Increases in bioactive compounds, calcium, and ion leakage	Psychrophilic and mesophilic bacteria: 1.0–2.6	[69]
Asparagus	40 kHz, 360 W, +2% acetic acid +50 mg/kg gibberellic acid	Low weight loss; increased soluble solids, phenols, ascorbic acid, and chlorophyll; inhibited PAL and POD	Yeast and mold, total number of colonies: 2	[85]
Cabbage	32 kHz, 10 W/L, 600 s + surfactant	-	*E. coli*: 2.91	[74]
	40 kHz, 400 W/L, 180 s, +40 °C + SAEW + wash water	Overall visual quality, browning, and off-odor under acceptable limits	Yeast and mold: 2.5–3.24, *E. coli*: 2.6–3.32, total bacteria count: 2.54–3.97, *L. monocytogenes*: 2.8–3.11, *Enterobacteriaceae*: 3.66	[75,86,87]
Cucumber	37 kHz, 13.57 W/L, + PA	Color, moisture content, and texture maintained	*C. sakazakii*: 0.60, 3.51	[76]
	20 kHz, 400 W, 226 W/cm^2, 300–600 s	Improved cell wall integrity; retained ascorbic acid; reduction in weight loss, firmness, and soluble solids	Yeast and mold: 0.41–0.84, total number of colonies: 0.49–1.02	[65]
Kiwifruit	30 kHz, 368 W/cm^2, + NaOCl	Reduced vitamin C, soluble solids, and firmness	Total bacteria count: 2.35, yeast and mold: 1.36	[88]
	40 kHz, 350 W, 600 s + nano-ZnO coating + MAP	Improved texture; low CO_2 and ethylene production	-	[89]
Lettuce	32 kHz, 10 W/L, 600 s + surfactant	-	*S.* Typhimurium: 2.7, *E. coli*: 2.11	[74]
	40 kHz, 400 W/L, 180 s, + SAEW + wash water	-	Yeast and mold: 2.25, total bacteria count: 2.6, *E. coli*: 2.5–2.8, *L. monocytogenes*: 2.6	[75]
	40 kHz, 30 W/L, 300 s, + organic acids	Maintained color and textural properties	*E. coli*: 2.75, *S.* Typhimurium: 3.18, *L. monocytogenes*: 2.87	[77]
	37 kHz, 30 W/L, 3600 s, + BPW + 24 kJ/m^2, 1200 s	-	*E coli*: 2.3, *S. aureus*: 1.7, *S.* Enteritidis: 5.72, *L. innocua*: 1.88, HAdV: 1.75–2.85	[19,43]
	26 kHz, 200 W, 90 μm + essential oils	Increased electrolyte leakage	*S. enterica*: 1.68–3.08, *E. coli*: 0.76–2.65	[15,66]
	25 kHz, 26 W/L, 60 s	Delayed browning; enhanced bioactive content	-	[72]
	42 kHz, 0.5 W/ml, + Cl_2 + surfactants	Higher values of color change and electrolyte leakage	*E. coli*: 2.61, *L. innocua*: 2.23, *P. fluorescens*: 1.10–2.10	[78]
	37 kHz, 13.57 W/L, + NaClO	pH and sugar content maintained	*C. sakazakii*: 0.58–4.44	[79]
	20 kHz, 131.25 W/L + NNEW	-	*E. coli*: 4.4, *S.* Typhimurium: 4.3	[80]

Table 2. *Cont.*

Fruit/Vegetable	Ultrasound Parameters	Main Findings		Sources
		Quality Attributes	Microbial Log Reduction and Disease Incidence	
Melons	40 kHz, 120 s, + 1% lactic acid	-	*E. coli*: 2.5, *S. enterica* Enteritidis: 3.1	[81]
Parsley	32 kHz, 10 W/L, 600 s + surfactant	-	*E. coli*: 0.84	[74]
Peach	40 kHz, 600 s, 8.8 W/L + SA	Higher phenol content and enzyme activities	*P. expansum*: growth inhibition	[82]
Pepper	35 kHz, 21.4 W/L, 120 s, 65 °C	Improved firmness; color preserved	*L innocua*: 7.4	[56]
	40 kHz, 120 s, +1% lactic acid	-	*E. coli*: 2.9, *S. enterica* Enteritidis: 2.8	[81]
	40 kHz, 400 W/L, SAEW+ mild heat	Color and firmness preserved	*L. monocytogens*, *S. enterica*: 1.0–3.0	[21]
Plum	40 kHz, 100 W, 600 s, +40 mg/L ClO_2	Reduced respiration rate; maintained acidity, flavonoid, and ascorbic acid	Aerobic mesophilic and aerobic psychotropic bacteria, yeast and mold: reduced counts	[8]
Sesame leaf	40 kHz, 400 W/L, 180 s, + SAEW + wash water	-	Yeast and mold: 1.76, *E. coli*: 2.33, *L. monocytogenes*: 2.4	[75]
Spinach	40 kHz, 400 W/L, 180 s, + SAEW + wash water	-	Yeast and mold: 1.97, *E. coli*: 2.41, *L. monocytogenes*: 2.49	[75]
	25 kHz, 79.41 W/L, 60 s	-	*E. coli*: 4.45, total aerobic count: 2.27, yeast and mold: 1.77	[63]
Strawberry	32 kHz, 10 W/L, 600 s + surfactant	-	*E. coli*: 1.96	[74]
	20 kHz, 30 W/L, 300 s, + O_3 + ClO_2	Maintained quality and physicochemical properties	*B. cinera*: growth and decay incidence inhibition	[10]
	35 kHz, 21.4 W/L, 120 s, 65 °C	Improved firmness; anthocyanin and color retained	Total mesophiles: 8.24	[56]
	40 kHz, 250–350 W, 600 s	Improved firmness, soluble solids, acidity, and vitamin C	Total bacteria: 1.9–2.42, yeast and mold: 1.35–2.45	[90]
	40 kHz, 24 W/L, 600 s + SAEW	Maintained physicochemical properties	Total aerobic bacteria, yeast and mold: 1.29	[83]
	37 kHz, 30 W/L, 3600 s, + BPW + 24 kJ/m^2, 1200 s	-	*E coli*: 3.04, *S. aureus*: 2.42, *S.* Enteritidis: 5.52, *L. innocua*: 6.12, HAdV: 2.73–3.98	[19,43]
	33 kHz, 60 W	Maintained physicochemical and antioxidant properties	Total bacteria count: 2.0, yeast and mold: 1.22	[71]

Table 2. *Cont.*

Fruit/ Vegetable	Ultrasound Parameters	Main Findings		Sources
		Quality Attributes	Microbial Log Reduction and Disease Incidence	
Tomato	20 kHz, 131.25 W/L, + NNEW	-	*E. coli*, *S.* Typhimurium: no detection	[80]
	45 kHz, 1140 s	Enhanced phenols; delayed color development	Total microbial count, yeast and mold: reduced count	[91]
	40 kHz, 24 W/L, 600 s + SAEW	Soluble solids; acidity and vitamin C maintained	Total aerobic bacteria: 1.77, yeast and mold: 1.50	[83]
	45 kHz, 600 s + peracetic acid	-	*S. enterica*: 3.90, yeast and mold: 3.4, aerobic mesophile: 4.44	[84]
	37 kHz, 30 W/L, 600 s, + BPW + 24 kJ/m^2, 1200 s	-	HAdV: 1.20–2.61	[43]
Watercress	35 kHz, 21.4 W/L, 120 s, 65 °C	Improved firmness; pronounced color difference	Total coliforms: 8.24	[56]
Zucchini	20 kHz + 0.165–8.2 kJ/m^2 + blanching (60–90 °C)	-	*Enterococcus faecalis*, *Deinococcus radiodurans*: 1–3	[26]

VI: vacuum impregnation, BPW: buffered peptone water, PAL: phenylalanine ammonia lyase, PA: peroxyacetic acid, SA: salicylic acid, POD: peroxidase, SAEW: slightly acidic electrolyzed water, NNEW: near neutral electrolyzed water, MAP: modified atmosphere packaging, HAdV: human adenovirus.

The systems for nano-ZnO coating application on kiwifruit after US treatment have also been evaluated [89]. Although individual US treatment of kiwifruit was somewhat effective, sequential application of US at 40 kHz, 350 W, and 1.2 g/L nano-ZnO coating was more effective in senescence delay, mass and firmness loss reduction, and storage life extension. The addition of mild heat to US treatment have also been explored, where a more pronounced effect of US (35 kHz, 120 W, and 21.4 W/L) was reported at 65 °C on red bell pepper, watercress, and strawberry, with improved quality retention and enhanced reductions of 7.43, 8.24, and 8.10 log CFU/g for *L. innocua*, total mesophiles, and total coliforms, respectively [56]. Similarly, the efficacy of US at 40 kHz, 400 W/L combined with SAEW was also enhanced at 40 and 60 °C, respectively, for fresh cut kale and bell pepper [21,86,87]. Browning and off-odor were under acceptable limits for fresh cut kale, while hardness and color parameters for control and treated samples were not significantly different for bell pepper. Enhanced bactericidal efficacy led to 3.32, 3.11, 3.00, 3.97, 3.66, 3.62, and 3.24 log CFU/g reductions in *E. coli*, *L. monocytogenes*, S. *Typhimurium*, total bacterial counts, *Enterobacteriaceae*, *Pseudomonas* spp., and yeast and mold counts, respectively.

4. Combination of UV-C Radiation and US Technology

Some studies have begun to examine the development of hurdle techniques involving UV-C radiation and US technology, either in sequential or simultaneous application. Sequential application involves treatment with one technology immediately followed by others and has shown improved microbial reductions and minimized quality changes in zucchini squash [26]. Alternative blanching treatments of UV-C radiation (1.17–8.2 kJ/m^2) followed by thermosonication (20 kHz, 90 °C) induced 3 log reductions in *Enterococcus faecalis* and *Deinococcus radiodurans*, compared with 1 log reduction achieved from individual treatment with UV-C radiation. Birmpa et al. [19] demonstrated that sequential application of US at 30 W/L followed by UV-C radiation at 24 kJ/m^2 evoked better log reductions (2–3 log) of human adenovirus (hAdVs) with less processing time, compared with individual results of UV-C radiation (2.13, 0.92, 1.25 log) and US (0.85, 0.36, 0.53 log) on lettuce, cherry tomato, and strawberry, respectively.

Simultaneous application, on the other hand, applies both technologies at the same time with enhanced results in disinfection efficacy and functional properties. A constant US intensity of 13.87 W/L from a 40 kHz, 1 kW transducer and UV-C radiation dosage of 0.72–10.76 kJ/m^2 revealed increasing reduction in the population of total aerobic bacteria on tomato and up to 2.33 log reduction, while yeast and mold were undetected at higher dosage levels [22]. Enhanced biosynthesis and extractability of phytochemicals were also reported in tomato, with 36, 60, 30, and 90% increases in antioxidant activities, vitamin C, total phenols, and lycopene contents, respectively [1,92]. The combined effects of UV-C radiation and US cavitation were observed to disrupt the structural integrity of macromolecules and chemical bonds of tomato phytochemicals, which stimulated their accumulation and facilitated their extraction. Collectively, these few studies present the potential of a hurdle technology involving UV-C radiation and US technology for the postharvest decontamination of fruits and vegetables, with this area of focus warranting further exploration.

The theory elucidating the mechanism of the synergy between US technology and UV-C radiation can be explained from their individual mechanisms of operation. It is possible that the resultant mechanical effect of the collapse of transient cavitation from US technology, which includes high heating and cooling rate (10^9 K/s), shock wave formation, high temperature (5000 K), and pressure (1000 atm), ruptures microbial cell envelopes [93]. The ruptured cell envelopes then facilitate access of UV-C radiation to light-sensitive DNA of the microorganisms within a short processing time. The formation of cyclobutane thymine dimmers from the action of UV-C radiation combined with chemical effects of OH^- and H^+ species formation from US technology eventually offers a more potent bactericidal effect that is capable of depleting microbial growth, leading to the eventual death of microorganisms [42,93]. However, this theory needs to be properly tested from more studies and relevant tools.

5. Future Perspectives and Industrial Relevance

The low penetrating capacity of UV-C radiation into opaque substances, high energy consumption, and long treatment time associated with US technology has limited their individual usage. Thus, research in the applications of UV-C radiation and US technology for postharvest decontamination of fruits and vegetables is focused on hurdle technology of combination applications, which have shown enhanced decontamination efficiency. The hurdle technology has shown potential of overcoming limitations and harnessing advantages of each individual technology. As studies reported on this hurdle technology are relatively few, it is necessary to have more validation towards achieving industrial-scale applications. Scale-up studies to larger proportions would reveal important considerations to ensure uniformity of US cavitation and UV-C exposure on food surfaces due to variations in ultrasonic field and non-penetrative ability of UV-C radiation, which have been observed as the most important challenges to the commercialization of US technology and UV-C radiation for fresh produce decontamination [63,94]. The use of agitation in free-flowing US baths and application of mixed-mode US frequencies can reduce standing-wave effects and creation of different cavitation bubble size to ensure uniform cavitation [20,63]. Wide variations in UV-C radiation intensity can be reduced so that energy reaches the samples more effectively [26]. The radiochromic film dosimetry approach could also be introduced to UV-C treatment to measure or evaluate exposure of radiation dose. Film dosimetry is reportedly associated with measuring and verifying the dosage level on food substances during irradiation [94,95]. Once non-uniformity of irradiation or wide variations in dose is established, a system to rotate irradiated samples as described by Yan et al. [26] can be introduced to ensure UV-C dosage uniformity on the samples for improved efficiency, especially in high loading industrial applications. For combination approaches of US technology and UV-C radiation, we can incorporate sanitizer solutions as the washing medium in US baths lined with UV-C radiation lamps to enhance efficiency and reduce cross-contamination. The heat generated from simultaneous combination energies, which may cause dehydration and damage the appearance of fresh produce, may be overcome by incorporating a water circulatory system that simultaneously introduces and drains water in the tank to help control temperature rise and ensure minimal cross-contamination. Identification of important factors influencing the decontamination process can be enhanced through better understanding of microbes' behavior where more accurate inactivation of target microorganisms can assist optimization of the process parameters for cost-effective practical applications.

6. Conclusions

The research on the postharvest decontamination of fruits and vegetables is rapidly evolving due to increased safety awareness and consumer demands for minimally processed foods with fresh-like characteristics. UV-C radiation and US technology have presented encouraging results in postharvest decontamination of fresh produce with minimal effects on quality characteristics. However, long exposure time, high US energy consumption, and limited penetration capability of UV-C radiation have plagued research in these areas. The potential for synergistic or complementary effects for improved efficiency have also been established when UV-C radiation or US technology is applied in combination with other antimicrobial techniques. The hurdle technology involving UV-C radiation and US technology offer substitute benefits towards ensuring overall safety and wholesomeness of fruits and vegetables in comparison with individual applications or their combinations with conventional approaches. More studies are required on this hurdle technology, while scale-up and optimization of the process will provide opportunities for industry-led collaborations and chart the path towards effective commercialization.

Author Contributions: Conceptualization, O.J.E. and N.L.C.; investigation, O.J.E.; resources, O.J.E. and N.L.C.; data curation, O.J.E.; writing—original draft preparation, O.J.E.; writing—review and editing, N.L.C.; supervision, N.L.C., Y.A.Y., and R.S.; project administration, N.L.C.; funding acquisition, O.J.E and N.L.C. All authors have read and agreed to the published version of the manuscript.

Funding: This research was funded by Universiti Putra Malaysia, grant number GP-I/2014/9439000, and the first author was funded by Niger Delta Development Commission, Nigeria (NDDC/DEHSS/2014PGFS/AKS/014), for postgraduate study.

Acknowledgments: The authors are grateful to Daniel Onwude for providing some insights and discussion of the manuscript.

Conflicts of Interest: The authors declare no conflict of interest.

References

1. Esua, O.J.; Chin, N.L.; Yusof, Y.A.; Sukor, R. Effects of simultaneous UV-C radiation and ultrasonic energy postharvest treatment on bioactive compounds and antioxidant activity of tomatoes during storage. *Food Chem.* **2019**, *270*, 113–122. [CrossRef] [PubMed]
2. Dyshlyuk, L.; Babich, O.; Prosekov, A.; Ivanova, S.; Pavsky, V.; Chaplygina, T. The effect of postharvest ultraviolet irradiation on the content of antioxidant compounds and the activity of antioxidant enzymes in tomato. *Heliyon* **2020**, *6*, e03288. [CrossRef] [PubMed]
3. Gustavsson, J.; Cederberg, C.; Sonesson, U.; van Otterdijk, R.; Meybeck, A. Global Food Losses and Food Waste: Extent, Causes and Prevention. Food and Agricultural Organisation of the United Nations, 2011. Available online: http://www.fao.org/fileadmin/user_upload/suistainability/pdf/Global_Food_Losses_and_Food_Waste.pdf (accessed on 2 March 2016).
4. Scallan, E.; Hoekstra, R.M.; Angulo, F.J.; Tauxe, R.V.; Widdowson, M.-A.; Roy, S.L.; Jones, J.L.; Griffin, P.M. Foodborne illness acquired in the United States—major pathogens. *Emerg. Infect. Dis.* **2011**, *17*, 7–15. [CrossRef] [PubMed]
5. Kirk, M.D.; Pires, S.M.; Black, R.E.; Caipo, M.; Crump, J.A.; Devleesschauwer, B.; Dopfer, D.; Fazil, A.; Fischer-Walker, C.L.; Hald, T.; et al. World Health Organization estimates of the global and regional disease burden of 22 foodborne bacterial, protozoal, and viral diseases, 2010: A data synthesis. *PLoS Med.* **2015**, *12*, e1001921. [CrossRef]
6. Esua, O.J.; Chin, N.L.; Yusof, Y.A.; Sukor, R. Antioxidant bioactive compounds and spoilage microorganisms of wax apple (*Syzygium samarangense*) during room temperature storage. *Int. J. Fruit Sci.* **2017**, *17*, 188–201. [CrossRef]
7. Food and Agricultural Organisation of the United Nations. Available online: http://www.fao.org/3/CA1397EN/ca1397en.pdf (accessed on 15 January 2019).
8. Chen, Z.; Zhu, C. Combined effects of aqueous chlorine dioxide and ultrasonic treatments on postharvest storage quality of plum fruit (*Prunus salicina* L.). *Postharvest Biol. Technol.* **2011**, *61*, 117–123. [CrossRef]
9. Phungamngoen, C.; Chiewchan, N.; Devahastin, S. Effects of various pretreatments and drying methods on *Salmonella* resistance and physical properties of cabbage. *J. Food Eng.* **2013**, *115*, 237–244. [CrossRef]
10. Aday, M.S.; Caner, C. Individual and combined effects of ultrasound, ozone and chlorinedioxide on strawberry storage life. *LWT Food Sci. Technol.* **2014**, *57*, 344–351. [CrossRef]
11. Martinez-Hernandez, G.B.; Navarro-Rico, J.; Gomez, P.A.; Oton, M.; Artes, F.; Artes-Hernandez, F. Combined sustainable sanitising treatments to reduce *Escherichia coli* and *Salmonella* Enteritidis growth on fresh-cut kalian-hybrid broccoli. *Food Control* **2015**, *47*, 312–317. [CrossRef]
12. Olosunde, W.A.; Aremu, A.K.; Onwude, D.I. Development of a solar powered evaporative cooling storage system for tropical fruits and vegetables. *J. Food Process. Preserv.* **2016**, *40*, 279–290. [CrossRef]
13. Tawema, P.; Han, J.; Vu, K.D.; Salmieri, S.; Lacroix, M. Antimicrobial effects of combined UV-C or gamma radiation with natural antimicrobial formulations against *Listeria monocytogenes, Escherichia coli* O157:H7, and total yeasts/molds in fresh cut cauliflower. *LWT Food Sci. Technol.* **2016**, *65*, 451–456. [CrossRef]
14. Dziedzic, E.; Blaszczyk, J.; Bieniasz, M.; Dziadek, K.; Kopec, A. Effect of modified (MAP) and controlled atmosphere (CA) storage on the quality and bioactive compounds of blue honeysuckle fruits (*Lonicera caerulea* L.). *Sci. Hortic.* **2020**, *265*, 109226. [CrossRef]
15. Millan-Sango, D.; Garroni, E.; Farrugia, C.; Van Impe, J.F.M.; Valdramidis, V.P. Determination of the efficacy of ultrasound combined with essential oils on the decontamination of *Salmonella* inoculated lettuce leaves. *LWT Food Sci. Technol.* **2016**, *73*, 80–87. [CrossRef]

16. Park, S.-H.; Kang, J.-W.; Kang, D.-H. Inactivation of foodborne pathogens on fresh produce by combined treatment with UV-C radiation and chlorine dioxide gas, and mechanisms of synergistic inactivation. *Food Control* **2018**, *92*, 331–340. [CrossRef]
17. Lu, C.; Ding, J.; Park, H.K.; Feng, H. High intensity ultrasound as a physical elicitor affects secondary metabolites and antioxidant capacity of tomato fruits. *Food Control* **2020**, *113*, 107176. [CrossRef]
18. Darras, A.I.; Tsikaloudakis, G.; Lycoskoufis, I.; Dimitriadis, C.; Karamousantas, D. Low doses of UV-C irradiation affects growth, fruit yield and photosynthetic activity of tomato plants. *Sci. Hortic.* **2020**, *267*, 109357. [CrossRef]
19. Birmpa, A.; Bellou, M.; Kokkinos, P.; Vantarakis, A. Effect of nonthermal, conventional, and combined disinfection technologies on the stability of human adenoviruses as fecal contaminants on surfaces of fresh ready-to-eat products. *J. Food Prot.* **2016**, *79*, 454–462. [CrossRef]
20. Mustapha, A.T.; Zhou, C.; Wahia, H.; Amanor-Atiemoh, R.; Otu, P.; Qudus, A.; Fakayode, O.A.; Ma, H. Sonozonation: Enhancing the antimicrobial efficiency of aqueous ozone washing techniques on cherry tomato. *Ultrason. Sonochem.* **2020**, *64*, 105059. [CrossRef] [PubMed]
21. Luo, K.; Oh, D.-H. Inactivation kinetics of *Listeria monocytogenes* and *Salmonella enterica* serovar Typhimurium on fresh-cut bell pepper treated with slightly acidic electrolyzed water combined with ultrasound and mild heat. *Food Microbiol.* **2016**, *53*, 165–171. [CrossRef] [PubMed]
22. Esua, O.J.; Chin, N.L.; Yusof, Y.A. Simultaneous UV-C and ultrasonic energy treatment for disinfection of tomatoes and its antioxidant properties. *J. Adv. Agric. Technol.* **2018**, *5*, 209–214. [CrossRef]
23. Collazo, C.; Noguera, V.; Aguilo-Aguayo, I.; Abadias, M.; Colas-Meda, P.; Nicolau, I.; Vinas, I. Assessing water-assisted UV-C light and its combination with peroxyacetic acid and *Pseudomonas graminis* CPA-7 for the inactivation and inhibition of *Listeria monocytogenes* and *Salmonella enterica* in fresh-cut 'Iceberg' lettuce and baby spinach leaves. *Int. J. Food Microbiol.* **2019**, *297*, 11–20. [CrossRef]
24. Langcake, P.; Pryce, R.J. The production of resveratrol and the viniferins by grapevines in response to ultraviolet irradiation. *Phytochemistry* **1977**, *16*, 1193–1196. [CrossRef]
25. Artes, F.; Allende, A. Processing lines and alternative preservation techniques to prolong the shelf-life of minimally fresh processed leafy vegetables. *Europ. J. Hort. Sci.* **2005**, *70*, 231–245.
26. Yan, R.; Mattheis, J.; Gurtler, J.; Sites, J.; Fan, X. UV-C inactivation of *Escherichia coli* and dose uniformity on apricot fruit in a commercial setting. *Postharvest Biol. Technol.* **2014**, *95*, 46–49. [CrossRef]
27. Taze, B.H.; Unluturk, S. Effect of postharvest UV-C treatment on the microbial quality of 'Salak' apricot. *Sci. Hortic.* **2018**, *233*, 370–377. [CrossRef]
28. Mukhopadhyay, S.; Ukuku, D.O.; Juneja, V.; Fan, X. Effects of UV-C treatment on inactivation of *Salmonella enterica* and *Escherichia coli* O157:H7 on grape tomato surface and stem scars, microbial loads, and quality. *Food Control* **2014**, *44*, 110–117. [CrossRef]
29. Lim, W.; Harrison, M.A. Effectiveness of UV light as a means to reduce Salmonella contamination on tomatoes and food contact surfaces. *Food Control* **2016**, *66*, 166–173. [CrossRef]
30. Neves, F.I.G.; Silva, C.L.M.; Vieira, M.C. Combined pre-treatments effects on zucchini (*Cucurbita pepo* L.) squash microbial load reduction. *Int. J. Food Microbiol.* **2019**, *305*, 108257. [CrossRef]
31. Bermudez-Aguirre, D.; Barbosa-Canovas, G.V. Disinfection of selected vegetables under non-thermal treatments: Chlorine, acid citric, ultraviolet light and ozone. *Food Control* **2013**, *29*, 82–90. [CrossRef]
32. Cote, S.; Rodoni, L.; Miceli, E.; Concellon, A.; Civello, P.M.; Vincente, A.R. Effect of radiation intensity on the outcome of postharvest UV-C treatments. *Postharvest Biol. Technol.* **2013**, *83*, 83–89. [CrossRef]
33. Adhikari, A.; Syamaladevi, R.M.; Killinger, K.; Sablani, S.S. Ultraviolet-C light inactivation of *Escherichia coli* O157:H7 and *Listeria monocytogenes* on organic fruit surfaces. *Int. J. Food Microbiol.* **2015**, *210*, 136–142. [CrossRef]
34. Syamaladevi, R.M.; Lu, X.; Sablani, S.S.; Insan, S.K.; Adhikari, A.; Killinger, K.; Rasco, B.; Dhingra, A.; Bandyopadhyay, A.; Annapure, U. Inactivation of *Eschrichia coli* population on fruit surfaces using UVC light: Influence on fruit surface characteristics. *Food Bioproc. Tech.* **2013**, *6*, 2959–2973. [CrossRef]
35. Nigro, F.; Ippolito, A.; Lattanzio, V.; Di Venere, D.; Salerno, M. Effect of ultraviolet –C light on postharvest decay of strawberry. *J. Plant. Pathol.* **2000**, *82*, 29–37.
36. Alegria, C.; Pinheiro, J.; Duthoit, M.; Goncalves, E.M.; Moldao-Martins, M.; Abreu, M. Fresh-cut carrot (cv. Nantes) quality as affected by abiotic stress (heat shock and UV-C irradiation) pre-treatments. *LWT Food Sci. Technol.* **2012**, *48*, 197–203. [CrossRef]

37. Yun, J.; Yan, R.; Fan, X.; Gurtler, J.; Phillips, J. Fate of *E. coli* O157:H7, *Salmonella* spp. and potential surrogate bacteria on apricot fruit, following exposure to UV-C light. *Int. J. Food Microbiol.* **2013**, *166*, 356–363. [CrossRef]
38. Graca, A.; Salazar, M.; Quintas, C.; Nunesaday, C. Low dose UV-C illumination as an eco-innovative disinfection system on minimally processed apples. *Postharvest Biol. Technol.* **2013**, *85*, 1–7. [CrossRef]
39. George, D.S.; Razali, Z.; Santhirasegaram, V.; Somasundram, C. Effect of postharvest ultraviolet-C treatment on the proteome changes in fresh cut mango (*Mangifera indica* L. cv. Chokanan). *J. Sci. Food Agric.* **2016**, *98*, 2851–2860. [CrossRef] [PubMed]
40. Wang, D.; Chen, L.; Ma, Y.; Zhang, M.; Zhao, Y.; Zhao, X. Effect of UV-C treatment on the quality of fresh-cut lotus (*Nelumbo nucifera* Gaertn.) root. *Food Chem.* **2019**, *278*, 659–664. [CrossRef]
41. Pinheiro, J.; Alegria, C.; Abreu, M.; Goncalves, E.M.; Silva, C.L.M. Use of UV-C postharvest treatment for extending fresh whole tomato (*Solanum lycopersicum*, cv. Zinac) shelf-life. *J. Food Sci. Technol.* **2015**, *52*, 5066–5074. [CrossRef]
42. Sommers, C.H.; Sites, J.E.; Musgrove, M. Ultraviolet light (254 nm) inactivation of pathogens on foods and stainless steel surfaces. *J. Food Saf.* **2010**, *30*, 470–479. [CrossRef]
43. Birmpa, A.; Sfika, V.; Vantarakis, A. Ultraviolet light and ultrasound as non-thermal treatments for the inactivation of microorganisms in fresh ready-to-eat foods. *Int. J. Food Microbiol.* **2013**, *167*, 96–102. [CrossRef] [PubMed]
44. Gomez, P.L.; Alzamora, S.M.; Castadhikariro, M.A.; Salvatori, D.M. Effect of ultraviolet-C light dose on quality of cut-apple: Microorganism, color and compression behavior. *J. Food Eng.* **2010**, *98*, 60–70. [CrossRef]
45. Perkins-Veazie, P.; Collins, J.K.; Howard, L. Blueberry fruit response to postharvest application of ultraviolet radiation. *Postharvest Biol. Technol.* **2008**, *47*, 280–285. [CrossRef]
46. Kim, C.; Hung, Y.C. Inactivation of *E. coli* O157:H7 on blueberries by electrolyzed water, ultraviolet light, and ozone. *J. Food Sci.* **2012**, *77*, 206–211. [CrossRef] [PubMed]
47. Martinez-Hernandez, G.B.; Artes-Hernandez, F.; Gomez, P.A.; Formica, A.C.; Artes, F. Combination of electrolysed water, UV-C and super-atmospheric O_2 packaging for improving fresh-cut broccoli quality. *Postharvest Biol. Technol.* **2013**, *76*, 125–134. [CrossRef]
48. Jemni, M.; Gomez, P.A.; Souza, M.; Chaira, N.; Ferchichi, A.; Oton, M.; Artes, F. Combined effect of UV-C, ozone and electrolysed water for keeping overall quality of date palm. *LWT Food Sci. Technol.* **2014**, *59*, 649–655. [CrossRef]
49. Freitas, P.M.; Lopez-Galvez, F.; Tudela, J.A.; Gil, M.I.; Allende, A. Postharvest treatment of table grapes with ultraviolet-C and chitosan coating preserves quality and increases stilbene content. *Postharvest Biol. Technol.* **2015**, *105*, 51–57. [CrossRef]
50. Santo, D.; Graca, A.; Nunes, C.; Quintas, C. *Escherichia coli* and *Cronobacter sakazakii* in Tommy Atkins minimally processed mangos: Survival, growth and effect of UV-C and electrolyzed water. *Food Microbiol.* **2018**, *70*, 49–54. [CrossRef]
51. Manzocco, L.; Da Pieve, S.; Maifreni, M. Impact of UV-C light on safety and quality of fresh-cut melon. *Innov. Food Sci. Emerg. Technol.* **2011**, *12*, 13–17. [CrossRef]
52. Artes-Hernandez, F.; Robles, P.A.; Gomez, P.A.; Tomas-Callejas, A.; Artes, F. Low UV-C illumination for keeping overall quality of fresh-cut watermelon. *Postharvest Biol. Technol.* **2010**, *55*, 114–120. [CrossRef]
53. Syamaladevi, R.M.; Lupien, S.L.; Bhunia, K.; Sablani, S.S.; Dugan, F.; Rasco, B.; Killinger, K.; Dhinger, A.; Ross, C. UV-C light inactivation kinetics of *Penicillium expansum* on pear surfaces: Influence on physicochemical and sensory quality during storage. *Postharvest Biol. Technol.* **2014**, *87*, 27–32. [CrossRef]
54. Schenk, M.; Guerrero, S.; Alzamora, S.M. Response of some microorganisms to ultraviolet treatment on fresh-cut pear. *Food Bioproc. Tech.* **2008**, *1*, 384–392. [CrossRef]
55. Graca, A.; Santo, D.; Quintas, C.; Nunes, C. Growth of *Escherichia coli*, *Salmonella enterica* and *Listeria* spp., and their inactivation using ultraviolet energy and electrolyzed water, on 'Rocha' fresh-cut pears. *Food Control* **2017**, *77*, 41–49. [CrossRef]
56. Alexandre, E.M.C.; Santos-Pedro, D.M.; Brandao, T.R.S.; Silva, C.L.M. Study on thermosonication and ultraviolet radiation processes as an alternative to blanching for some fruits and vegetables. *Food Bioproc. Tech.* **2011**, *4*, 1012–1019. [CrossRef]

57. Manzocco, L.; Plazzotta, S.; Maifreni, M.; Calligaris, S.; Anese, M.; Nicoli, M.C. Impact of UV-C light on storage quality of fresh-cut pineapple in two different packages. *LWT Food Sci. Technol.* **2016**, *65*, 1138–1143. [CrossRef]
58. Kim, H.-G.; Song, K.B. Combined treatment with chlorine dioxide gas, fumaric acid, and ultraviolet-C light for inactivating *Escherichia coli* O157:H7 and *Listeria monocytogenes* inoculated on plums. *Food Control* **2017**, *71*, 371–375. [CrossRef]
59. Lopez-Rubira, V.; Conesa, A.; Allende, A.; Artes, F. Shelf life and overall quality of minimally processed pomegranate arils modified atmosphere packaged and treated with UV-C. *Postharvest Biol. Technol.* **2005**, *37*, 174–185. [CrossRef]
60. Artes-Hernandez, F.; Escalona, V.H.; Robles, P.A.; Martinez-Hernandez, G.B.; Artes, F. Effect of UV-C radiation on quality of minimally processed spinach leaves. *J. Sci. Food Agric.* **2009**, *89*, 414–421. [CrossRef]
61. Choi, D.S.; Park, S.H.; Choi, S.R.; Kim, J.S.; Chun, H.H. The combined effects of ultraviolet-C irradiation and modified atmosphere packaging for inactivating *Salmonella enterica* serovar Typhimurium and extending the shelf life of cherry tomatoes during cold storage. *Food Packag Shelf Life* **2015**, *3*, 19–30. [CrossRef]
62. Mansourbahmani, S.; Ghareyazie, B.; Kalatejari, S.; Mohammadi, R.S.; Zarinnia, V. Effect of post-harvest UV-C irradiation and calcium chloride on enzymatic activity and decay of tomato (*Lycopersicon esculentum* L.) fruit during storage. *J. Integr. Agric.* **2017**, *16*, 2093–2100. [CrossRef]
63. Zhou, B.; Feng, H.; Pearlstein, A.J. Continuous-flow ultrasonic washing system for fresh produce surface decontamination. *Innov. Food Sci. Emerg. Technol.* **2012**, *16*, 427–435. [CrossRef]
64. Mason, T.J. *Chemistry with Ultrasound*; Elsevier: London, UK, 1990; pp. 1–26.
65. Fan, K.; Zhang, M.; Jiang, F. Ultrasound treatment to modified atmospheric packaged fresh-cut cucumber: Influence on microbial inhibition and storage quality. *Ultrason. Sonochem.* **2019**, *54*, 162–170. [CrossRef] [PubMed]
66. Millan-Sango, D.; McElhatton, A.; Valdramidis, V.P. Determination of the efficacy of ultrasound in combination with essential oil of oregano for the decontamination of *Escherichia coli* on inoculated lettuce leaves. *Food Res. Int.* **2015**, *67*, 145–154. [CrossRef]
67. Aday, M.S.; Temizkan, R.; Buyukcan, M.B.; Caner, C. An innovative technique for extending shelf life of strawberry: Ultrasound. *LWT Food Sci. Technol.* **2013**, *52*, 93–101. [CrossRef]
68. Wang, W.; Ma, X.; Zou, M.; Jiang, P.; Hu, W.; Li, J.; Zhi, Z.; Chen, J.; Li, S.; Ding, T.; et al. Effects of ultrasound on spoilage microorganisms, quality, and antioxidant capacity of postharvest cherry tomatoes. *J. Food Sci.* **2015**, *80*, 2117–2126. [CrossRef]
69. Yilmaz, F.M.; Bilek, S.E. Ultrasound-assisted vacuum impregnation on the fortification of fresh-cut apple with calcium and black carrot phenolics. *Ultrason. Sonochem.* **2018**, *48*, 509–516. [CrossRef]
70. Wiktor, A.; Sledz, M.; Nowacka, M.; Rybak, K.; Witrowa-Rajchert, D. The influence of immersion and contact ultrasound treatment on selected properties of the apple tissue. *Appl. Acoust.* **2016**, *103*, 136–142. [CrossRef]
71. Gani, A.; Baba, W.N.; Ahmad, M.; Shah, U.; Khan, A.A.; Wani, I.A.; Masoodi, F.A.; Gani, A. Effect of ultrasound treatment on physico-chemical, nutraceutical and microbial quality of strawberry. *LWT Food Sci. Technol.* **2016**, *66*, 496–502. [CrossRef]
72. Yu, J.; Engeseth, N.J.; Feng, H. High intensity ultrasound as an abiotic elicitor—effects on antioxidant capacity and overall quality of romaine lettuce. *Food Bioproc. Tech.* **2016**, *9*, 262–273. [CrossRef]
73. Huang, T.-S.; Xu, C.; Walker, K.; West, P.; Zhang, S.; Weese, J. Decontamination efficacy of combined chlorine dioxide with ultrasonication on apples and lettuce. *J. Food Sci.* **2006**, *71*, 134–139. [CrossRef]
74. Seymour, I.J.; Burfoot, D.; Smith, R.L.; Cox, L.A.; Lockwood, A. Ultrasound decontamination of minimally processed fruits and vegetables. *Int. J. Food Sci. Technol.* **2002**, *37*, 547–557. [CrossRef]
75. Forghani, F.; Oh, D.-H. Hurdle enhancement of slightly acidic electrolyzed water antimicrobial efficacy on Chinese cabbage, lettuce, sesame leaf and spinach using ultrasonication and water wash. *Food Microbiol.* **2013**, *36*, 40–45. [CrossRef] [PubMed]
76. Bang, H.-J.; Park, S.Y.; Kim, S.E.; Rahman, M.M.F.; Ha, S.-D. Synergistic effects of combined ultrasound and peroxyacetic acid treatments against *Cronobacter sakazakii* biofilms on fresh cucumber. *LWT Food Sci. Technol.* **2017**, *84*, 91–98. [CrossRef]
77. Sagong, H.-G.; Lee, S.-Y.; Chang, P.-S.; Heu, S.; Ryu, S.; Choi, Y.-J.; Kang, D.-H. Combined effect of ultrasound and organic acids to reduce *Escherichia coli* O157:H7, *Salmonella* Typhimurium, and *Listeria monocytogenes* on organic fresh lettuce. *Int. J. Food Microbiol.* **2011**, *145*, 287–292. [CrossRef]

78. Huang, K.; Wrenn, S.; Tikekar, R.; Nitin, N. Efficacy of decontamination and a reduced risk of cross-contamination during ultrasound-assisted washing of fresh produce. *J. Food Eng.* **2018**, *224*, 95–104. [CrossRef]
79. Park, S.Y.; Mizan, M.F.R.; Ha, S.-D. Inactivation of *Cronobacter sakazakii* in head lettuce by using a combination of ultrasound and sodium hypochlorite. *Food Control* **2016**, *60*, 582–587. [CrossRef]
80. Afari, G.K.; Hung, Y.-C.; King, H.C.; Hu, A. Reduction of *Escherichia coli* O157:H7 and *Salmonella* Typhimurium DT104 on fresh produce using an automated washer with near neutral electrolyzed (NEO) water and ultrasound. *Food Control* **2016**, *63*, 246–254. [CrossRef]
81. de Sao Jose, J.F.B.; de Medeiros, H.S.; Bernardes, P.C.; de Andrade, N.J. Removal of *Salmonella enterica* Enteritidis and *Escherichia coli* from green peppers and melons by ultrasound and organic acids. *Int. J. Food Microbiol.* **2014**, *190*, 9–13. [CrossRef] [PubMed]
82. Yang, Z.; Cao, S.; Cai, Y.; Zheng, Y. Combination of salicylic acid and ultrasound to control postharvest blue mold caused by *Penicillium expansum* in peach fruit. *Innov. Food Sci. Emerg. Technol.* **2011**, *12*, 310–314. [CrossRef]
83. Ding, T.; Ge, Z.; Shi, J.; Xu, Y.-T.; Jones, C.L.; Liu, D.-H. Impact of slightly acidic electrolyzed water (SAEW) and ultrasound on microbial loads and quality of fresh fruits. *LWT Food Sci. Technol.* **2015**, *60*, 1195–1199. [CrossRef]
84. de Sao Jose, J.F.B.; Vanetti, M.C.D. Effect of ultrasound and commercial sanitizers in removing natural contaminants and *Salmonella enterica* Typhimurium on cherry tomatoes. *Food Control* **2012**, *24*, 95–99. [CrossRef]
85. Wang, J.; Fan, L. Effect of ultrasound treatment on microbial inhibition and quality maintenance of green asparagus during cold storage. *Ultrason. Sonochem.* **2019**, *58*, 104631. [CrossRef] [PubMed]
86. Mansur, A.R.; Oh, D.-H. Combined effect of thermosonication and slightly acidic electrolyzed water to reduce foodborne pathogens and spoilage microorganisms on fresh-cut kale. *J. Food Sci.* **2015**, *80*, 1277–1284. [CrossRef] [PubMed]
87. Mansur, A.R.; Oh, D.-H. Combined effects of thermosonication and slightly acidic electrolyzed water on the microbial quality and shelf life extension of fresh-cut kale during refrigeration storage. *Food Microbiol.* **2015**, *51*, 154–162. [CrossRef] [PubMed]
88. Vivek, K.; Subbarao, K.V.; Srivastava, B. Optimization of postharvest ultrasonic treatment of kiwifruit using RSM. *Ultrason. Sonochem.* **2016**, *32*, 328–335. [CrossRef]
89. Meng, X.; Zhang, M.; Adhikari, B. The effects of ultrasound treatment and nano-zinc oxide coating on the physiological activities of fresh-cut kiwifruit. *Food Bioproc. Tech.* **2014**, *7*, 126–132. [CrossRef]
90. Cao, S.; Hu, Z.; Pang, B. Optimization of postharvest ultrasonic treatment of strawberry fruit. *Postharvest Biol. Technol.* **2010**, *55*, 150–153. [CrossRef]
91. Pinheiro, J.; Alegria, C.; Abreu, M.; Goncalves, E.M.; Silva, C.L.M. Influence of postharvest ultrasounds treatments on tomato (*Solanum lycopersicum*, cv. Zinac) quality and microbial load during storage. *Ultrason. Sonochem.* **2015**, *27*, 552–559. [CrossRef]
92. Esua, O.J.; Chin, N.L.; Yusof, Y.A.; Sukor, R. Combination of ultrasound and ultraviolet-C irradiation on kinetics of color, firmness, weight loss, and total phenolic content changes in tomatoes during storage. *J. Food Process. Preserv.* **2019**, *43*, 14161. [CrossRef]
93. Feng, H.; Yang, W.; Hielscher, T. Power ultrasound. *Food Sci. Technol. Int.* **2008**, *14*, 433–436. [CrossRef]
94. Yan, R.; Yun, J.; Gurtler, J.; Fan, X. Radiochromic film dosimetry for UV-C treatments of apple fruit. *Postharvest Biol. Technol.* **2017**, *127*, 14–20. [CrossRef]
95. Fan, X.; Thayer, D.W. Quality of irradiated alfalfa sprouts. *J. Food Protect.* **2001**, *64*, 1574–1578. [CrossRef]

Publisher's Note: MDPI stays neutral with regard to jurisdictional claims in published maps and institutional affiliations.

processes

MDPI

Article

Impact of Process Parameters and Bulk Properties on Quality of Dried Hops

Sharvari Raut [1,*], Gardis J. E. von Gersdorff [1], Jakob Münsterer [2], Klaus Kammhuber [3], Oliver Hensel [1] and Barbara Sturm [1,4,5]

1. Department of Agricultural and Biosystems Engineering, University of Kassel, Nordbahnhofstr. 1a, D-37213 Witzenhausen, Germany; g.gersdorff@uni-kassel.de (G.J.E.v.G.); agrartechnik@uni-kassel.de (O.H.); Barbara.sturm@uni-kassel.de (B.S.)
2. Bavarian State Research Center for Agriculture (LfL), Hop Competence Centre Production Technologies, Hüll, 85354 Wolnzach, Germany; Jakob.Muensterer@lfl.bayern.de
3. Bavarian State Research Center for Agriculture (LfL), Hop Research Centre Hüll, 85354 Wolnzach, Germany; Klaus.Kammhuber@LfL.bayern.de
4. Leibniz Institute for Agricultural Engineering and Bioeconomy (ATB), Max-Eyth Allee 100, 14469 Potsdam, Germany
5. Albrecht Daniel Thaer-Institute of Agricultural and Horticultural Sciences, Humboldt University of Berlin, Hinter der Reinhardtstr. 6–8, 10115 Berlin, Germany

* Correspondence: sharvari.raut@uni-kassel.de

Received: 26 October 2020; Accepted: 18 November 2020; Published: 20 November 2020

Abstract: Hops are critical to the brewing industry. In commercial hop drying, a large bulk of hops is dried in multistage kilns for several hours. This affects the drying behavior and alters the amount and chemical composition of the hop oils. To understand these changes, hops of the var. Hallertauer Tradition were dried in bulks of 15, 25 and 35 kg/m^2 at 60 °C and 0.35 m/s. Additionally, bulks of 25 kg/m^2 were also dried at 65 °C and 0.45 m/s to assess the effect of change in temperature and velocity, respectively. The results obtained show that bulk weights significantly influence the drying behavior. Classification based on the cone size reveals 45.4% medium cones, 41.2% small cones and 8.6% large cones. The highest ΔE value of 6.3 and specific energy consumption (113,476 kJ/kg$_{H2O}$) were observed for the 15 kg/m^2 bulk. Increasing the temperature from 60 °C to 65 °C increased the oil yield losses by about 7% and myrcene losses by 22%. The results obtained show that it is important to define and consider optimum bulk and process parameters, to optimize the hop drying process to improve the process efficiency as well the product quality.

Keywords: bulk weight; hop cones size distribution; chemical analysis; energy consumption

1. Introduction

Hop (Humulus lupulus) is a perennial climbing plant that is used in the brewing industry to impart a bitter taste and provide a "hoppy" aroma to beer [1]. About 97% of the world's hop crop is used for beer making [2]. Initially, hops were primarily used to provide stability to beer and increase shelf life. However, with extensive research and knowledge acquired over time, hops have also increasingly been used for other purposes such as creation of foam characteristics [3]. The bittering acids, tannins, chemical and aroma compounds are situated in the lupulin glands of the hop cone [1,4]. As of 2016, hops can be spilt into three categories namely aroma hops, bitter hops, and dual-purpose hops. Aroma hops are responsible for contributing fruity, piney, citrusy, and tropical aromas to the beer while bittering hops are associated with its bitterness [5,6]. Dual-purpose hops are described as bitter hops with special aroma characteristics [6]. However, freshly harvested hop cones have a high water content, which in turn promotes microbial spoilage and thus, cannot be stored for an extended period of time [7]. Therefore, to ensure minimal post-harvest losses and consistent product quality, hop cones need to be processed immediately after harvesting. In the Hallertau region, 22 hop varieties are grown [6]. As harvest lasts for a maximum of eight weeks a year, and with harvesting period for each variety lasting for less than two weeks, the efficient processing of harvested hops is of great economic importance to growers [8].

To extend shelf life and increase stability of hops, convection drying is used as a preservation technique by hop growers. Hop drying is a complicated process due to both the physical and chemical characteristics of the hop cone. The anatomy of the hop cone consists of strig as its main axis, followed by bracteole which is made up of sheets of fine petals adhering and covering the strig and contain the lupulin glands and finally the bract as the outer structure of the hop cone which consists the hop leaves. Compared to the mass, the bracteole and bract account for relatively larger surface area of the cone. The strig, however has an extremely small surface area. During the drying process, the strig is shielded from the drying air due to the exterior bract and thus leading to high-water content retention. Furthermore, the cone size and shape, strig shape, arrangement of the cone leaves and density of the capillaries in the bracteole differ in individual hop varieties and thus further complicate the drying process [9]. According to [9], at the end of the drying process if a hop cone has a water content of 8–9%, the water content within the bracteole is 4–7%, while that within the strig is up to 20–30%. The larger amount of water within the strig is released to the bracteole and bracts after the drying process during ventilation in the conditioning chamber [9]. This stage allows for the establishment of constant moisture equilibrium between the strig, the bracteole and bract [10]. Therefore, the stage of conditioning is of utmost important in commercial drying of hop cones. Optimum process parameters during hop drying are of paramount importance as over drying of hops (moisture content ≤ 7%) can increase lupulin losses due to the shattering of hop cones [11]. Conversely, stopping the drying process at moisture contents ≥ 13% would initiate microbial spoilage. Furthermore, the temperature of the drying process also influences quality parameters such as color, oil yield and oil composition of the produce.

Traditionally, hops were dried in a single layer (3–5 cm). The drying time required was largely dependent on the weather conditions and usually ranged between two to ten days [7]. With improvement in technology and understanding of the main characteristics for hop drying, this traditional method was replaced by multistage kiln and belt driers. In belt driers, the hops are dried in three layers in a continuous process. The fresh hops are fed onto the uppermost belt, while the dried hops remain on the lowest belt until the final moisture content is achieved. For optimum drying using belt driers, it is important to evenly feed fresh hops over the entire belt width. Unlike the continuous process of belt driers, multistage kiln drying is a batch process [10]. The drier consists of three to four superimposed perforated trays, through which hot air is passed (bottom to top) to reduce the moisture content within hops. Freshly harvested hop cones are filled within the topmost tray and as the drying progresses, the bulk of hop cones is transferred to the trays below. Once the drying process is ended, the dried hop cones are transferred into a conditioning chamber. Depending on the variety in focus and the drying system, the drying time with multi stage kilns can vary between five to eight hours [7]. One of the biggest problems with kiln driers is potential uneven drying of the hop cones due to the formation of so called "nests". Nest generation is the formation of compacted zones within the hop stack and can lead to varying air resistances from the beginning of the drying process. Parameters such as air velocity, bulk height and density as well as air temperature are the influencing process parameters that need to be considered and optimized to avoid the formation of nests [9].

With hops being so important to the brewing industry, the need for optimisation of the drying process and understanding of the changes within the product during drying is imminent. Currently, to determine moisture content, temperature and humidity sensors are placed within the hop stack. However, this technique lacks in providing information of the overall stack and thus, leads to not detecting potential variations of moisture content within the hop stack [10]. With advancement in technology, smart drying systems that combine non-invasive measurement techniques with process control have also been developed. Implementation of non–invasive techniques such as computer-aided vision (CAV), thermography (TI), laser backscattering (LB), and hyperspectral imaging (HSI) have demonstrated the ability to analyse the physical and chemical properties, colour, shrinkage, texture, shape, water content, porosity, defects and firmness of the product [12]. A recent investigation performed also used color and HSI imaging as non-invasive techniques to monitor quality parameters such as color of hops of different bulk weights, provide spectral correlations and build models for prediction of selected chemical components. This study has shown promising results for novel real time monitoring during the drying of hops [3]. Applications of thermal imaging in kilns have also been used to understand the drying behavior of the hop cones [9,13] and further potentially combine them with control systems for real time product based control of the drying process.

Drying is an energy intensive process and in case of hop drying, fresh air is heated to 63 °C–68 °C using fuel oil or gas burners [7]. On average, 44 litres of fuel oil is currently required per 100 kg of dried hops. With increasing fuel prices, alternative energy sources and heat recovery systems are becoming of larger interest [14]. In the context of hop drying, options such as preheating of air, use of waste heat from power generator, use of combined solar-building heat and heat recovery from exhaust air are some of the considered alternatives. However, the economic efficiency of these alternative energy sources largely depends on the acquisition costs, operating costs, and equipment life expectancy [9]. Another option to reducing heating oil consumption was investigated by performing experiments with inclusion of a heat recovery system and optimisation of drying process parameters. The results for the performed experiments reveal a potential of 20% savings in heating oil [7,9]. The results from these experiments successfully led to the installation of heat recovery systems on industrial hop kilns and further led to the development of a pilot plant scale dryer that considers better control of the process parameters and gentler kilning of the hop cones [7].

The experiments conducted within this study were performed on the aforementioned pilot plant dryer and aimed to understand the effect of varying bulk weights and hop size distributions on the drying behavior of hops and the associated changes in quality parameters such as color, hop oil yield and hop oil content. As energy is a crucial aspect of the drying process, specific energy requirement for the different bulk weights were also calculated.

2. Materials and Methods

Investigations were performed at the Bavarian State Research Center for Agriculture (LfL), Hop Research Centre Hüll, Wolnzach. Organic aroma hops of the variety Hallertauer Tradition were used for investigation. Based on the harvest dates for this variety in 2018, the experiments were conducted for a week starting from end of August to beginning of September. Hop vines were freshly harvested prior to each investigation and further mechanically de-vined using a hop cone harvester (Hopfenpflückmaschine Wolf WWE 220, Giesenfeld, Germany).

2.1. Dryer

The pilot plant scale dryer described by [3,7] was used for conducting the experiments within this study (Figure 1). The dryer consisted of two drying layers and a slip case which is also considered as a drying stage. However, in case of this experimental investigation, the second drying layer was not used, and the slip case was only used to remove the hop cones. Thus, the overall drying process was conducted using the top drying layer itself. This means that the hop cones were not tipped into the second drying layer but were dried continuously on the top layer and tipped directly into the slip case at the end of drying. Temperature and humidity of the air entering and exiting the top drying layer was measured using Testo 174 H data loggers with an accuracy of ±0.5 °C for temperature and ±3% for relative humidity (Testo SE & Co. KGaA, Lenzkirch, Germany).

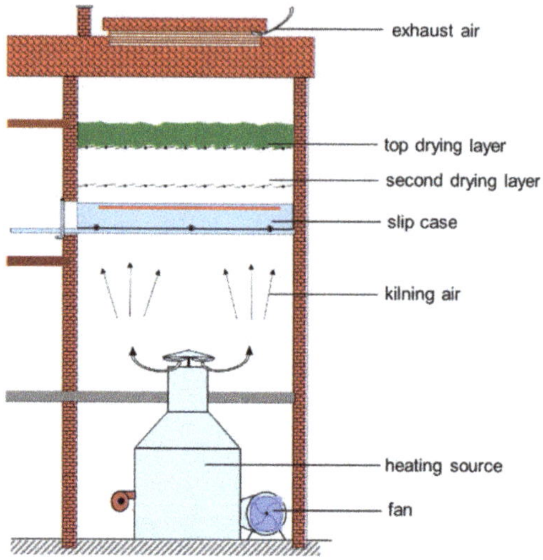

Figure 1. Schematic of hops dryer (adapted and revised from [9]).

Experimental investigations were conducted for hops with bulk weights of 15, 25 and 35 kg/m^2 at 60 °C and 0.35 m/s, respectively. The 25 kg/m^2 bulk was considered as the control bulk and changes to the velocity to 0.45 m/s and temperature of 65 °C were only performed at this bulk weight. Thus, allowing for better comparison for further analysis. The drying time for each bulk weight was estimated based on the set final moisture content of ≅ 7–8% (equivalent moisture ratio [MR] ≅ 0.10–0.12). For reproducibility and repeatability, three repetitions were performed for all experiments. Table 1 provides a summary of the experimental plan for the investigations conducted.

Table 1. Experimental plan for drying of Hallertauer Tradition.

Temperature [°C]-Velocity [m/s]	Bulk Weight [kg/m^2]		
	15	25	35
60-0.35	X	X	X
60-0.45		X	
65-0.35		X	

2.2. Sampling

For further analysis, the samples were extracted from the front section of the dryer at defined time intervals. At each time interval, a small set of approximately 50 g was extracted from the dryer. To dissuade the formation of artificial holes due to sample extraction, the hop cones were redistributed throughout the bulk. The initial moisture content of hops for the different bulk weight varied between 78% and 80%. For moisture content analysis, the samples were collected at a 20 min interval from the start for the first hour, followed by a 30 min interval for the next hour, and a 60 min interval thereafter. End time for the drying process varied depending on the weight of the bulk. Table 2 provides details of the sampling intervals for the different experiments.

Table 2. Sampling intervals for different experiments.

Interval	60 °C-0.35 m/s			60 °C-0.45 m/s	65 °C-0.35 m/s
	15 kg	25 kg	35 kg	25 kg	25 kg
	Time Section in Minutes				
20 min	0–60	0–60	0–60	0–60	0–60
30 min	60–120	60–120	60–120	60–120	60–120
60 min	120–180	120–180	120–180	120–180	120–180
End	180	210	240	200	180

The samples extracted at different intervals were further placed in an oven at 105 °C for 24 h in accordance with the official Association of Official Analytical Collaboration (AOAC) method [15] to obtain the final moisture content of the samples. The moisture content on dry basis (MC_{db}) was calculated using Equation (1).

$$MC_{db} = \frac{m_t}{m_{dm}} - 1 \quad (1)$$

where m_t is the total mass of a sample at a given time and m_{dm} is the absolute dry matter in a sample. In addition to MC_{db}, the moisture ratio (MR) of the samples was also calculated as it allows to compare the different samples with each other by normalizing the moisture contents to their initial moisture contents and thus displaying them in values ranging from one to zero. MR was calculated using Equation (2).

$$MR = \frac{MC_{db,1} - MC_e}{MC_{db,\,0} - MC_e} \quad (2)$$

where $MC_{db,1}$ is the dry basis MC at a given time, $MC_{db,0}$ the dry basis MC before drying and MC_e is the equilibrium moisture content. As the influence of MC_e is relatively small (≤2%) [16], the MR equation is further simplified to Equation (3).

$$MR = \frac{MC_{db,1}}{MC_{db,\,0}} \quad (3)$$

Sampling intervals similar to those described for moisture content analysis were used for color acquisition. A detailed explanation on the color measurements is provided in Section 2.3.

Hop cone size varies both between and within the varieties. This variation is evidently observed from Figure 2 wherein hop cone sizes for two varieties namely Hallertauer Tradition and Hallertauer Herkules is shown.

(a) (b)

Figure 2. Hop cone size variation in varieties (**a**) Hallertauer Tradition, (**b**) Hallertauer Herkules [9].

Therefore, to understand the distribution of the hop cone size, 50 g of fresh hops were extracted from the bulk and sorted based on their length and diameter. Cones smaller than 2 cm in length were classified as small, between 2 cm–3.2 cm as medium and larger than 3.4 cm as large. The classified cones were further weighed to establish their ratios within the 50 g mix. The investigation of hop cone size distribution was limited to fresh hop cones (i.e., 0 min) as the aim of this study was to investigate the effect of hop cone size on the drying process.

For distillation of essential oils, samples were collected prior to drying (fresh) and at the end (dried) of each trial. Distillation of essential oils was carried out only for the control bulk i.e., 25 kg/m^2 bulk dried at air velocity of 0.35 m/s and at 60 °C and 65 °C, respectively, to assess the effect of temperature on the overall hop oil yield and hop oil composition. Finally, to mimic the commercial drying process, it was ensured that the bulk was thoroughly mixed using a rake prior to sample collection.

2.3. Imaging

During the drying process, hop cones undergo significant color change. To understand the development of color, hop samples were placed in a photo box (Life of Photo 60 cm LED Light Cube, Weiwa Foto, Gummersbach, Germany). The photo box includes an array of 160 LED lamps as the illumination and a window for the camera. To capture the images, an RGB Camera (model no. 61BUC02, The Imaging Source Europe GmbH, Bremen, Germany) was placed at a distance of 29 cm from the surface of the product [17] using the camera window of the photo box. A schematic of the imaging system used is presented in Figure 3.

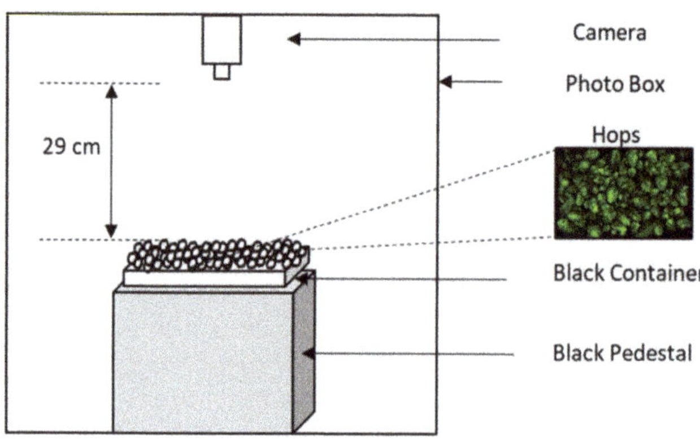

Figure 3. Schematic of RGB camera imaging system for hops.

The acquisition of the images was realized using the IC Capture software (the Imaging Source Europe GmbH, Bremen, Germany). The captured images were further processed in MATLAB® (The MathWorks, Inc., Natick, MA, USA) to convert RGB values to CIE L*a*b* values. For this purpose, RGB images were converted to gray and further into binary images (0, 1) using appropriate threshold factors. In the binary images, the values 0 and 1 were assigned to the background and the sample, respectively. To remove any undesirable objects, the function bwareaopen was applied. The images were then masked to ensure that all values now belong to the image and no values are being assigned to the background. Finally, the function rgb2lab was used to convert the RGB values into CIE L*a*b* values for the images captured [18]. The function rgb2lab within MATLAB® uses a two-step process. In the first step, RGB values are converted into XYZ space using linear matrix in Equation (4) [19].

$$\begin{bmatrix} X \\ Y \\ Z \end{bmatrix} = \begin{bmatrix} 0.412453 & 0.357580 & 0.180423 \\ 0.212671 & 0.019334 & 0.072169 \\ 0.019334 & 0.119193 & 0.950227 \end{bmatrix} \begin{bmatrix} R \\ G \\ B \end{bmatrix} \quad (4)$$

The XYZ values thus obtained are further normalized to the D65 white points. In the final step, the XYZ values are converted to CIE L*a*b* values using Equations (5)–(7) [20].

$$L^* = 116 \times \left(\frac{Y}{Y_n}\right)^{\frac{1}{3}} - 16 \text{ if } \left(\frac{Y}{Y_n}\right) > 0.008856 \qquad 116 \times \left(\frac{Y}{Y_n}\right) \text{ if } \left(\frac{Y}{Y_n}\right) > 0.008856 \tag{5}$$

$$a^* = 500 \times \left[\left(\frac{X}{X_n}\right)^{\frac{1}{3}} - \left(\frac{Y}{Y_n}\right)^{\frac{1}{3}}\right] \tag{6}$$

$$a^* = 200 \times \left[\left(\frac{Y}{Y_n}\right)^{\frac{1}{3}} - \left(\frac{Z}{Z_n}\right)^{\frac{1}{3}}\right] \tag{7}$$

where L*, a* and b* represents lightness, red/green and yellow/blue, respectively. Xn, Yn, and Zn are the values of the blank references.

To compare the non-invasive method to a gold standard color measurement method, a CR-400 Chromameter Konica Minolta Sensing Europe B.V., München, Germany) in combination with the SpectraMagic NX Software (Konica Minolta Sensing Europe B.V., München, Germany) was also used. For calibration, a ColourChecker Classic chart (X-rite Gmbh, Planegg-Martinsried, Germany) was used to compare the values obtained from the chromameter and the RGB camera for the 24 color patches.

Finally, the values obtained from each of the repetitions were further averaged to calculate the total color difference (ΔE_{Lab}) using Equations (8)–(11):

$$\Delta E_{Lab} = \sqrt{(\Delta L^*)^2 + (\Delta a^*)^2 + (\Delta b^*)^2} \tag{8}$$

$$\Delta L^* = L_1^* - L_0^* \tag{9}$$

$$\Delta a^* = a_1^* - a_0^* \tag{10}$$

$$\Delta b^* = b_1^* - b_0^* \tag{11}$$

where L^*_1, a^*_1, b^*_1 is the lightness, red/green and blue/yellow at the specific time interval and L^*_0, a^*_0, b^*_0 is the initial lightness, red/green and blue/yellow [21].

2.4. Gas Chromatography

Gas Chromatography (GC) analysis was performed using a Dani GC with flame ionisation detection (FID; DANI Instruments SpA, Milan, Italy) with a defined temperature program. The gas flow was divided in a ratio of 1:25 (spilt) with helium being used as the carrier gas. The injector and detector temperature were set to 200 °C. The GC unit was also calibrated using the internal standard developed by the Hops Research Center, Wolnzach. Myrcene, linalool, ß-caryophyllene and humulene—representing the most valuable oils related to Hallertauer Tradition—were the main components analysed in the GC unit. Mathematical calculation for quantity of hops required for distillation of essential oils was determined using the method described in [17].

2.5. Energy Consumption

To calculate the specific energy consumption per bulk weight it was essential to first calculate the amount of dry matter at the end of the drying time. To this end, Equations (12) and (13) were used to calculate the dry matter in the bulk:

$$DM_i = \frac{100 - MC_f}{100} \tag{12}$$

$$DM_f = DM_i \times W_i \tag{13}$$

where DM_i is the amount of dry matter per kg, MC_f is the final moisture content, DM_f is the amount of dry matter in the bulk (kg) and W_i is the initial weight of bulk.

The specific energy requirement per bulk weight was calculated using Equations (14) and (15) [22,23]:

$$E_t = A \times \vartheta \times \Delta T \times \rho_a \times C_{p,a} \times t \tag{14}$$

$$E_s = \frac{E_t}{W_i - DM_f} \tag{15}$$

where E_t is the total energy consumed (kJ), A is the area of the drying chamber, v is the velocity of the air entering the drying chamber (m/s), ΔT is the temperature difference between the inlet air and surrounding air (K), ρ_a is the density of air (kg/m^3), $C_{p,a}$ is the specific heat capacity of air (kJ/kg K), t is the run time (s), E_s is the specific energy required (kJ/kg) and W_i is the initial weight of the bulk.

3. Results and Discussion

3.1. Drying Behavior

Figure 4 presents the temperature and the humidity profile of the air exiting the drying chamber. From the extracted results, it is observed that the humidity for all bulks other than that for 15 kg/m^2 is close to 100% at the initial phase of drying. As the drying continues, the temperature of the air exiting (top layer of the hop bulk) begins to increase and the humidity to decrease. The time required for the temperature to reach the set drying temperature significantly depends on the temperature as well as the bulk weight. The lower the temperature and the higher the bulk weight the longer the time required for drying and vice versa.

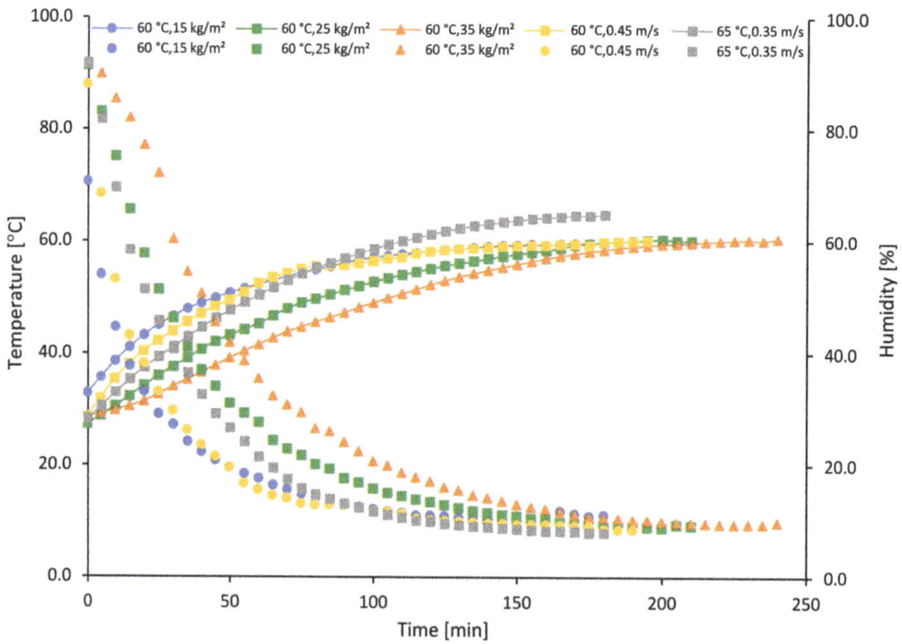

Figure 4. Temperature (symbol and line) and humidity (symbol) profile for different bulk weight and different process settings.

Figure 5 presents the drying curves for different bulk weights of hops dried at 60 °C and 0.35 m/s air velocity. The bulk weight had a significant influence on the drying time required to attain the set moisture ratio of 0.12. Hops of 15 kg/m^2 bulk weight required 180 min while 25 kg/m^2 and 35 kg/m^2 bulks required 210 min and 240 min, respectively. Similar results were also obtained by [3] for hops with bulk weights of 12 kg, 20 kg and 40 kg dried at 60 °C and 0.35 m/s. Overall, the development of the drying curve for all three bulks is characteristic to the curves obtained from convective drying process; with rapid losses in moisture content during the initial phase of drying and eventual reduction as it transitions from the first to the second phase of drying. In the case of hops, this development of drying behavior can be associated to two factors: structure of hop cone and the variety in focus. As mentioned previously, the structure of the hop cone plays a crucial role in the drying behavior. The strig which contains high water content and is surrounded by the bracteole and bract of the hop cone, only begins to dry once the bracteole (leaves of the hop cone) has reached a specific moisture content. It is believed that the development of high suction tension acts as the driving force to draw the moisture from the strig [3] and thus, resulting in specific drying behavior. Furthermore, the weight percentage ratio between the strig and the bracteole also depends on the variety in focus. In case of Hallertauer Tradition, this ratio varies from 9–10% [24]. In addition to the variety, harvesting conditions such as environmental temperature, humidity, location of harvest, soil quality, hop cone size and shape also affect the overall hop cone drying behavior [3]. The effect of hop size, shape and distribution on the overall drying behavior has been further explained in Section 3.2.

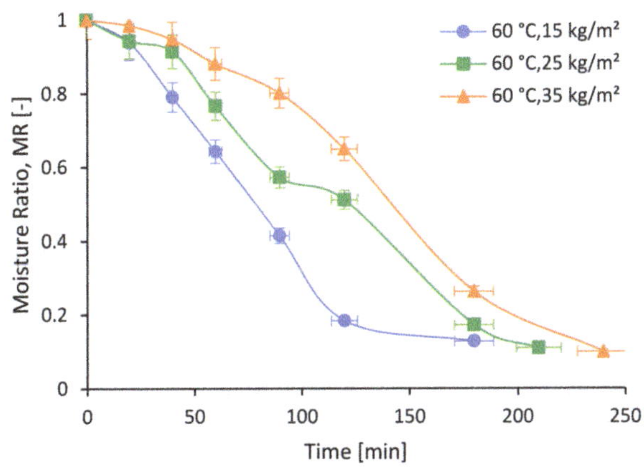

Figure 5. Moisture ratio (MR) vs. time (min) for hops with three different bulk weights dried at 60 °C and 0.35 m/s.

Figure 6 presents the results obtained from changing the velocity (Figure 6a) and changing the temperature (Figure 6b) for 25 kg/m^2 bulk of hops.

No significant impact on the drying time between air velocity of 0.35 m/s (210 min) and 0.45 m/s (200 min) was observed. This is contradictory to several investigations performed through the years that identify air velocity to be an influencing factor for reducing the drying time [20,25]. For most of these studies, a thin layer convective drying of products is investigated as compared to a bulk of 25 kg/m^2. The insignificant influence of air velocity on the bulk could be due to the air being completely saturated with water vapor prior to reaching the top of the bulk. Thus, the deficit in saturation of the drying air acts as the limiting factor as compared to the air velocity [26]. This further explains the similarity in drying behavior for the first 90 min and then the observed rapid moisture losses for 0.45 m/s for the rest of the drying period. However, increasing the air velocity is also not an ideal solution to reduce the drying time as increasing the velocity leads to faster drying at the surface as compared to the center, in turn stiffening the surface and causing uneven shrinkage of the product [27]. Furthermore, increasing velocity also increases the overall energy demand for the drying process [9].

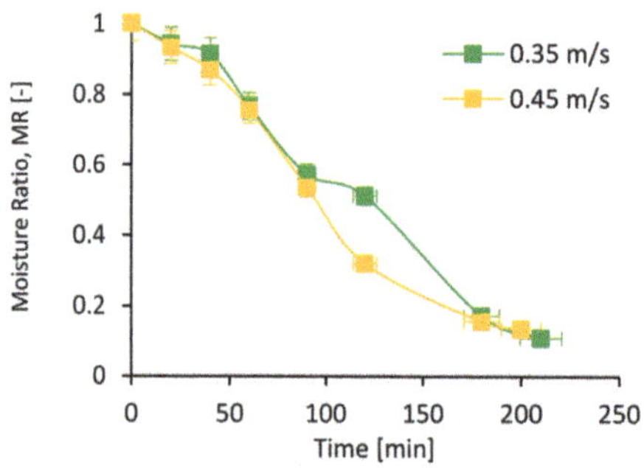

(a)

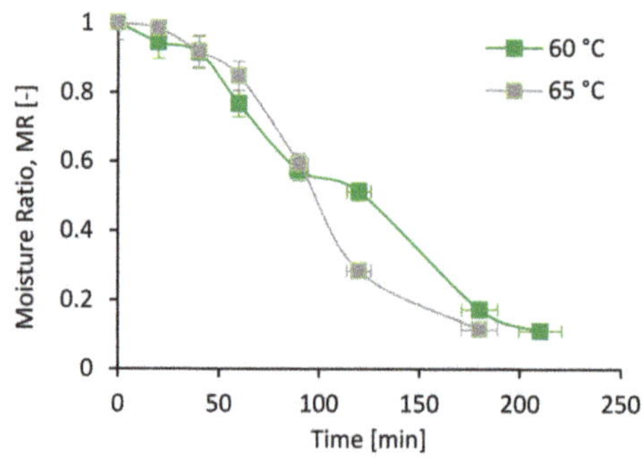

(b)

Figure 6. Moisture ratio (MR) vs. time (min) for hops dried at (**a**) 0.35 m/s and 0.45 m/s at 60 °C (**b**) 60 °C and 65 °C at 0.35 m/s.

Changing the drying temperature, the overall drying time is observed to have reduced from 210 min for 60 °C to 180 min for 65 °C. This is analogous to the several investigations performed on the influence of drying temperature on the drying time [20,23,28–30]. From the drying curve presented in Figure 6b, it is also observed that the moisture losses were rapid for 60 °C than 65 °C for the initial phase (up to 60 min), indicating that the 60 °C bulk dried faster than the 65 °C bulk. Within the 7-day investigation period, a severe thunderstorm on Day 3 led to harvesting of wet hops, thus increasing the overall moisture content of the hop cones. Furthermore, as the hop cones were harvested just prior to the beginning of each drying trial, it was observed that those harvested in the morning had a higher moisture content due to the morning dew as compared to those harvested after midday. The trial period within the experiments were randomized to ensure that each repetition was performed on a different day and at a different time. However, the difference in the environmental conditions and time for 65 °C trials could have led to the observed results. A study performed on the impact of environmental conditions on the hop cones during drying, shows that weather conditions such as heavy rains, morning dew and sunny days significantly influence the drying process and the overall product quality [3].

3.2. Hop Cone Size

Results from the classification of average hop cone sizes within a hop stack is presented in Table 3.

Table 3. Average hop cone size distribution within a hop stack.

Cone Size	Average Cone Length [cm]	Average Cone Diameter [cm]	Average Weight of One Cone [g]	Average Hop Weight [g]
Small	1.6	0.9	0.3	20.6
Medium	2.7	1.3	0.7	22.7
Large	3.9	1.3	1.1	4.3
Total Weight				47.57

For the Hallertauer Tradition, the distribution of cone size lies between small to medium with small hop cones accounting for 41.2%, medium for 45.4% and large for 8.6% of the total measured weight. A total of 50 g of hops were weighed for classification of which 4% were lost due to the cones either being broken or shattered. In addition to the size, a linear increase in the average weight of one cone is also observed. A study conducted on drying of coarse lignite particles of varying particle size indicates that larger particle sizes require longer drying times as compare to the small particles [31]. Particles smaller in size have a larger specific surface area which results in faster heating rate and better transfer of heat from the drying medium to the center. Additionally, the moisture transfer from inside to the surface is also faster within smaller particles, while smaller diameter also provides less resistance to the internal heat transfer [32], thus, leading to faster drying processes. However, bulks with a large number of small particles such as that investigated within the current study, tend to increase resistance within the bulk due to a decrease of average particle size and an increase of void fraction, and thus the associated increase in pressure drop. Investigations performed on grain drying showed that the grain surface characteristics can potentially tend to increase the airflow resistance [33]. For locust beans at low moisture content, the cotyledons tend to become hard and smooth which in turn increases the frictional losses due to high resistance to airflow [34].

Furthermore, the degree of compaction of the hop cones also increases with bulk height. This results in a rapid decrease in the air flow rate and the drying capacity while simultaneously increasing the retention time of the hop in the top layer. As drying progresses, the cones tend to become solid and maintain their individual shape and their position in relation to each other in the top layer. This results in the creation of air channels and formation of zones with uneven air flow [9]. In the current investigation, the proportion of small size cones is significantly high and hence the degree of compaction within the stack is also high. Although no air channels formation was observed in 25 or 35 kg/m^2 stack, relatively many air channels were observed to have formed in the 15 kg/m^2 bulk. Figure 7 shows the formation of air channels within the bulk.

Figure 7. Air channel formation in a 15 kg/m^2 bulk of hops.

The overall bulk height of a 15 kg/m^2 is not high. Therefore, bulks with low heights such as those for 15 kg/m^2 combined with high numbers of small sized cones and set air flow rate can also lead to the formation of air channels. A study performed on pistachio nuts also revealed a linear increase in air flow resistance with increasing depth of the pistachio column. The study also adds that the air flow rate significantly influenced the pressure drop within the column as compared to other parameters such as moisture content, particle interaction and filling method [35]. The inclusion of a higher number of small cones resulted in faster drying but also led to the formation of air pockets or so called "nests" within the drying chamber which in turn led to uneven drying [10]. Investigations performed on storage of hops revealed that hops stored for a longer amount of time undergo oxidation which lead to formation of compounds that interfere with the aromatic smell associated with hops [17]. Bulks with unevenly distributed moisture content can also undergo a similar degradation and thus leading to significant losses of both physical and chemical attributes of the product. Therefore, in commercial hop drying units, to avoid formation of large air pockets, low bulk weights such as 15 kg/m^2 are not usually considered. Furthermore, in the commercial process, the air velocity is also lowered as the drying progress so as to ensure even drying of hop cones and deter the formation of air channels. Moreover, during extraction of samples at specified intervals, a number of small hop cones were observed to have exited the drying chamber. This could be due to two reasons. Firstly, as mentioned earlier, small cones have large surface area and hence dry faster. Secondly, the pressure difference within the drying chamber and the surrounding environment could have also led to the dried small hop cones to exit the dryer.

As the size of the hop cone is an uncontrolled parameter, a solution to optimize the drying performance, and reduce the air resistance, is to fill low bulks followed by tipping at shorter interval. Thus, aide in loosening the compacted layers within the bulk and reducing air resistance which increases the air flow and the drying performance [9].

3.3. Chemical Analysis

3.3.1. Oil Content

The variety Hallertauer Tradition is a product of cross breeding between the variety Hallertauer Tradition Gold and a male breeding line. It has a slightly flowery, herby-spicy aroma with an amount of oil ranging between 0.005 L/kg–0.007 L/kg [6,9]. Within the experimental investigation conducted, the amount of oil extracted for fresh hops ranged between 0.0025–0.0027 L/kg while those for dried hops ranged between 0.0023–0.0024 L/kg. Table 4 summarizes the amount of oil extracted and the standard deviations for the investigations conducted.

Table 4. Amount of oil extracted in L/kg with standard deviations (SD) for prior and after drying process.

Temperature [°C]	Amount of Oil [L/kg hops]	
	Fresh	Dried
60	0.0024	0.0023
SD	0.0006	0.0005
65	0.0027	0.0024
SD	0.0003	0.0004

As compared to the dried samples, significant deviation is observed within the fresh samples which could be due to the natural heterogeneity of hops within and between the different bulks. Figure 8 presents the amount of oil extracted prior to drying and after drying at 60 °C and 65 °C, respectively.

The overall amount of oil extracted for fresh hops was lower than that reported in literature (0.007 L/kg) [6]. Factors such as location, environmental conditions at the location and harvest time have been shown to have a significant effect on the total amount of oil [36,37]. In case of the Hallertauer Tradition, a five-week investigation conducted for three continuous years for essential oil content reveals that the amount of oil extracted is lower at the initial harvest time (about 0.005 L/kg) and increases over time (about 0.02 L/kg) until it reaches an optimum point (roughly Week 3). Harvesting after this period has also shown a decrease in the overall amount of oil [38]. The investigation within this study was performed in the first week of the harvesting period for Hallertauer Tradition. The early harvest combined with associated environmental conditions could have led to the lower amounts of oil extraction.

Drying temperature also had a significant effect on the overall hop oil content. Within the current investigation it was observed that increasing the drying temperature decreased the total amount of oil extracted after drying. An investigation performed on the Mandarina Bavaria variety of hops revealed losses in the total oil extracted after drying as compared to prior to drying [17]. Furthermore, drying of Laurel (*Laurus nobilis* L.) leaves at either 45 °C or 65 °C accounted for losses close to 83% [39]. An experimental study performed on drying the aerial parts of sweet wormwood (*Artemisia annua* L.) also revealed a decrease in the yield of oil with an increase in drying temperature [40].

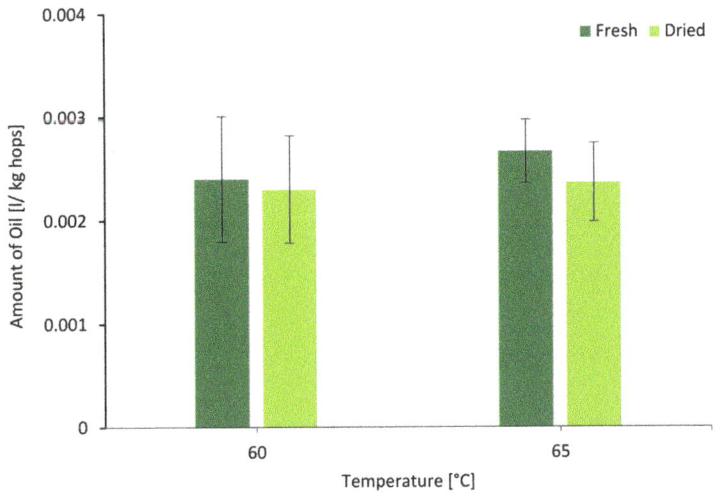

Figure 8. Amount of oil in L/kg hops extracted for samples extracted at 0 min and 210 min for hop cones dried at 60 °C and 65 °C at 0.35 m/s.

3.3.2. Gas Chromatography of Essential Oils

The effect of drying is quite significant on the overall hop oil content with losses accounting for 30–40% [41] and are mainly associated to the high water vapor volatility [42]. Table 5 provides a summary with standard deviation values of the essential oils namely myrcene, linalool, β-caryophyllene and humulene analyzed through gas chromatography for 25 kg/m^2 bulk dried at 60 °C and 65 °C, respectively.

Table 5. Essential oils with standard deviation (SD) prior to and after drying process extracted using GC-FID.

Temperature [°C]	Essential Oils [g/kg hops]							
	Myrcene		Linalool		β-Caryophyllene		Humulene	
	Fresh	Dried	Fresh	Dried	Fresh	Dried	Fresh	Dried
60	4.52	4.17	0.09	0.09	0.72	0.73	2.35	2.45
SD	0.90	1.30	0.01	0.08	0.21	0.35	0.71	1.16
65	7.30	5.09	0.14	0.10	0.76	0.75	2.47	2.54
SD	4.84	0.81	0.02	0.004	0.08	0.06	0.33	0.20

In line with the results on hop oil yield (Section 3.3.1), large deviations in the essential oils is observed. As mentioned previously, hops are a heterogeneous product and the aroma/essential oil content significantly varies within and between the bulks. Furthermore, as mentioned previously the environmental conditions during harvest (dry, dew or rain wet) between the repetitions also varied significantly. Factors such as environmental conditions and harvest time, influences the amount on essential oil components within hops [43]. With harvest lasting for less than two weeks, variations in harvesting conditions during this period can further add to the large variations in oil components [44]. In additions to these factors, the overall sample size was small and hence large deviations were observed. Thus, making it a poor estimator for representation of deviations within the samples. Figure 9 represents the myrcene content prior to and after drying for 60 °C and 65 °C respectively. Losses in myrcene content after drying are observed in both cases. The losses are relatively higher for 65 °C as compared to 60 °C.

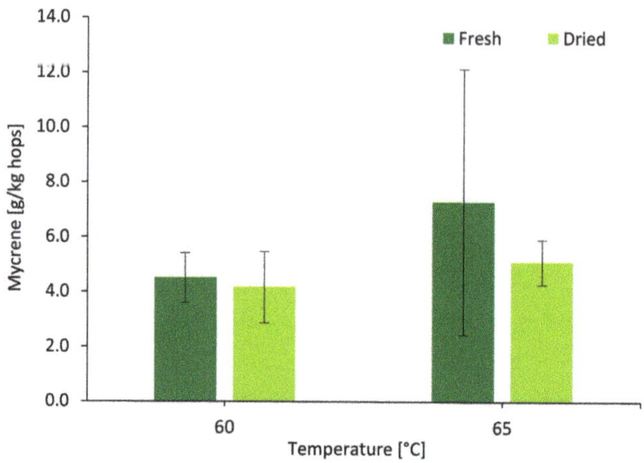

Figure 9. Results obtained from gas chromatography analysis for myrcene in g/kg hops at 0 min and 210 min for hop cones dried at 60 °C and 65 °C at 0.35 m/s.

Myrcene is a terpene hydrocarbon and is one of the most important and sensorially dominant component of fresh hop cones [41]. It is a key contributor to the hop aroma and encompasses 17–37% of the total hop oil [45]. During the drying process, myrcene undergoes oxidation and polymerization which leads to cyclic reactions that form products such as terpenoids which include linalool, and geraniol [46]. Furthermore, myrcene is highly volatile at high temperature. With hop drying temperatures ranging between 55 °C and 65 °C, significant losses of myrcene are expected. Studies conducted on essential oil losses show that myrcene losses range between 25% and 30% [44,47]. Within the present study myrcene losses of 7.74% and 30.27% were observed for 60 °C and 65 °C, respectively. Experimental investigation conducted on Laurel (*Laurus nobilis* L.) leaves also reveal losses in monoterpene hydrocarbons such as myrcene due to volatility with steam or potential chemical rearrangement of compounds [39].

Linalool is one of the key contributors to the hoppy flavor within beer. It provides a citrusy, flowery and fruity aroma to the beer [4,48]. During the drying process linalool content also undergoes significant changes but unlike myrcene, the losses associated to the linalool are much lower [47]. The results from GC analysis (Figure 10) reveal that after drying at 60 °C, no significant losses in the linalool content were observed. However on increasing the temperature to 65 °C, the losses were much higher as compared to 60 °C. Investigation conducted by [47] show that increasing the temperature increases the losses in linalool content. Too high temperatures such as 80 °C or 90 °C have shown to increase the linalool content. This increase is associated to presence of glycosidically bound linalool that are still active at high temperature and indirectly increase the linalool content. Low temperatures such as 50 °C have shown to retain maximum linalool content as compared to those at 65 °C [47]. Similar losses in linalool content were observed for the Mandarina Bavaria variety dried at 60 °C [17].

β-caryophyllene and humulene belong to the sesquiterpenes group and are both hydrophobic in nature; thus have limited solubility in aqueous solutions such as beer [46]. During the drying process, both compounds undergo oxidation and form various other products. Oxidation of β-caryophyllene leads to formation of caryophyllene oxide, 14-hydroxy-b-caryophyllene, caryolan-1-ol,4S-dihydrocaryophyllene-5-one, (3Z)-caryophylla-3, 8(13)-diene-5a-ol, caryophylla-4(12), and 8(13)-diene-5a/b-ol [46,49]. The dominant products formed are caryophyllene oxide and 14-hydroxy-b-caryophyllene that have musty, floral, spicy, cedar, woody odor. Humulene theoretically oxidizes to three humulene monoepoxides however only two, namely mono and diepoxides, are found to be present in hops and have hay-like, moldy, cedar like odor [46]. From the results obtained for β-caryophyllene (Figure 11) and humulene (Figure 12), the values after drying are observed to have remained either similar or increased as compared to fresh hops. As mentioned previously, the Hallertauer Tradition has a typical herby, spicy aroma and oxidation of β-caryophyllene and humulene could have led to the potential increase of the byproducts of these main compounds. Thus, explaining the corresponding increase in the β-caryophyllene and humulene content after drying.

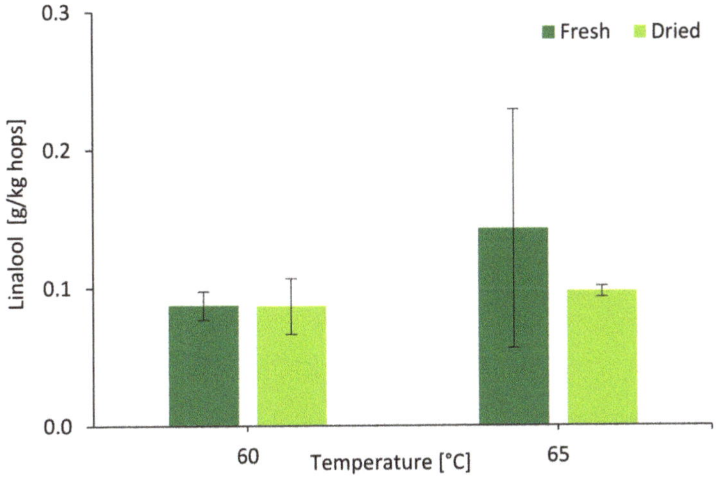

Figure 10. Results obtained from gas chromatography analysis for linalool in g/kg hops at 0 min and 210 min for hop cones dried at 60 °C and 65 °C at 0.35 m/s.

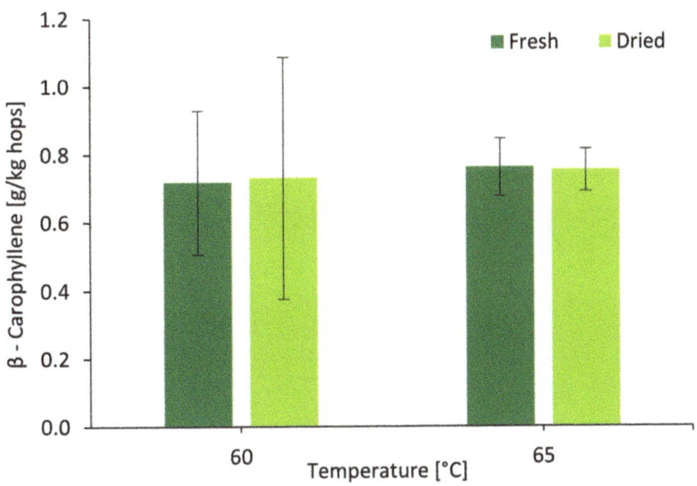

Figure 11. Results obtained from gas chromatography analysis for β-caryophyllene in g/kg hops at 0 min and 210 min for hop cones dried at 60 °C and 65 °C at 0.35 m/s.

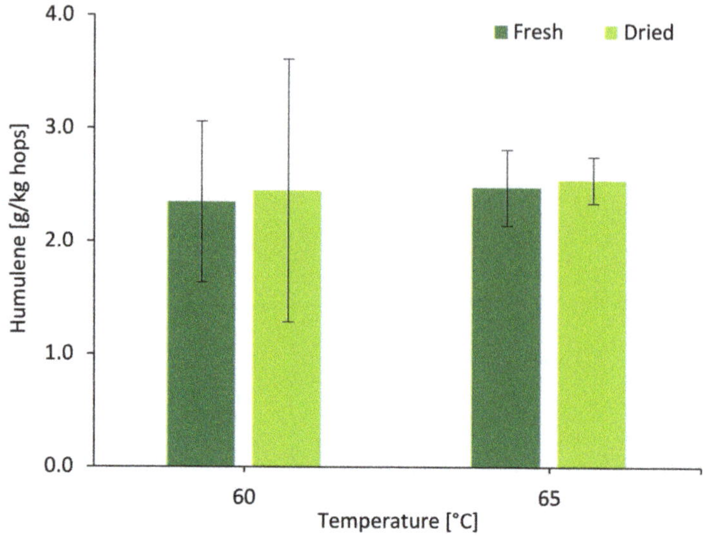

Figure 12. Results obtained from gas chromatography analysis for humulene in g/kg hops at 0 min and 210 min for hop cones dried at 60 °C and 65 °C at 0.35 m/s.

3.4. Color Development

Figure 13 provides the results for the CIE L*a*b* values measured using a chromameter and the values obtained from processing the acquired RGB images. R^2 values of 0.96, 0.90 and 0.95 were obtained for L*, a* and b* respectively. However, significant differences in the actual values were observed especially in case of L* values. The illumination system used for the RGB imaging was observed to have lower intensity as compared to the illumination within the chromameter and could potentially have affected the lower values from the RGB camera system.

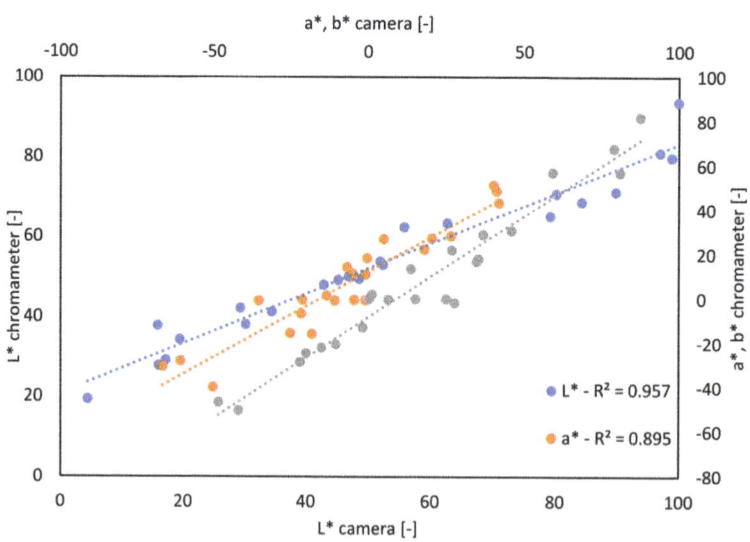

Figure 13. Correlation between CIE L*a*b* values between chromameter and RGB camera system.

The color of fresh hop cones varies between slight dark green to yellow green on the exteriors depending on the variety and harvest time. During the drying process, these hop cones undergo discoloration due to influencing factors such as drying time, temperature, and air velocity. The results obtained for the change in color during the drying process under various conditions is presented in Figure 14.

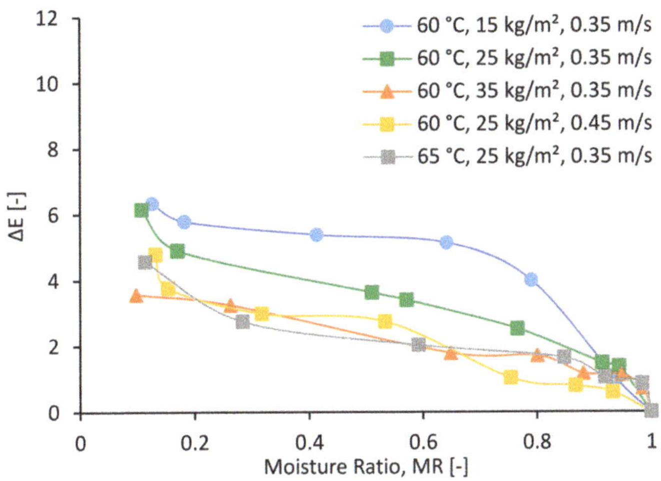

Figure 14. ΔE as a function of moisture ratio (MR) for hops dried under different process settings and different bulk weights.

The highest ΔE values were observed for hops with bulk weight of 15 kg/m^2 dried at 60 °C and the lowest ΔE values for hops with bulk weight of 35 kg/m^2 dried again at 60 °C. The associated discoloration is caused during the initial drying phase when the hop cones are colder than the surrounding air. Thus, leading to recondensation of water on the top layers of the hop stack [8,50]. Furthermore, the formation of compact zones or nests in the 15 kg/m^2 bulk led to uneven drying of hop cones which in turn increased the required specific drying time. As drying time is also an influencing factor for color changes, the longer drying time combined with lower bulk weight, could have led to the observed discoloration in the 15 kg/m^2 bulk. The overall results for the total change in color is in contradiction to the results obtained by [3]. According to investigation performed on Mandarina Bavaria with bulk weight of 12, 20 and 40 kg at 60 °C, the highest color changes were observed for 40 kg bulk and lowest color changes for 12 kg bulk [3]. The contradiction between the two studies could potentially be due to two factors. Firstly, the RGB system set up within this study was outside the drying chamber as compared to the system setup of [3] wherein the camera was placed above the bulk within the drying chamber. Secondly, the study conducted by [3] considers the variety Mandarina Bavaria while the current study focuses on the variety Hallertauer Tradition. As each variety has slight differences in color as well as the initial moisture content values, it is possible that opposite results are obtained for experimental investigations performed using the same equipment but slight parameter variations. Furthermore, the varying cone size and the natural heterogeneity in hop cone color could have also influenced the color measurements.

Comparing the results for 25 kg/m^2 bulk dried at 60 °C and 65 °C, it is observed that the ΔE values for samples dried at 65 °C are lower than those dried at 60 °C. Drying time and drying temperature are two of the influencing factors with degradation of color [51]. An increase in either of two factors can lead to higher degradation of color. The exposure time for the hop cones dried at 65 °C was lower than that of those dried at 60 °C. Investigations on 3 mm thick carrot slices have shown maximum degradation at 50 °C where the exposure time is higher and are concurrent to the findings from the current study [30]. Increasing the velocity has also shown to reduce the color change within the hops. Similar results for apple drying under increased velocity were also obtained [20].

Drying temperature has also shown to influence the lupulin color within the hops. Under optimum conditions, the color tends to remain lemon yellow while at higher temperature and longer exposure time the color degrades to brown [52].

3.5. Energy Consumption

As shown in Figure 15, the highest specific energy consumption was required by the 15 kg/m^2 bulk (113,476 kJ/ kg_{H_2O}), followed by the 25 kg/m^2 bulk (91,955 kJ/ kg_{H_2O}) and finally the 35 kg/m^2 bulk (84,794 kJ/ kg_{H_2O}). The 15 kg/m^2 bulk required 180 min for drying, while the 25 kg/m^2 and 35 kg/m^2 bulk required 210 min and 240 min, respectively. During the drying process for the 15 kg/m^2 bulk, air pockets were observed to have formed after the first 60 min. These air pockets might have caused the hops cones to clump in one section and thus requiring a much longer time to dry. For 25 kg/m^2 and 35 kg/m^2, the overall bulk weight and the height of the bulk dissuaded the formation of air pockets during the drying process and hence required comparatively lower energy to dry the hop cones.

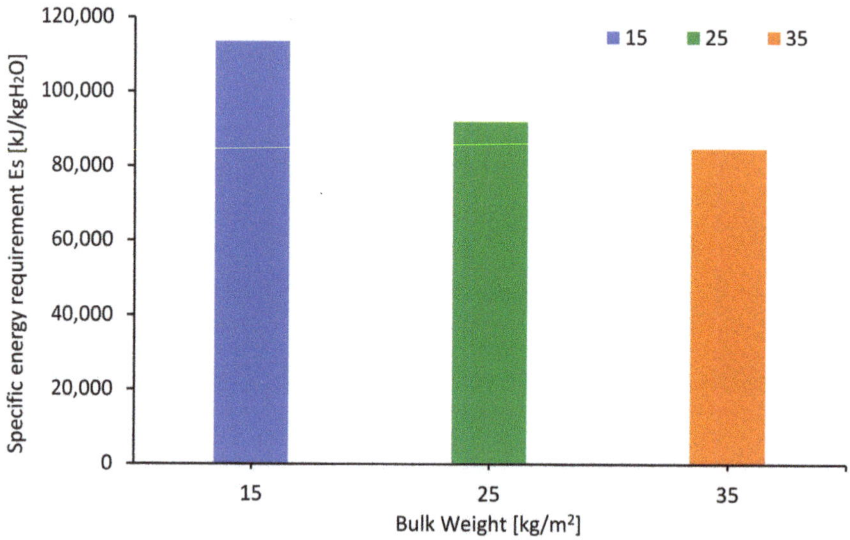

Figure 15. Specific energy requirement for three different hop bulks dried at 60 °C and 0.35 m/s.

On increasing the velocity of the 25 kg/m^2 bulk to 0.45 m/s, the overall energy consumption is also observed to increase as compared to 0.35 m/s. The difference between the drying time is insignificant and, thus, does not drastically affect the consumption values. The factor influencing the increased energy consumption in this case is the velocity. Results of energy consumption with varying velocity is presented in Figure 16a.

Lower energy consumptions were observed for 65 °C (Figure 16b). The decrease in drying time for higher temperatures compensates for the overall energy demand. The effect of temperature was observed to have a significant influence on the specific energy consumption and is in agreement with previously conducted investigations by [23,53].

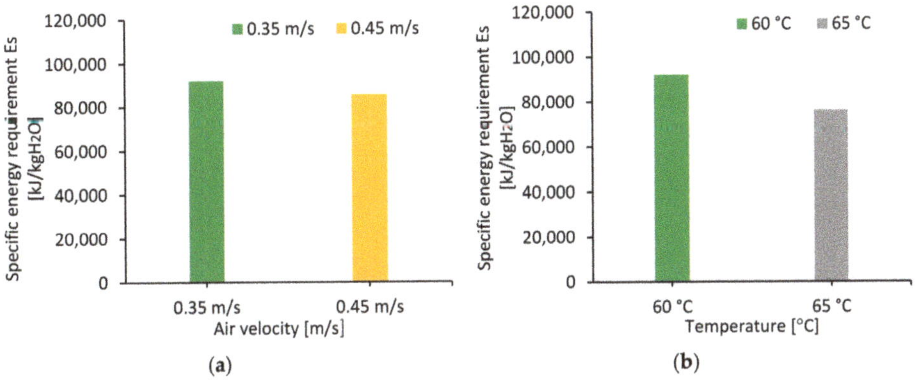

Figure 16. Specific energy requirements for 25 kg/m^2 bulk dried at (**a**) 0.35 m/s and 0.45 m/s at 60 °C (**b**) 60 °C and 65 °C at 0.35 m/s.

4. General Discussion

This study investigated the effects of varying bulk weights and process settings on the drying behavior and product quality attributes such as color, hop oil yield, hop oil components and energy consumption. The study also extended to analyze the influence of hop cone size on the overall drying behavior and its extended effects on quality.

Results obtained from the drying curves show that low temperature in combination with high bulk weight lead to an increase in the overall drying time. Increasing the velocity or the temperature does not significantly influence the drying behavior. Factors such as the environmental conditions during harvest as well as during the drying process, the hop variety in focus and the cone size distribution influence the overall drying behavior. Results obtained from classifying hop cones by size reveals that for the variety Hallertauer Tradition, the majority of the cones lie between small to medium size. The results reveal that such distribution of cone can lead to increase in the air resistance, void fraction and hence increase the pressure drop. In case of 15 kg/m^2 bulk, this further led to the formation of air channel which affect the moisture content distribution within the hop stack. Thus, the 15 kg/m^2 bulk requiring a comparatively longer specific drying time than expected. Hop cone size is a natural parameter and varies according to the variety in focus. In order to reduce the resistances, considering bulk density rather than bulk weights/height has been observed to be a better solution [3,9,54].

In terms of color changes, the highest discoloration of $\Delta E = 6.83$ was obtained for the 15 kg/m^2 bulk, followed by the 25 kg/m^2 at $\Delta E = 6.16$. The lowest discoloration was observed for the 35 kg/m^2 bulk at $\Delta E = 3.58$. The higher discoloration of the 15 kg/m^2 bulk can be associated to the formation of nests or compact zones which increased the required amount of drying time. As temperature and drying time are influencing factors for color changes, an increase in either of these factors, can cause significant effects on the color quality of the product. Temperature and time are also influencing factors for the specific energy consumption, with the 15 kg/m^2 bulk consuming the high amount of energy per unit of product. In order to optimize and reduce the energy costs, it is important to consider alternative solutions that can be implemented with the current drying techniques [9].

Drying temperature is not only an influencing factor affecting color, but also affects the yield and chemical compositions within hops. The drying process has a significant effect on the hop oil yield and composition, and thus a decrease in the values after completion of drying is expected. On increasing the temperature from 60 °C to 65 °C, increased losses for yield were observed. This is analogous to the results of previous studies [3,17]. In terms of composition, the expected trend of losses was observed for myrcene and linalool, however, an increase in β-caryophyllene and humulene was observed at higher temperature of 65 °C. The increase in β-caryophyllene and humulene was previously observed in storage of hops for 24 h prior to drying wherein it was believed the formation of other by products led to an overall increase in these components [17]. As no storage tests were conducted within the current study, the results obtained for β-caryophyllene and humulene are in contradiction. However, hops in general have a large heterogeneity in the aroma components which fluctuate based on various conditions such as hop variety, hop garden, soil, environmental factors at harvest, etc. [44]. The increase in the β-caryophyllene and humulene content could potentially also be associated to the other influencing factors.

5. Conclusions

The investigations conducted within the current study demonstrated that process and product quality in hop drying depend on multiple factors such as harvest time, drying conditions, and bulk characteristics (weight and cone size distribution). An optimization of the process would require consideration of multiple potential target conflicts, i.e., energy demand vs product quality, reduction of drying time with bulk reduction vs additional labor demand and machine times caused by additional harvesting schedules.

The study, therefore, indicates the need for development of systems that consider various factors such as hop variety, harvest maturity, environmental conditions at harvest (dry, rain wet) and bulk quality (e.g., cone size distribution), for optimum process settings. This includes the necessity for the development of non-invasive measurement systems in combination with integration of adaptive control systems. With advances in optical sensors, computational models and with the advent of internet of things (IoT) integrated processing systems, it will be possible to further develop tailored systems which consider all above aspects. In addition to integration of measurement and control systems, the study also highlights the need for further work with bulk density rather than bulk weight/height in combination with pressure drop measurements as this will not only aide in optimizing the process parameters but also help in reducing the energy consumption and, thus, the associated processing costs.

Author Contributions: Conceptualization, S.R., G.J.E.v.G., J.M. and B.S.; data curation, S.R., G.J.E.v.G. and B.S.; formal analysis, S.R. and G.J.E.v.G.; funding acquisition, J.M., O.H. and B.S.; investigation, S.R., G.J.E.v.G., J.M. and K.K.; methodology, S.R., G.J.E.v.G. and B.S.; project administration, J.M., O.H. and B.S.; resources, J.M. and B.S.; software, S.R.; supervision, J.M., O.H. and B.S.; validation, S.R.; visualization, S.R. and B.S.; writing—original draft, S.R. and G.J.E.v.G.; writing—review and editing, S.R., G.J.E.v.G., J.M., K.K., O.H. and B.S. All authors have read and agreed to the published version of the manuscript.

Funding: This project was funded by transnational funding bodies, being partners of the CORE Organic Cofund, and the cofund from the European Commission. This study is part of the SusOrgPlus project within the framework of the CORE Organic Cofund Programme and is supported by funds of the Federal Ministry of Food and Agriculture (BMEL) based on a decision of the Parliament of the Federal Republic of Germany via the Federal Office for Agriculture and Food (BLE) under the BÖLN programme (Project number: BLE–2817OE005).

Acknowledgments: The authors would like to thank Sebastian Wambach for his support during the experimental investigations.

Conflicts of Interest: The authors declare no conflict of interest.

References

1. Schieberle, P.; Steinhaus, M. Characterization of the Odor-Active Constituents in Fresh and Processed Hops (variety Spalter Select). In *Gas Chromatography-Olfactometry*; Leland, J.V., Schieberle, P., Buettner, A., Acree, T.E., Eds.; American Chemical Society: Washington, DC, USA, 2001; pp. 23–32. ISBN 0-8412-3713-1.
2. Schönberger, C.; Kostelecky, T. 125th Anniversary Review: The Role of Hops in Brewing. *Inst. Brew. Distill.* **2011**, *117*, 259–267. [CrossRef]
3. Sturm, B.; Raut, S.; Kulig, B.; Münsterer, J.; Kammhuber, K.; Hensel, O.; Crichton, S.O. In-process investigation of the dynamics in drying behavior and quality development of hops using visual and environmental sensors combined with chemometrics. *Comput. Electron. Agric.* **2020**, *175*, 105547. [CrossRef]

4. Steinhaus, M.; Schieberle, P. Comparison of the most odor-active compounds in fresh and dried hop cones (Humulus lupulus L. variety spalter select) based on GC-olfactometry and odor dilution techniques. *J. Agric. Food Chem.* **2000**, *48*, 1776–1783. [CrossRef] [PubMed]
5. Rubottom, L. Hop Kilning Temperature Sensitivity of Dextrin-Reducing Enzymes in Hops. Master's Thesis, Oregon State University, Corvallis, OR, USA, 2020.
6. Biendl, M.; Forster, A.; Gahr, A.; Lutz, A.; Schimdt, R.; Schönberger, C.; Schüll, F. *Pocket Guide 2016 Hops from Germany*; German Hop Growers Association: Wolnzach, Germany, 2016.
7. Hofmann, R.; Weber, S.; Rettberg, N.; Thörner, S.; Garbe, L.A.; Folz, R. Optimization of the Hop Kilning Process to Improve Energy Effi ciency and Recover Hop Oils. *Brew. Sci.* **2013**, *66*, 23–30.
8. Crichton, S.; Münsterer, J.; Kammhuber, K.; Sturm, B. Measurement of hop moisture content and chromaticity during drying with VNIR hyperspectral imaging. In Proceedings of the International Conference on Agricultural Engineering, CIGR—AgEng, Aarhus, Denmark, 26–29 June 2016.
9. Münsterer, J. Trocknung und Konditionierung von Hopfen, 2020. Available online: https://www.lfl.bayern.de/mam/cms07/publikationen/daten/informationen/trocknung-konditionierung-hopfen_lfl-information.pdf (accessed on 7 October 2020).
10. Münsterer, J. *Optimale Trocknung und Konditionierung von Hopfen*; Bayerische Landesanstalt für Landwirtschaft (LfL): Freising, Germany, 2006.
11. Heřmánek, P.; Rybka, A.; Honzík, I. Determination of moisture ratio in parts of the hop cone during the drying process in belt dryer. *Agron. Res.* **2018**, *16*, 723–727. [CrossRef]
12. Nicolaï, B.M.; Defraeye, T.; de Ketelaere, B.; Herremans, E.; Hertog, M.L.A.T.M.; Saeys, W.; Torricelli, A.; Vandendriessche, T.; Verboven, P. Nondestructive measurement of fruit and vegetable quality. *Annu. Rev. Food Sci. Technol.* **2014**, *5*, 285–312. [CrossRef]
13. LfL. *Annual Report 2018—Special Crop: Hops, 2019*. Available online: https://www.lfl.bayern.de/mam/cms07/ipz/dateien/hopfen_jahresbericht_englisch_2019.pdf (accessed on 27 May 2020).
14. Münsterer, J. Kosteneinsparung bei der Hopfentrocknung durch alternative Energiequellen und Wärmerückgewinnung. Available online: https://www.lfl.bayern.de/mam/cms07/ipz/dateien/hopfen_alternative_energiequellen.pdf (accessed on 7 October 2020).
15. AOAC International. *Official Methods of Analysis of AOAC International*, 20th ed.; AOAC International: Rockville, MD, USA, 2016; ISBN 0935584870.
16. Diamante, L.M.; Munro, P.A. Mathematical modelling of the thin layer solar drying of sweet potato slices. *Solar Energy* **1993**, *51*, 271–276. [CrossRef]
17. Raut, S.; von Gersdorff, G.J.E.; Münsterer, J.; Kammhuber, K.; Hensel, O.; Sturm, B. Influence of Pre-Drying Storage Time on Essential Oil Components in Dried Hops (Humulus lupulus L.). *J. Sci. Food Agric.* **2020**. [CrossRef]
18. Nasirahmadi, A.; Behroozi-Khazaei, N. Identification of bean varieties according to color features using artificial neural network. *Span. J. Agric. Res.* **2013**, *11*, 670. [CrossRef]
19. Ruzon, M. RGB2lab. Mathworks. 2009. Available online: https://www.mathworks.com/matlabcentral/fileexchange/24009-rgb2lab (accessed on 13 November 2020).
20. Sturm, B.; Hofacker, W.C.; Hensel, O. Optimizing the Drying Parameters for Hot-Air–Dried Apples. *Dry. Technol.* **2012**, *30*, 1570–1582. [CrossRef]
21. Martin, A. Chapter 4. Colour Management in the Graphic Technologies. In *Graphic Design and Print Production Fundamentals*; BCcampus: Vancouver, BC, Canada, 2017.
22. Aghbashlo, M.; kianmehr, M.H.; Samimi-Akhijahani, H. Influence of drying conditions on the effective moisture diffusivity, energy of activation and energy consumption during the thin-layer drying of berberis fruit (Berberidaceae). *Energy Convers. Manag.* **2008**, *49*, 2865–2871. [CrossRef]
23. Ndisya, J.; Mbuge, D.; Kulig, B.; Gitau, A.; Hensel, O.; Sturm, B. Hot air drying of purple-speckled Cocoyam (Colocasia esculenta (L.) Schott) slices: Optimisation of drying conditions for improved product quality and energy savings. *Therm. Sci. Eng. Prog.* **2020**, *18*, 100557. [CrossRef]
24. Münsterer, J. *Neue Erkenntnisse zur Leistungssteigerung und Energieeffizienz von Hopfentrocknungsanlagen*; Bayerische Landesanstalt für Landwirtschaft (LfL): Wolnzach, Germany, 2015.
25. Contreras, C.; Martín-Esparza, M.E.; Chiralt, A.; Martínez-Navarrete, N. Influence of microwave application on convective drying: Effects on drying kinetics, and optical and mechanical properties of apple and strawberry. *J. Food Eng.* **2008**, *88*, 55–64. [CrossRef]

26. Müller, J. Convective drying of medicinal, aromatic and spice plants: A review. *Stewart Postharvest Rev.* **2007**, *3*, 1–6. [CrossRef]
27. Ratti, C. Shrinkage during drying of foodstuffs. *J. Food Eng.* **1994**, *23*, 91–105. [CrossRef]
28. Sturm, B.; Nunez Vega, A.-M.; Hofacker, W. Influence of process control strategies on drying kinetics, colour and shrinkage of air dried apples. *Appl. Therm. Eng.* **2014**, *62*, 455–460. [CrossRef]
29. Von Gersdorff, G.J.; Shrestha, L.; Raut, S.; Retz, S.K.; Hensel, O.; Sturm, B. Impact of processing temperature on drying behavior and quality changes in organic beef. In Proceedings of the 21st International Drying Symposium, València, Spain, 11–14 September 2018; Cárcel, J.A., Ed.; Universitat: València, Spain, 2018. ISBN 9788490486887.
30. Md Saleh, R.; Kulig, B.; Hensel, O.; Sturm, B. Investigation of dynamic quality changes and optimization of drying parameters of carrots (Daucus carota var. laguna). *J. Food Process. Eng.* **2019**, *48*, 1. [CrossRef]
31. Pusat, S.; Akkoyunlu, M.T.; Erdem, H.H.; Teke, I. Effects of Bed Height and Particle Size on Drying of a Turkish Lignite. *Int. J. Coal Prep. Util.* **2015**, *35*, 196–205. [CrossRef]
32. Tahmasebi, A.; Yu, J.; Han, Y.; Li, X. A study of chemical structure changes of Chinese lignite during fluidized-bed drying in nitrogen and air. *Fuel Process. Technol.* **2012**, *101*, 85–93. [CrossRef]
33. Górnicki, K.; Kaleta, A. Resistance of Bulk Grain to Airflow—A Review. Part I: Equations for Airflow Resistance. *Ann. Wars. Univ. Life Sci.* **2015**, 31–41. [CrossRef]
34. Jekayinfa, S.O. Effect of Airflow Rate, Moisture Content and Pressure Drop on the Airflow Resistance of Locust Bean Seed. *Agric. Eng. Int. CIGR Ej.* **2006**, *VIII*, 1–9.
35. Kashaninejad, M.; Maghsoudlou, Y.; Khomeiri, M.; Tabil, L.G. Resistance to airflow through bulk pistachio nuts (Kalleghochi variety) as affected by moisture content, airflow rate, bed depth and fill method. *Powder Technol.* **2010**, *203*, 359–364. [CrossRef]
36. Kammhuber, K. Analytical Aroma Characterization of the New Hüller "Special Flavor Hops". In Proceedings of the Scientific Commission. International Hop Growers' Convention, Kiev, Ukraine, 4–9 June 2013; Seigner, E., Ed.; Hop Research Center Hüll—Bavarian State Research Center for Agriculture (LfL): Wolnzach, Germany, 2013; pp. 37–39.
37. Rodolfi, M.; Chiancone, B.; Liberatore, C.M.; Fabbri, A.; Cirlini, M.; Ganino, T. Changes in chemical profile of Cascade hop cones according to the growing area. Journal of the Science of Food and Agriculture. *J. Sci. Food Agric.* **2019**, *99*, 6011–6019. [CrossRef] [PubMed]
38. Bayerische Landesanstalt für Landwirtschaft. *Hopfen für alle Biere der Welt*; Bavarian State Research Center for Agriculture (LfL): Wolnzach, Germany, 2011.
39. Sellami, I.H.; Wannes, W.A.; Bettaieb, I.; Berrima, S.; Chahed, T.; Marzouk, B.; Limam, F. Qualitative and quantitative changes in the essential oil of Laurus nobilis L. leaves as affected by different drying methods. *Food Chem.* **2011**, *126*, 691–697. [CrossRef]
40. Prusinowska, R.; Śmigielski, K. Losses of essential oils and antioxidants during the drying of herbs and spices. A review. *Eng. Sci. Technol.* **2015**. [CrossRef]
41. Aberl, A. Methodenentwicklung der Headspace-Trap-GC-MS Analyse zur schnellen quantitativen Bestimmung flüchtiger Verbindungen in Hopfen und Bier—Mit Korrelationsanalyse der Hopfenölkomponenten. Ph.D. Thesis, Technischen Universitaet Munchen, München, Germany, 2016.
42. Münsterer, J. Flavour-Hopfen optimal trocknen (in German). *Brauwelt* **2017**, *36*, 1063–1066.
43. Kammhuber, K.; Lutz, A.; Münsterer, J. *Einfluss von Erntezeitpunkt und Trocknung auf den Brauwert der Flavor-Hopfensorten Hallertau Blanc, Mandarina Bavaria und Polaris*; Bavarian State Research Center for Agriculture (LfL): Wolnzach, Germany, 2017. (In German)
44. Kammhuber, K. *Was Beeinflusst das Hopfenaroma—Eine Analytische und Sensorische Annäherung*; Bavarian State Research Center for Agriculture (LfL): Wolnzach, Germany, 2018. (In German)
45. Krottenthaler, M. *Entwicklung moderner Technologien zur Optimierung der Würze- und Bierqualität*; Habilitationsschrift; Hochschule Weihenstephan-Triesdorf: Freising, Germany, 2007. (In German)
46. Rettberg, N.; Biendl, M.; Garbe, L.-A. Hop Aroma and Hoppy Beer Flavor: Chemical Backgrounds and Analytical Tools—A Review. *J. Am. Soc. Brew. Chem.* **2018**, *76*, 1–20. [CrossRef]
47. Kaltner, D. Untersuchungen zur Ausbildung des Hopfenaromas und technologische Maßnahmen zur Erzeugung Hopfenaromatischer Biere. Ph.D. Thesis, Technische Universität München, Muenchen, Geramy, 2000. (In German).

48. Hanke, S. Linalool—A Key Contributor to Hop Aroma. Global Emerging Issues. 2009. Available online: https://www.mbaa.com/brewresources/Documents/Linalool.pdf (accessed on 4 October 2020).
49. Eyres, G.T.; Marriott, P.J.; Dufour, J.-P. Comparison of odor-active compounds in the spicy fraction of hop (Humulus lupulus L.) essential oil from four different varieties. *J. Agric. Food Chem.* **2007**, *55*, 6252–6261. [CrossRef]
50. Münsterer, J. Neueste Erkenntnisse zur Leistungssteigerung und Energieeffizienz bei der Trocknung von Hopfen. In Proceedings of the 5th Congress of the International Hopgrowers' Convention IHGC, Bad Gögging, Germany, 31 July–3 August 2015. (In German).
51. Saxena, A.; Maity, T.; Raju, P.S.; Bawa, A.S. Degradation Kinetics of Colour and Total Carotenoids in Jackfruit (Artocarpus heterophyllus) Bulb Slices During Hot Air Drying. *Food Bioprocess Technol.* **2012**, *5*, 672–679. [CrossRef]
52. Rybáček, V. (Ed.) *Hop Production*; Elsevier: Amsterdam, The Netherlands; New York, NY, USA, 1991; ISBN 0444987703.
53. Koyuncu, T.; Pinar, Y.; Lule, F. Convective drying characteristics of azarole red (Crataegus monogyna Jacq.) and yellow (Crataegus aronia Bosc.) fruits. *J. Food Eng.* **2007**, *78*, 1471–1475. [CrossRef]
54. Sturm, B.; Münsterer, J.; Kammhuber, K.; Crichton, S. Impact of bulk weight on drying behaviour and hop quality after drying. In Proceedings of the International Conference on Agricultural Engineering, CIGR—AgEng, Aarhus, Denmark, 26–29 June 2016.

processes

MDPI

Article

Evaluation of Postharvest Processing of Hazelnut Kernel Oil Extraction Using Uniaxial Pressure and Organic Solvent

Gürkan Alp Kağan Gürdil [1,*], Abraham Kabutey [1], Kemal Çağatay Selvi [1], Čestmír Mizera [1], David Herák [1] and Adéla Fraňková [2]

1 Department of Mechanical Engineering, Faculty of Engineering, Czech University of Life Sciences Prague, 16521 Prague, Czech Republic; kabutey@tf.czu.cz (A.K.); kcselvi@omu.edu.tr (K.Ç.S.); mizera@tf.czu.cz (Č.M.); herak@tf.czu.cz (D.H.)

2 Department of Food Science, Faculty of Agrobiology, Food and Natural Resources, Czech University of Life Sciences Prague, 16521 Prague, Czech Republic; frankovaa@af.czu.cz

* Correspondence: ggurdil@omu.edu.tr; Tel.: +90-362-312-1919

Received: 8 July 2020; Accepted: 6 August 2020; Published: 8 August 2020

Abstract: Uniaxial loading and organic solvent are small-scale oil expression methods used to evaluate the mechanical behavior, oil content, and oil efficiency of oil-bearing materials aimed at designing a low-cost mechanical pressing system. Bulk kernels of pressing height 40 mm were heated from 40 to 60 °C and compressed at maximum force of 60 kN and speeds from 4 to 8 mm/min. Relaxation times between 3 and 12 min were applied to assess the kernel oil efficiency. The kernel oil point was identified at deformation levels between 15 and 25 mm at a speed of 4 mm/min using a litmus test. The kernel oil was analyzed for peroxide value and free fatty acid. Kernel oil content was determined by Soxhlet extraction. Increased speed caused a serration effect on the force–deformation curve leading to lower oil yield. Lower and upper oil point forces at 6.21 ± 0.58 and 10.61 ± 0.71 kN were observed to be useful for predicting the pressure for maximum output oil. The peroxide value and free fatty acid content of kernel oil decreased with increasing temperature, indicating its quality usage. The relaxation time of 12 min after compression increased kernel oil efficiency of 15.6%. In designing new presses, there is a need to consider compression and relaxation processes to reduce the residual kernel cake oil.

Keywords: bulk hazelnut kernels; mechanical properties; heating temperature; oil efficiency; relaxation process

1. Introduction

Hazelnut (*Corylus avellana* L.) is a fruit consisting of a hard shell with a kernel inside it that is edible [1]. The edible kernel is covered by a removable thin fibrous pellicle, with the internal tissue of the cotyledons consisting of parenchyma cells separated by very small intercellular spaces [2,3]. The tree crop belongs to the family Betulaceae and is grown in the Mediterranean Sea regions, mainly in Turkey, Spain, and Italy [4,5]. Turkey is the main producer and exporter with 420,000 tons/year (contributing around 70% of the total global production), followed by Italy with 120,572 tons/year, while Spain produces approximately 15,300 tons/year [6–8]. Portugal, France, United States (Oregon and Washington), New Zealand, China, Azerbaijan, Chile, Iran, Georgia, Kirgizstan, Poland, and Croatia are other producing countries [9–11].

Hazelnuts are rich in antioxidant compounds such as tocopherols and polyphenols, unsaturated fatty acids, flavan-3-ols (flavanols) and proanthocyanidins, essential amino acids, dietary fibers, vitamins (E and B), and complex and essential minerals (Ca, Mg, P, K), which have a wide range of health benefits, including reducing oxidative stress and risk of cancer, stroke, inflammation, and other neurodegenerative diseases [7,12–14]. Hazelnuts are usually used in dairy, bakery, coffee, spreads, confectionery products, and salads due to their nutritious quality and unique flavor [14,15]. In addition, the hazelnut oil is used in cooking, cosmetics, and pharmaceutical industries [16,17]. Hazelnut is primarily composed of oil content (58–65%), protein (11–16%), fat content (60%), and carbohydrates (15–18%) [18,19].

The roasting process is performed to remove the pellicles of kernels, inactivate enzymes, destroy microorganisms, and reduce moisture [11,20]. In addition, roasting is used to improve the sensory features (color, crispy texture, appearance, and flavor) of the product [11,21,22]. Postharvest practices, such as handling, processing, and storage, could affect the biochemical composition and sensory properties of hazelnuts [23–25]. Against this background, knowledge of the physical and mechanical properties of the nuts is useful for the design and development of systems for handling, processing, packaging, storage, and transportation [26–30]. The physical properties of the nuts—mostly the average length, width, thickness, geometric mean diameter, sphericity, unit mass, volume, bulk density, true density, porosity, projected area, terminal velocity, and static and dynamic coefficients of friction—have been studied as a function of moisture content [30–33] where they tend to increase and/or decrease with increases in moisture content. The mechanical properties, on the other hand, include the rupture force, deformation at rupture point, deformation ratio at rupture point, hardness, and energy [1,34,35]. According to [30,36,37], the rupture force is the minimum force required to break the sample. The deformation at rupture point is the deformation at loading direction which can be used for the determination of the gap size between the surfaces to compress the fruit or nut for dehulling. Deformation ratio at the rupture point is the axial strain at the rupture point of the sample. Hardness is the ratio of rupture force and deformation at rupture point. The energy for rupture is the energy needed to rupture the sample, which is characterized by the area under the force–deformation curve. It is important to note that moisture content, speed, heating temperature, heating time, pressure, varieties, storage conditions, and vessel diameter thus affect the mechanical properties and force–deformation curves of oil-bearing material [38].

Based on the available literature, there is a lack of information on the postharvest processing of bulk hazelnut kernels in relation to processing factors such as speed, heating temperature, and relaxation time. To aid the design and development of optimal oil processing technology, it is necessary to enhance the knowledge of the uniaxial compression process. The uniaxial compression involves placing the bulk oil-bearing material in a pressing vessel with holes of about 0.5 mm around the bottom to allow the oil discharge while retaining the seed/kernel cake under a given load and speed using a universal testing machine [39,40]. The output data based on the preset pressing variables are usually the force–deformation and relaxation curves, which can further be analyzed to determine the mechanical properties and oil expression efficiency of the oil-bearing material under consideration [36,41,42]. The kernel oil extraction efficiency is determined based on the ratio of kernel oil recovered to that of the kernel oil content determined by Soxhlet extraction. The Soxhlet/solvent extraction technique involves repeated solvent distillation through a solid sample to remove the oil or analyte of interest [43]. Consequently, the study aimed to examine the effects of heating temperature, speed and relaxation time on hazelnut kernel oil processing in uniaxial pressure and to evaluate the chemical properties (peroxide value and free fatty acid) of the kernel oil in relation to heating temperatures.

2. Materials and Methods

2.1. Samples and Moisture Content Determination

Samples of bulk hazelnut kernels (Figure 1a) were from Ordu province of northeast Turkey, called the East Black Sea region of Turkey (40°58.8′ N, 37°52.2′ E). The Cakildak variety was used. The moisture content of 3.46 ± 0.11% (w.b.) was determined using the standard procedure where samples were randomly selected and dried at a temperature of 105 °C for 24 h [44]. An electronic balance (Kern 440–35, Kern & Sohn GmbH, Balingen, Germany) having an accuracy of 0.01 g was used to measure the mass (g) of the samples before and after oven drying. To determine the moisture content value, Equation (1) was applied [45].

$$MC = \left[\left(\frac{m_b - m_a}{m_b}\right) \cdot 100\right] \quad (1)$$

where MC is the moisture content % (w.b.) and m_a and m_b are the masses of the samples before and after oven drying (g).

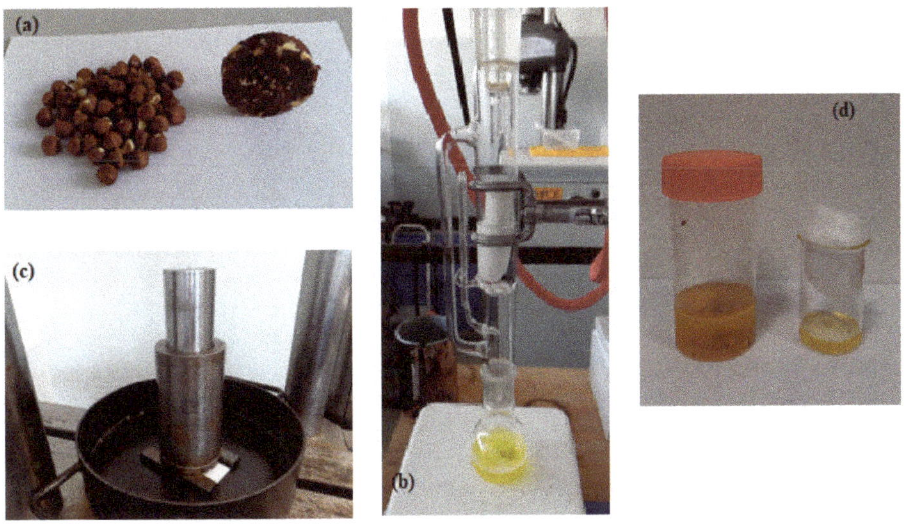

Figure 1. Set up of compression test and Soxhlet extraction: (**a**) hazelnut kernels and seed cake after compression; (**b**) Soxhlet extraction setup for kernel oil content determination; (**c**) a pan holding the pressing vessel with a plunger placed on two separated metal bars with a piece of white paper for oil point identification similar for the compression test; and (**d**) hazelnut kernel oil from the compression test and Soxhlet extraction (covered with cotton wool).

2.2. Determination of Percentage Kernel Oil Content

The percentage kernel oil content (Figure 1b) was determined to be (64.67 ± 0.01%) using the Soxhlet extraction method [46–48]. Based on the procedure, the sample mass of 8 g was ground and packed into a thimble. The thimble was placed in a Soxhlet extractor attached to a 150 mL round-bottom flask containing 100 mL of petroleum ether. The oil extraction process was allowed for 24 h involving several extraction cycles. The residual water and solvent in the extracted oil were removed by drying at 105 °C for 5 h. The kernel oil content was calculated based on the dry weight of the sample [45].

2.3. Heating of Samples

Samples were heated at temperatures between 40 and 60 °C using the oven device (MEMMERT GmbH + Co. KG, Schwabach, Germany) at a heating time of 30 min. The samples laboratory temperature of 25 °C served as the control.

2.4. Compression Tests

The universal compression machine (Tempos, ZDM 50, Czech Republic) together with the pressing vessel of diameter, D (60 mm) with a plunger (Figure 1c) were used to record the compression and relaxation curves of bulk hazelnut kernels at constant initial pressing height, H, of 40 mm (mass 58.31 ± 0.23 g and volume 11.31·10^{-5} m^3). The preset load of 60 kN was determined from a preliminary test where the load was initially set at 100 kN and speed 5 mm/min, which indicated an undulation pattern on the force–deformation curve (ejection of kernel cake through the holes of the pressing vessel). Four separate experiments were performed. The first experiment varied the speed from 4 to 8 mm/min at force 60 kN without sample pretreatment. The second experiment considered pretreatment of the samples at temperatures between 40 and 60 °C for 30 min and compressed at a force of 60 kN and speed of 4 mm/min. The third experiment varied the relaxation time from 0 to 12 min for the samples with a heating temperature of 60 °C, speed of 4 mm/min, and force of 60 kN to determine maximum oil output. The speed and temperature values were selected based on the results of the preliminary study. The fourth experiment determined the first leakage of the oil (oil point) of the samples before and after pretreatment based on the litmus test procedure [39]. For identifying the lower and upper oil points with the corresponding strain, force, and energy values, the sample deformation levels between 15 and 25 mm were examined at a speed of 4 mm/min.

2.5. Determination of Peroxide Value (PV) and Free Fatty Acid (FFA)

The hazelnut kernel oil obtained from the compression test (Figure 1d) was evaluated based on peroxide value and free fatty acid. For peroxide value determination, an oil sample of mass of 5 g was measured into a volumetric flask and dissolved in 30 mL of a chloroform and glacial acetic acid mixture of ratio 2:3. Saturated potassium iodine (KI), 1 mL, was added to the solution and left in a dark place for 5 min. Distilled water (40 mL) was added and the solution was titrated with 0.1 M $Na_2S_2O_3$ until the yellow color ceased to be visible. Subsequently, 1 mL of 1% starch was added, and the solution was titrated with $Na_2S_2O_3$ until complete discoloration. For FFA determination, 5 g of the oil sample was measured into a titration flask, then a 100 mL of neutralized ethanol (warmed up to 60–65 °C) and 2 mL of 1% phenolphthalein were added. Titration was performed directly with ethanolic KOH (0.1 g/mol) until the color was light pink. Peroxide value was expressed as (µg/g of active oxygen) and free fatty acid was expressed as (mg KOH/g) [1,48,49].

2.6. Compression Tests Calculated Parameters

The deformation values were obtained directly from the compression tests. The oil yield (%) and oil expression efficiency (%) were calculated according to Equations (2) and (3), as given by [50,51].

$$OY = \left[\left(\frac{m_o}{m_s}\right)\cdot 100\right] \tag{2}$$

where OY is the oil yield (%), m_o is the mass of oil obtained as the difference of mass of seed cake and initial mass of the sample m_s (g).

$$OEE = \left[\left(\frac{OY}{O_s}\right)\cdot 100\right] \tag{3}$$

where OEE is the oil expression efficiency (%) and O_s is the percentage of oil content (%). The energy (kJ) was calculated based on the area under the force–deformation curve, which is given by Equation (4) [52,53].

$$E = \sum_{n=0}^{n=i-1}\left[\left(\frac{F_{n+1}+F_n}{2}\right)\cdot(x_{n+1}-x_n)\right] \tag{4}$$

where E is the deformation energy (kJ), $F_{n+1}+F_n$ and $x_{n+1}-x_n$ are the compressive force (kN) and deformation (mm), n is the number of data points, and i is the number of sections in which the axis deformation was divided.

2.7. Statistical Analyses

All experiments were repeated twice based on prior knowledge of the statistical techniques applied and the reliability of the results of the preliminary experiment. The data were statistically analyzed using STATISTICA software (Version 13) [54] by employing the ANOVA (San Francisco, CA, USA), correlation, and regression techniques.

3. Results

3.1. Effects of Speed, Temperature, and Relaxation Time on Responses

The effects of speed, temperature, and relaxation time on deformation, oil yield, oil expression efficiency, and energy of hazelnut kernels are presented in Tables 1–3, respectively. The speed and temperature did not greatly influence the deformation values. The lower speed of 4 mm/min combined with a higher temperature of 60 °C increased the oil expression efficiency by 4.68%. A higher speed rather increased the energy for recovering the kernel oil compared to a higher temperature. This could be attributed to the softening of the kernels and occurrence of the serration effect, which increases the area under the force–deformation curve, which is defined as the energy. Lower speed and higher temperature recorded an equal amount of energy of 0.21 kJ. The relaxation time of 12 min after compression increased the oil expression efficiency, meaning that the residual kernel oil of 15.6% was recovered from the kernel cake after compression.

Table 1. Effect of speed on deformation, oil yield, oil expression efficiency, and energy.

* Speed (mm/min)	Deformation (mm)	Oil Yield (%)	Oil Expression Efficiency (%)	Energy (kJ)
4	27.75 ± 2.79	39.76 ± 0.02	61.48 ± 0.04	0.21 ± 0.01
8	27.04 ± 1.17	37.00 ± 0.51	57.22 ± 0.79	0.23 ± 0.01
12	28.06 ± 0.74	33.83 ± 0.88	52.31 ± 1.37	0.27 ± 0.04

* Sample compressive force of 60 kN.

Table 2. Effect of heating temperature on deformation, oil yield, oil expression efficiency, and energy.

* Heating Temperature (°C)	Deformation (mm)	Oil Yield (%)	Oil Expression Efficiency (%)	Energy (kJ)
25	27.76 ± 0.24	36.21 ± 0.04	55.99 ± 0.06	0.22 ± 0.001
40	30.61 ± 0.50	39.83 ± 0.64	61.59 ± 0.98	0.20 ± 0.001
50	30.41 ± 2.41	42.35 ± 0.43	65.48 ± 0.67	0.21 ± 0.003
60	29.96 ± 2.11	42.79 ± 0.24	66.16 ± 0.37	0.21 ± 0.001

* Sample compressive force of 60 kN and speed of 4 mm/min.

Table 3. Effect of relaxation time on sample heating temperature of 60 °C.

** Relaxation Time (min)	Oil Yield (%)	Oil Expression Efficiency (%)
0	36.21 ± 0.04	55.99 ± 0.06
3	41.73 ± 0.34	64.53 ± 0.52
6	45.08 ± 0.03	69.71 ± 0.05
9	46.04 ± 0.96	71.21 ± 1.48
12	46.31 ± 0.34	71.59 ± 0.53

** Compressive force of 60 kN and speed of 4 mm/min.

3.2. Force–Deformation Curves, Strain, Oil Point Force, and Oil Point Energy

The force–deformation curves for different speeds and heating temperatures are illustrated in Figures 2 and 3. The strain, oil point force, and oil point energy are also given in Tables 4 and 5. The maximum applied force of 60 kN was based on a preliminary experiment where a higher force using a pressing vessel diameter of 60 mm caused a serration effect characterized by the ejection of kernel cake through the holes of the pressing vessel. The strain is the ratio of deformation to the initial height of the sample and is useful for identifying the oil point force with the corresponding energy. The lower and upper oil points were observed at deformation values of 20 and 25 mm at a speed of

4 mm/min with the corresponding strain values of 0.5 and 0.625 (−) for both heat-treated kernels and those without heat treatment. The oil point force and energy of the samples before and after heating were determined from the deformation levels between 15 and 25 mm (Figure 4). The lower oil point denotes the minimum force required for the first drop of kernel oil whiles the upper oil point describes the maximum force needed to recover maximum kernel oil.

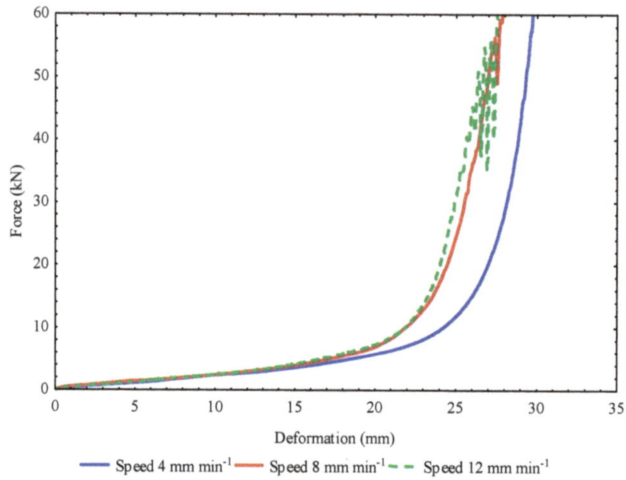

Figure 2. Force–deformation curves of bulk hazelnut kernels for different speeds.

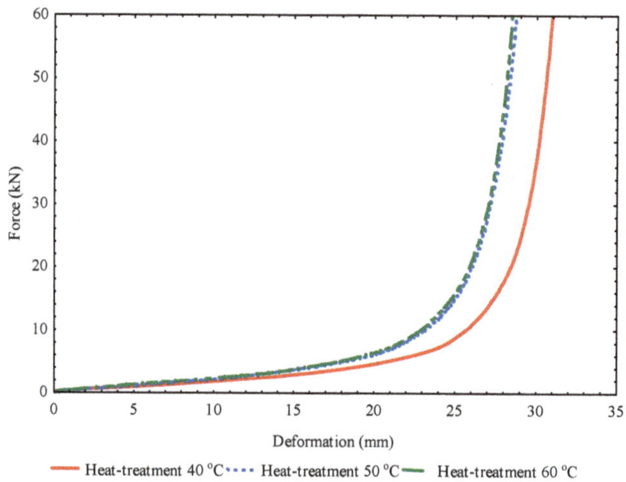

Figure 3. Force–deformation curves of bulk hazelnut kernels for heating temperatures.

Table 4. Strain and oil point force of deformation levels of kernels at a speed of 4 mm/min.

* Deformation Levels (mm)	Strain (−)	** Oil Point Force (kN)	** Oil Point Energy (kJ)
15	0.375	3.58 ± 1.13	0.03 ± 0.01
20	0.5 [a]	6.46 ± 0.08	0.06 ± 0.002
25	0.625 [b]	10.79 ± 2.62	0.09 ± 0.01

* For initial pressing height, H = 40 mm; [a] Lower oil point; [b] Upper oil point; ** (Control at 25 °C).

Table 5. Strain and oil point force of deformation levels of kernels at a speed of 4 mm/min.

* Deformation Levels (mm)	Strain (–)	*** Oil Point Force (kN)	*** Oil Point Energy (kJ)
15	0.375	3.22 ± 0.21	0.02 ± 0.001
20	0.5 [a]	6.21 ± 0.58	0.05 ± 0.01
25	0.625 [b]	10.61 ± 0.71	0.08 ± 0.01

* For initial pressing height, H = 40 mm; [a] Lower oil point; [b] Upper oil point; *** Heating temperature of 60 °C.

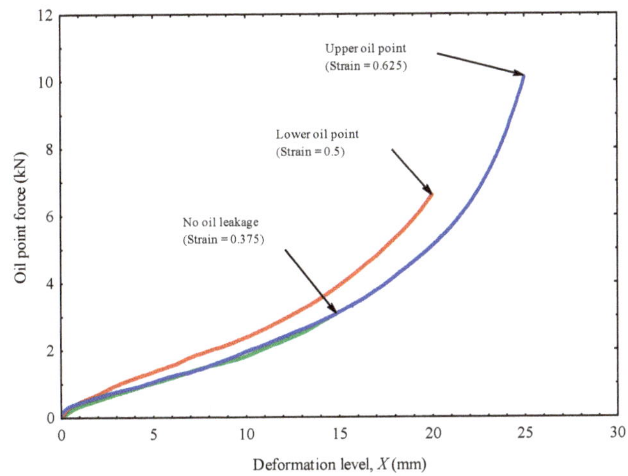

Figure 4. Oil point identification at deformation levels of bulk hazelnut kernels.

3.3. Peroxide Value and Free Fatty Acid

In Figure 5 is illustrated the box plot of values for peroxide and free fatty acid of kernel oil at different heating temperatures. The peroxide values of hazelnut kernel oil increased from 40 to 50 °C and then decreased at 60 °C but the free fatty acid values decreased linearly with heating temperatures.

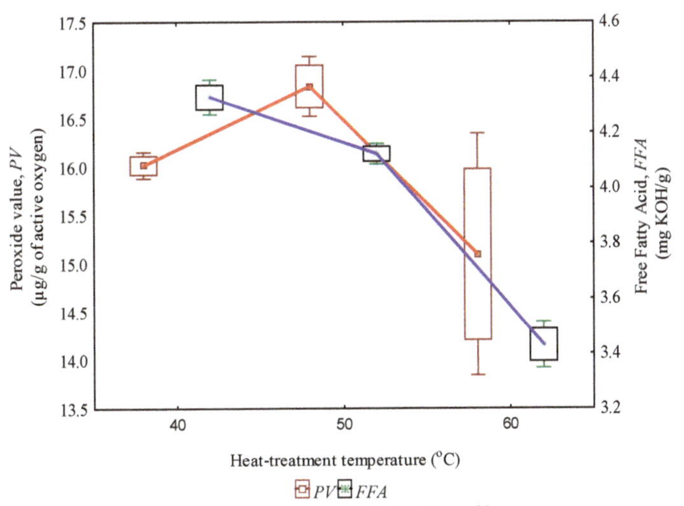

Figure 5. Box plot of peroxide value and free fatty acid of hazelnut kernel oil under heat treatment temperature.

3.4. ANOVA, Correlation, and Regression Results

The results of ANOVA, correlation, and regression of the effects of speed, heating temperature, and relaxation time are given in Tables 6–10. The ANOVA results show that the increase in speed and heating temperature did not significantly ($p > 0.05$) decrease or increase the deformation values of hazelnut kernels. Oil yield and oil expression efficiency significantly ($p < 0.05$) decreased with increased speed while they increased significantly ($p < 0.05$) with heating temperature. The speed and heating temperature combined did not indicate a significant effect ($p > 0.05$) on energy. The correlation results also showed no relationship between deformation and speed as well as deformation and temperature. Oil yield and oil expression efficiency correlated with speed and heating temperature effects with high correlation efficiency between 97 and 99% respectively. Energy showed no correlation with the effects of speed and heating temperature ($p > 0.05$). The oil point force, oil expression efficiency, and energy were statistically significant ($p < 0.05$) with a coefficient of determination (R^2) of 87%. Peroxide value was not significant ($p > 0.05$) with heating temperature but free fatty acid proved significant ($p < 0.05$) with a correlation efficiency of 95%. The regression equations representing oil yield, oil expression efficiency, and free fatty acid as functions of speed, heating temperature, and relaxation time effects, respectively, indicated significance ($p < 0.05$). The energy and peroxide value against speed and heating temperature revealed a non-significant ($p > 0.05$) effect.

Table 6. ANOVA and correlation results of the effect of speed and heat treatment temperature on the dependent variables.

Dependent Variables	R	R^2	F-Value	*p*-Value
Deformation (mm)	0.09^a 0.49^b	0.10^a 0.49^b	0.16^a 1.29^b	$>0.05^{a,b}$
Oil yield (%)	-0.99^a 0.89^b	0.97^a 0.99^b	50.65^a 111.89^b	$<0.05^{a,b}$
Oil expression efficiency (%)	-0.99^a 0.89^b	0.97^a 0.99^b	50.65^a 111.89^b	$<0.05^{a,b}$
Energy (kJ)	0.78^a 0.22^b	0.63^a 0.94^b	2.53^a 19.27^b	$>0.05^{a,b}$

Compressive force of 60 kN and speed 4 mm/min; [a] Speed; [b] heating temperature; Significant (*p*-value < 0.05 or F-value > *p*-value); Non-significant (*p*-value > 0.05 or F-value < *p*-value).

Table 7. ANOVA and correlation results of the effect of relaxation time on oil expression efficiency.

Dependent Variables	R	R^2	F-Value	*p*-Value
* Oil yield (%)	0.91	0.99	157.07	<0.05
* Oil expression efficiency (%)	0.91	0.99		

* Heating temperature of 60 °C; compressive force of 60 kN; speed of 4 mm/min.

Table 8. ANOVA and correlation results of oil point force, oil expression efficiency and energy.

Dependent Variables	R	R^2	F-Value	F-Value
* Oil point force (kN)	0.92^a 0.99^b	0.87^a 0.98^b	9.73^a 96.03^b	$<0.05^{a,b}$
* Oil point expression efficiency (%)	0.98^a 0.99^b	0.99^a 0.99^b	267.45^a 1549.86^b	
* Oil point energy (kJ)	0.97^a 0.99^b	0.95^a 0.97^b	29.13^a 56.87^b	

* Speed of 4 mm/min; [a] Control at 25 °C; [b] Heating temperature of 60 °C.

Table 9. ANOVA and correlation results of *PV* and *FFA* in relation to heating temperature.

Dependent variables	R	R^2	F-Value	*p*-Value
Peroxide value, *PV* (μg/g of active oxygen)	−0.43	0.64	2.70	>0.05
Free Fatty Acid, *FFA* (mg KOH/g)	−0.95	0.99	108.37	<0.05

Table 10. Regression results of the effects of speed, heating temperature, and relaxation time.

Dependent Variables	Speed, S_p (mm/min)			
	Equation	R^2	F-Value	*p*-Value
Oil yield (%)	$42.79 - 0.74 \times S_p$	0.97	127.60	<0.05
Oil expression efficiency (%)	$66.17 - 0.15 \times S_p$	0.97	127.60	<0.05
Energy (kJ)	$0.18 + 0.01 \times S_p$	0.60	6.09	>0.05
Dependent variables	**Heating temperature, T_p (°C)**			
	Equation	R^2	**F-value**	***p*-value**
Oil yield (%)	$34.26 + 0.15 \times T_p$	0.81	16.78	<0.05
Oil expression efficiency (%)	$52.98 + 0.23 \times T_p$	0.81	16.78	<0.05
Energy (kJ)	$0.20 + 0.0001 \times T_p$		0.21	>0.05
Dependent variables	**Relaxation time R_t (min)**			
	Equation	R^2	**F-value**	***p*-value**
Oil yield (%)	$38.18 + 0.82 \times R_t$	0.82	37.57	<0.05
Oil expression efficiency (%)	$59.03 + 1.26 \times R_t$	0.82	37.57	
Dependent variables	**Heating temperature, T_p (°C)**			
	Equation	R^2	**F-value**	***p*-value**
Peroxide value, *PV* (μg/g of active oxygen)	$18.31 - 0.05 \times T_p$	0.18	0.89	>0.05
Free Fatty Acid, *FFA* (mg KOH/g)	$6.21 - 0.04 \times T_p$	0.91	35.27	<0.05

T_p: Heating temperature (°C); S_p: Speed (mm/min); R_t: Relaxation time (min); Significant (*p*-value < 0.05 or F-value > *p*-value); Non-significant (*p*-value > 0.05 or F-value < *p*-value).

4. Discussion

The experimental and statistical results of the effects of speed, heating temperature, and relaxation time on deformation, oil yield, oil expression efficiency, force–deformation curves, peroxide value, and free fatty acid of postharvest processing of hazelnut kernels and oil under uniaxial compression have been described in the preceding sections. The results are herein elaborated.

A lower speed of 4 mm/min recovered maximum kernel oil compared to a higher speed of 12 mm/min, where the minimum oil yield was realized. Increases in speed increased the energy but decreased the oil yield and oil expression efficiency whereas increases in heating temperature showed a significant increase in oil yield and oil expression efficiency but not in energy. Similar results were reported by [55] for jatropha kernels heated at 80 °C. It has been reported that the increase in heat will not always increase the oil yield [56]. The oil yield, however, is dependent on the quality of the oil-bearing material, moisture content, and the oil extraction process. Theoretically, higher screw speed means more seed material throughput and higher oil content residual in the press cake due to less time being available for the oil to discharge from the solids. A lower energy input also means lower oil recovery efficiency, higher oil residue in the press cake, and higher seed material throughput, and higher pressure will lead to higher temperature generation and higher oil recovery efficiency [57–59].

The relaxation time for kernels heated at 60 °C and compressed at a force of 60 kN and speed of 4 mm/min increased the oil yield and oil expression efficiency. A significant increase in oil yield and oil expression efficiency occurred at 3 and 6 min of relaxation time in comparison with 9 and 12 min of the relaxation time. However, the percentage increase of 10.1% and 15.6% of oil yield and oil expression efficiency was observed at 12 min relaxation. The correlation efficiency between oil expression efficiency and relaxation time was 91%. The compression and relaxation processes are important for evaluating the oil yield and energy at the compression area and oil expression efficiency without energy at the relaxation phase, which is dependent on several factors such as the volume of kernels, moisture content, vessel diameter, force, and speed, among other factors [48]. Further, from the principles of viscoelasticity, the relaxation time shows how fast the material dissipates after receiving a sudden deformation, that is, when the maximum force is attained, and it can also be used to characterize the elastic and viscous components in the behavior of the material. The relaxation data can be used to develop rheological models useful in the design of food engineering processes by applying the concepts of momentum, energy, and mass balances [60,61].

The lower oil point force of 6.21 ± 0.58 kN and the upper oil point force of 10.61 ± 0.71 kN of hazelnut kernels were observed. Based on these values, a maximum force of 60 kN was applied to recover the maximum kernel oil. However, it was also observed that the oil point forces at both lower and upper oil points for heat-treated kernels were lower compared to kernels without heat treatment. This indicates that kernels without heat treatment are more resistant to breakage and, hence, higher energy than those with heat treatment. The oil point force decreased with increased heating temperature. The results are in agreement with [62] based on cashew kernels. The oil point pressure/force indicates the threshold pressure at which oil emerges from the seed or kernel during mechanical or uniaxial compression [63–65]. The effective pressure for the oil to discharge from oil-bearing cells is determined by first identifying the oil point pressure. The oil point pressure improves the oil expression efficiency and provides useful information for the design and performance evaluation of oil expellers [62,65]. Nevertheless, the oil point force is dependent on the initial moisture content, heat treatment, heating time, speed, vessel diameter, and kernel pressing height and varieties. This suggests that the optimal processing factors for identifying the oil point pressure/force with corresponding energy and for recovering higher percentage kernel oil need to be determined to help in the design of energy-efficient oil processing system.

Peroxide value and free fatty acid of hazelnut kernel oil decreased at the highest heating temperature of 60 °C. This indicates that the quality of the oil at the abovementioned heating temperature might be within the acceptable limit. The explanation is that higher temperature leads to the higher acid value, resulting in high free fatty acid. The lower the acid value, the lower the free fatty acid hence, the better the oil usage quality [55]. Peroxide value and free fatty acid are strongly correlated [66]. Generally, peroxide values in edible oils indicate lipid oxidation whilst free fatty acids indicate lipid hydrolysis [67–69]. Oxidation of edible oils reduces oil quality and human health benefits and, therefore, it is important to minimize lipid oxidation and rancidity [70]. Peroxide values depend on the cultivar, harvest time, delaying of storage after harvest, ecological conditions, cultural management during the growing season, and handling procedures after harvest [71]. On the other hand, enzymatic lipid hydrolysis produces free fatty acids which are driven by a reaction between moisture and oil [72,73].

In addition to the abovementioned facts, mechanical pressing of oil from oil-bearing materials involves the application of pressure by using either a hydraulic or screw press to recover the oil [52,64]. Improving the performance of the oil expression process and/or designing an optimum universal pressing system requires understanding of the mechanical and relaxation behaviors of the oil material under uniaxial compression. The results of the present study thus provide supplementary information to the available literature.

5. Conclusions

A laboratory-scale study was conducted to examine the mechanical behavior, oil content, and oil efficiency of hazelnut kernels with different speeds, heating temperatures, and relaxation times.

The quality of the kernel oil was assessed in terms of peroxide value and free fatty acid. The following findings were observed. For a pressing vessel of diameter 60 mm and sample pressing height of 40 mm, kernel oil efficiency was achieved at a maximum force of 60 kN and speed of 4 mm/min. The maximum force applied was based on a preliminary experiment, which showed that a greater force under the same processing conditions demonstrated the serration effect, which is not energy efficient for recovering the maximum kernel oil. The lower and upper oil points with strain values of 0.5 and 0.625 with the corresponding forces of 6.21 ± 0.58 and 10.61 ± 0.71 kN were in reference to the maximum force applied for recovering the kernel oil. Peroxide value, *PV* (µg/g of active oxygen), and free fatty acid, *FFA* (mg KOH/g), of hazelnut kernel oil decreased with heating temperature, indicating that the quality of the oil at 60 °C is acceptable for use. Regression models representing oil yield, oil expression efficiency, and free fatty acid were described based on the predictors of speed, heat treatment temperature, and relaxation time. The coefficients of the model predictors were statistically significant ($p < 0.05$) for describing the responses. The relaxation time of 12 min after compression increased the kernel oil expression efficiency of 15.6%, specifying that the compression and relaxation processes for each pressing chamber along the screw configuration need to be considered when designing a screw press to reduce the additional energy for recovering the residual kernel cake oil.

Author Contributions: Conceptualization, G.A.K.G.; A.K.; Č.M. and D.H.; Funding acquisition, D.H.; Methodology, G.A.K.G.; A.K.; Č.M. and A.F.; data curation, A.K.; Č.M.; K.Ç.S. and A.F.; Writing—original draft, G.A.K.G.; A.K. and K.Ç.S.; Writing—review and editing, G.A.K.G.; A.K., K.Ç.S.; Č.M. and D.H. All authors have read and agreed to the published version of the manuscript.

Funding: The research through the project "supporting the development of international mobility of research staff at CULS Prague, was funded by EU, Managing Authority of the Czech Operational Programme Research, Development and Education" Grant Number: CZ.02.2.69/0.0/0.0/16_027/0008366.

Conflicts of Interest: The authors declare no conflict of interest.

References

1. Guner, M.; Dursun, E.; Dursun, J.G. Mechanical behaviour of hazelnut under compression loading. *Biosyst. Eng.* **2003**, *85*, 485–491. [CrossRef]
2. Delprete, C.; Giacosa, S.; Raviola, E.; Rolle, L.; Sesana, R. Experimental characterization and numerical modeling of the compressive mechanical of hazelnut kernels. *J. Food Eng.* **2015**, *166*, 364–369. [CrossRef]
3. Young, W.J. *A Study of Nuts with Special Reference to Microscopic Identification*; U.S. Government Printing Office: Washington, DC, USA, 1912.
4. Demirkaya, E.; Dal, O.; Yuksel, A. Liquefaction of waste hazelnut shell by using sub-and supercritical solvents as a reaction medium. *J. Supercrit Fluids.* **2019**, *150*, 11–20. [CrossRef]
5. Baran, Y.; Gokce, H.S.; Durmaz, M. Physical and mechanical properties of cement containing regional hazelnut shell ash wastes. *J. Clean. Prod.* **2020**, *259*, 1–9. [CrossRef]
6. FAO. Produccion Avellana. 2016. Available online: http://www.fao.org/faostat/es/#data/QC/visualize (accessed on 20 July 2020).
7. Lainas, K.; Alasalvar, C.; Bolling, B.W. Effects of roasting on proanthocyanidin contents of Turkish Tombul hazelnut and its skin. *J. Funct. Foods.* **2016**, *23*, 647–653. [CrossRef]
8. Perez-Armada, L.; Rivas, S.; Gonzalez, B.; Moure, A. Extraction of phenolic compounds from hazelnut shells by green processes. *J. Food Eng.* **2019**, *255*, 1–8. [CrossRef]
9. Ciemniewska-Żytkiewicz, H.; Verardo, V.; Pasini, F.; Bryś, J.; Koczoń, P.; Caboni, M.F. Determination of lipids and phenolic fraction in two hazelnut (*Corylus avellana L.*) cultivars grown in Poland. *Food Chem.* **2015**, *168*, 615–622. [CrossRef] [PubMed]
10. Pelvan, E.; Alasalvar, C.; Uzman, S. Effects of roasting on the antioxidant status and phenolic profiles of commercial turkish hazelnut varieties (Corylus avellana L.). *J. Agric. Food Chem.* **2012**, *60*, 1218–1223. [CrossRef] [PubMed]
11. Marzocchi, S.; Pasini, F.; Verardo, V.; Ciemniewska-Zytkiewicz, H.; Caboni, M.F.; Romani, S. Effects of different roasting conditions on physical-chemical properties of Polish hazelnuts (*Corylus avellana* L. Var. Katalonski). *LWT Food Sci. Technol.* **2017**, *77*, 440–448. [CrossRef]

12. Köksal, A.I.; Artik, N.; Simsek, A.; Günes, N. Nutrient composition of hazelnut (*Corylus avellana* L.) varieties cultivated in Turkey. *Food Chem.* **2006**, *99*, 509–515. [CrossRef]
13. Del Rio, D.; Calani, L.; Dall'Asta, M.; Brighenti, F. Polyphenolic composition of hazelnut skin. *J. Agric. Food Chem.* **2011**, *59*, 9935–9941. [CrossRef] [PubMed]
14. Kara, H.; Orem, A.; Yulug, E.; Yucesan, F.B.; Kerimoglu, G.; Yaman, S.O.; Bodur, A.; Turedi, S.; Alasalvar, C. Hazelnut consumption improves testicular antioxidant function and semen quality in young and old male rats. *Food Chem.* **2019**, *294*, 1–8. [CrossRef] [PubMed]
15. Yuan, B.; Lu, M.; Eskridge, K.M.; Isom, L.D.; Hanna, A.M. Extraction, identification, and quantification of antioxidant phenolics from hazelnut (*Corylus avellana* L.) shells. *Food Chem.* **2018**, *244*, 7–15. [CrossRef] [PubMed]
16. Shahidi, F.; Alasalvar, C.; Liyana-Pathirana, C.M. Antioxidant phytochemicals in hazelnut kernel (*Corylus avellana* L.) and hazelnut byproducts. *J. Agric. Food Chem.* **2007**, *55*, 1212–1220. [CrossRef] [PubMed]
17. Eskandari, J.; Kermani, A.M.; Kouravand, S.; Zarafshan, P. Design, fabrication, and evaluation a laboratory dry-peeling system for hazelnut using infrared radiation. *LWT Food Sci. Technol.* **2018**, *90*, 570–576. [CrossRef]
18. Tas, N.G.; Yilmaz, C.; Gokmen, V. Investigation of serotonin, free and protein-bound tryptophan in Turkish hazelnut varieties and effect of roasting on serotonin content. *Food Res. Int.* **2019**, *120*, 865–871. [CrossRef]
19. Alasalvar, C.; Karamac, M.; Kosinska, A.; Rybarczyk, A.; Shahidi, F.; Amarowicz, R. Antioxidant activity of hazelnut skin phenolics. *J. Agric. Food Chem.* **2009**, *57*, 4645–4650. [CrossRef]
20. Ozdemir, M.; Seyan, F.G.; Bakan, A.K.; Ilter, S.; Ozay, G.; Devres, O. Analysis of internal browning of roasted hazelnuts. *Food Chem.* **2001**, *73*, 191–196. [CrossRef]
21. Burdack-Freitag, A.; Schieberle, P. Changes in the key odorants of Italian hazelnuts (*Corylus avellana* L. Var. Tonda Romana) induced by roasting. *J. Agric. Food Chem.* **2010**, *58*, 6351–6359. [CrossRef]
22. Belviso, S.; Dal Bello, B.; Giacosa, S.; Bertolino, M.; Ghirardello, D.; Giordano, M.; Rolle, L.; Gerbi, V.; Zeppa, G. Chemical, mechanical and sensory monitoring of hot air-and infrared-roasted hazelnuts (*Corylus avellana* L.) during nine months of storage. *Food Chem.* **2017**, *217*, 398–408. [CrossRef]
23. Guler, S.K.; Bostan, S.Z.; Con, A.H. Effects of gamma irradiation on chemical and sensory characteristics of natural hazelnut kernels. *Postharvest Biol. Technol.* **2017**, *123*, 12–21. [CrossRef]
24. Ghirardello, D.; Rolle, L.; Zeppa, G. Effects of storage conditions on hazelnut (*Corylus avellana L.*) textural characteristics. In Proceedings of the VII International Congress of Hazelnuts, Viterbo, Italy, 23 June 2008.
25. Ghirardello, D.; Contessa, C.; Valentini, N.; Zeppa, G.; Rolle, L.; Gerbi, V.; Botta, R. Effect of storage conditions on chemical and physical characteristics of hazelnut (*Corylus avellana* L.). *Postharvest Biol. Technol.* **2013**, *81*, 37–43. [CrossRef]
26. Akinoso, R.; Raji, A.O. Physical properties of fruit, nut and kernel of oil palm. *Int. Agrophys.* **2011**, *25*, 1–6.
27. Jahromi, M.K.; Rafiee, S.; Jafari, A.; Bousejin, M.R.G.; Mirasheh, R.; Mohtasebi, S.S. Some physical properties of date fruit (cv. Dairi). *Int. Agrophys.* **2008**, *22*, 221–224.
28. Ozturk, I.; Ercisli, S.; Kara, M. Chosen physical properties of olive cultivars (Olea europaea L.). *Int. Agrophys.* **2009**, *23*, 309–312.
29. Yurtlu, Y.B.; Yesiloglu, E.; Arslanoglu, F. Physical properties of bay laurel seeds. *Int. Agrophys.* **2010**, *24*, 325–328.
30. Ercisli, S.; Ozturk, I.; Kara, M.; Kalkan, F.; Seker, H.; Duyar, O.; Erturk, Y. Physical properties of hazelnuts. *Int. Agrophys.* **2011**, *25*, 115–121.
31. Aydin, C. Physical properties of hazelnuts. *Biosyst Eng.* **2002**, *82*, 297–303. [CrossRef]
32. Ozdemir, F.; Akinci, I. Physical and nutritional properties of four major commercial Turkish hazelnut varieties. *J. Food Eng.* **2004**, *63*, 341–347. [CrossRef]
33. Kibar, H.; Ozturk, T. The effect of moisture content on the physic-mechanical properties of some hazelnut varieties. *J. Stored Prod. Res.* **2009**, *45*, 14–18. [CrossRef]
34. Valentini, N.; Rolle, L.; Stevigny, C.; Zeppa, G. Mechanical behaviour of hazelnuts used for table consumption under compression loading. *J. Sci Food Agric.* **2006**, *86*, 1257–1262. [CrossRef]
35. Giacosa, S.; Belviso, S.; Bertolino, M.; Bello, B.D.; Gerbi, V.; Ghirardello, D.; Giordano, M.; Zeppa, G.; Rolle, L. Hazelnut kernels (*Corylus avellana* L.) mechanical and acoustic properties determination: Comparison of test speed, compression or shear axis, roasting, and storage condition effect. *J. Food Eng.* **2016**, *173*, 59–68. [CrossRef]

36. Sirisomboon, P.; Kitchaiya, P.; Pholpho, T.; Mahuttanyavanitch, W. Physical and mechanical properties of *Jatropha curcas* L. fruits, nuts and kernels. *Biosyst. Eng.* **2007**, *97*, 201–207. [CrossRef]
37. Karaj, S.; Muller, J. Determination of physical, mechanical and chemical properties of seeds and kernels of *Jatropha curcas* L. *Ind. Crop. Prod.* **2010**, *32*, 129–138. [CrossRef]
38. Divisova, M.; Herak, D.; Kabutey, A.; Sigalingging, R.; Svatoňová, T. Deformation curve characteristics of rapeseeds and sunflower seeds under compression loading. *Sci. Agric. Bohem.* **2014**, *45*, 180–186.
39. Herak, D.; Kabutey, A.; Hrabe, P. Oil point determination of Jatropha curcas L. bulk seeds under compression loading. *Biosyst. Eng.* **2013**, *116*, 470–477. [CrossRef]
40. Munson-Mcgee, S.H. D-optimal experimental designs for uniaxial expression. *J. Food Process. Eng.* **2014**, *37*, 248–256. [CrossRef]
41. Herak, D.; Kabutey, A.; Choteborsky, R.; Petru, M.; Sigalingging, R. Mathematical models describing the relaxation behaviour of Jatropha curcas L. bulk seeds under axial compression. *Biosyst. Eng.* **2015**, *131*, 77–83. [CrossRef]
42. Mizera, C.; Herak, D.; Hrabe, P.; Kabutey, A. Mathematical models describing the creep and stress relaxation behaviour of false banana's fiber (Ensete ventricosum). *J. Nat. Fibers.* **2019**. [CrossRef]
43. Demirbas, A. Oils from Hazelnut Shell and Hazelnut Kernel Husk for Biodiesel Production. *Energ. Sournce Part. A* **2008**, *30*, 1870–1875. [CrossRef]
44. ISI. *Indian Standard Methods for Analysis of Oilseeds, IS:3579*; Indian Standard Institute: New Delhi, India, 1966.
45. Blahovec, J. *Agromatereials Study Guide*; Czech University of Life Sciences: Prague, Czech Republic, 2008.
46. Niu, L.; Li, J.; Chen, M.-S.; Xu, Z.-F. Determination of oil contents in Sacha inchi (*Plukenetia volubilis*) seeds at different developmental stages by two methods: Soxhlet extraction and time-domain nuclear magnetic resonance. *Ind Crop. Prod.* **2014**, *56*, 187–190. [CrossRef]
47. Danlami, J.M.; Arsad, A.; Zaini, M.A.A. Characterization and process optimization of castor oil (*Ricinus communis* L.) extracted by the Soxhlet method using polar and non-polar solvents. *J. Taiwan Inst. Chem Eng.* **2015**, *47*, 99–104. [CrossRef]
48. Gürdil, G.A.K.; Kabutey, A.; Selvi, K.C.; Hrabě, P.; Herák, D.; Fraňková, A. Investigation of heating and freezing pretreatments on mechanical, chemical and spectral properties of bulk sunflower seeds and oil. *Processes* **2020**, *8*, 411. [CrossRef]
49. Chatepa, L.E.C.; Uluko, H.; Masamba, K. Comparison of oil quality extracted from selected conventional and non conventional sources of vegetable oil from Malawi. *Afr. J. Biotechnol.* **2019**, *18*, 171–180.
50. Perrier, A.; Delsart, C.; Boussetta, N.; Grimi, N.; Citeau, M.; Vorobiev, E. Effect of ultrasound and green solvents addition on the oil extraction efficiency from rapeseed flakes. *Ultrason. Sonochem.* **2017**, *39*, 58–65. [CrossRef]
51. Deli, S.; Masturah, F.; Aris, T.Y.; Nadiah, W.W.A. The effects of physical parameters of the screw press oil expeller on oil yield from Nigella sativa L. Seeds. *Int. Food Res. J.* **2011**, *18*, 1367–1373.
52. Demirel, C.; Kabutey, A.; Herak, D.; Gurdil, G.A.K. Comparison of deformation energy of particular oil-bearing crops. In Proceedings of the 58th ICMD, Prague, Czech Republic, 6–8 September 2017; pp. 60–65.
53. Gupta, R.K.; Das, S.K. Fracture resistance of sunflower seed and kernel to compressive loading. *J. Food Eng.* **2000**, *46*, 1–8. [CrossRef]
54. StatSoft Inc. *(1995). STATISTICA for Windows*; StatSoft Inc.: Tulsa, OK, USA, 2013.
55. Sirisomboon, P.; Kitchaiya, P. Physical properties of Jatropha curcas L. kernels after heat treatment. *Biosyst. Eng.* **2009**, *102*, 244–250. [CrossRef]
56. Baryeh, E.A. Effect of palm oil processing parameters on yield. *J. Food Eng.* **2001**, *48*, 1–6. [CrossRef]
57. Willems, P.; Kuipers, N.J.M.; De Haan, A.B. Hydraulic pressing of oilseeds: Experimental determination and modeling of yield and pressing rates. *J. Food Eng.* **2008**, *89*, 8–16. [CrossRef]
58. Willems, P.; Kuipers, N.J.M.; De Haan, A.B. A consolidation based extruder model to explore GAME process configurations. *J. Food Eng.* **2009**, *90*, 238–245. [CrossRef]
59. Karaj, S.; Muller, J. Optimizing mechanical oil extraction of Jatropha curcas L. seeds with respect to press capacity, oil recovery and energy efficiency. *Ind Crop. Prod.* **2011**, *34*, 1010–1016. [CrossRef]
60. Sahin, S.; Gulum, S. *Physical Properties of Foods*; Middle East Technical University: Ankara, Turkey, 2006.
61. Chakespari, A.G.; Rajabipour, A.; Mobli, H. Anisotropic relaxation and creep properties of apple (cv. Shafi Abadi and Golab Kohanz). *Adv. J. Food Sci Technol.* **2010**, *2*, 200–205.

62. Ogunsina, B.S.; Owolarafe, O.K.; Olatunde, G.A. Oil point pressure of cashew (*Anacardium occidentale*) kernels. *Int. Agrophys.* **2008**, *22*, 53–59.
63. Faborade, M.O.; Favier, J.F. Identification and significance of the oil-point in seed-oil expression. *J. Agric. Res.* **1996**, *65*, 335–345. [CrossRef]
64. Ajibola, O.O.; Okunade, D.A.; Owolarafe, O.K. Oil point pressure of soybean. *J. Food Process. Eng.* **2002**, *25*, 407–416. [CrossRef]
65. Mrema, G.C.; McNulty, P.B. Mathematical model of mechanical oil expression from oilseeds. *J. Agric. Eng. Res.* **1985**, *31*, 361–370. [CrossRef]
66. Christopoulos, M.V.; Tsantili, E. Oil composition in stored walnut cultivars—Quality and nutritional value. Eur. *J. Lipid Sci. Technol.* **2015**, *117*, 338–348. [CrossRef]
67. Maskan, M.; Karataş, Ş. Fatty acid oxidation of pistachio nuts stored under various atmospheric conditions and different temperatures. *J. Sci. Food Agric.* **1998**, *77*, 334–340. [CrossRef]
68. Buransompob, A.; Tang, J.; Mao, R.; Swanson, B.G. Rancidity of walnuts and almonds affected by short time heat treatments for insect control. *J. Food Process. Preserv.* **2003**, *27*, 445–464. [CrossRef]
69. Gecgel, U.; Gumus, T.; Tasan, M.; Daglioglu, O.; Arici, M. Determination of fatty acid composition of γ-irradiated hazelnuts, walnuts, almonds, and pistachios. *Radiat. Phys. Chem.* **2011**, *80*, 578–581. [CrossRef]
70. Shahidi, F.; Zhong, Y. Lipid oxidation and improving the oxidative stability. *Chem. Soc. Rev.* **2010**, *39*, 4067–4079. [CrossRef] [PubMed]
71. Gama, T.; Wallace, H.M.; Trueman, S.J.; Hosseini-Bai, S. Quality and shelf life of tree nuts: A review. *Sci. Hortic.* **2018**, *242*, 116–126. [CrossRef]
72. Phatanayindee, S.; Borompichaichartkul, C.; Srzednicki, G.; Craske, J.; Wootton, M. Changes of chemical and physical quality attributes of macadamia nuts during hybrid drying and processing. *Dry. Technol.* **2012**, *30*, 1870–1880. [CrossRef]
73. Ling, B.; Hou, L.; Li, R.; Wang, S. Thermal treatment and storage condition effects on walnut paste quality associated with enzyme inactivation. *Food Sci. Technol.* **2014**, *59*, 786–793. [CrossRef]

processes

MDPI

Article

Physico-Chemical Changes, Microbiological Properties, and Storage Shelf Life of Cow and Goat Milk from Industrial High-Pressure Processing

Sok Fern Tan [1], Nyuk Ling Chin [1,*], Tuan Poy Tee [2] and Sook Kuan Chooi [3]

1 Department of Process and Food Engineering, Faculty of Engineering, Universiti Putra Malaysia, UPM Serdang 43400, Selangor, Malaysia; ferncharlotte@hotmail.com

2 Department of Animal Science, Faculty of Agriculture, Universiti Putra Malaysia, UPM Serdang 43400, Selangor, Malaysia; ttpoy@upm.edu.my

3 Kara F&B Productions Sdn Bhd, 5, Jalan Signature Industrial Park 2, Signature Industrial Park, Semenyih 43500, Selangor, Malaysia; sharonchui@jussu.com.my

* Correspondence: chinnl@upm.edu.my; Tel.: +6-03-97696353

Received: 21 March 2020; Accepted: 11 April 2020; Published: 16 June 2020

Abstract: Industrial high-pressure processing (HPP) was conducted on cow and goat milk in comparison to conventional heat pasteurization. No significant changes were found in the physico-chemical properties of the treated milk except for pH, where pasteurized cow milk experienced a decrease while goat milk's pH increased for both pasteurized and HPP treated. HPP-treated cow and goat milk both achieved microbial shelf life of 22 days at 8 °C storage with no increase in *Bacillus cereus*, mesophilic aerobic spores, coliform, yeast and mold but slight increase in psychrotrophic bacteria and total plate count. Pasteurized goat milk was spoilt at the end of storage with exceeding count of psychrotrophic bacteria (9.0×10^8 CFU/mL) and total plate count (3.5×10^8 CFU/mL). HPP-treated cow milk exhibited higher physico-chemical stability than goat milk as evidenced by non-significant change of titratable acidity but goat milk experienced an increase of 0.04% averagely.

Keywords: milk quality; high pressure processing; pasteurization; milk storage; shelf life

1. Introduction

Milk is well-known as a major source of dietary energy, protein, and fat. It contributes 134 kcal of energy/capita per day on average [1]. Apart from cow milk, goat milk has also started to gain interest worldwide in recent years [2]. Goat milk contains lower amount of caseins by 89% compared to cow milk, hence gives a greater digestive utilization compared to cow milk [3]. However, the nature of milk and its chemical composition render it as one of the ideal culture media for microbial growth and multiplication [4]. Hence, the major challenge in milk industries is to ensure delivery of safe and healthy milk products of consistent quality to an ever-increasing demand of milk.

Thermal treatments such as pasteurization high temperature short time (HTST–72 °C/15 s), low temperature long time (LTLT–63 °C/30 min), and ultra-high temperature (UHT–135 °C/3 s) are usually used to treat raw milk in order to be rendered safe for consumer and increase its shelf life [5]. Studies have shown that microbial population counted is significant beyond 20 days of storage for pasteurized milk [6–8]. UHT-treated milk typically has up to 4–6 months of commercial shelf life at ambient temperature [9]. However, these conventional heat treatments have detrimental effects on the nutrient content of milk. Studies have proven that thermal treatments caused 5–15% of whey protein denaturation, 10–25% of water soluble vitamins destruction, disturbed structure of casein micelles, and also enhancement of allergenic properties of milk proteins [5,10,11].

High pressure processing (HPP) is a promising alternative to conventional thermal treatments as it is capable to inactivate foodborne pathogens while minimizing the loss of nutrients, such as water-soluble vitamins and maintaining the fresh-like characteristics of food products [12–14]. During HPP, pressure is allowed to remain in contact with the product for a specific holding time to activate the destruction of spore and microbes, and formation of non-covalent bonds of food components to gelatinize the enzyme [13,15,16]. Pressure regime of 200–600 MPa with holding time of 3–10 min is usually applied in HPP of milk. Generally, higher pressure resulted in higher rate of microbial destruction as well as longer shelf life. Pereda et al. [17] reported a shelf life of 14–18 days for milk treated at 200 MPa at storage temperature of 4 °C. Mussa and Ramaswamy [18] reported that pressure of 350 MPa renders a milk shelf life of 18 days at similar storage temperature. When milk was treated at 600 MPa, shelf life of 28 days was achieved at similar storage temperature [14]. Storage temperature of milk also plays a role in milk's shelf life. Rademacher and Kessler [19] reported that milk subjected to pressure treatment of 400 MPa for 15 min or 600 MPa for 3 min could only achieve 10 days of shelf life when stored at 10 °C, which is 18 days shorter than that of storing at 4 °C.

While studies about the HPP effects on nutrient content like fat, protein, and moisture are very limited, several studies have proven that HPP is able to preserve vitamins in milk better than pasteurization and UHT [20,21]. From the microbiological perspective, a more pronounced microbial inactivation was obtained by increasing HPP pressure levels and holding time. Alexandros et al. [14] reported that HPP above 550 MPa and 3 min reduced the number of *E. coli*, *Salmonella* spp., and *L. monocytogenes* to below limit of detection while pressure regime of 400 MPa for 1 min did not result in statistically significant differences in reduction levels. Yang et al. [22] reported that holding time of 30 min at 300 MPa is sufficient to inactivate *Salmonella* spp., *E. coli*, *Shigella*, and *S. aureus*.

Most HPP studies conducted on milk are based in countries with seasonal climates. For a hot tropical country like Malaysia, HPP for milk is a new technology. Because of the high consumption of pasteurized UHT milk, this study was conducted to investigate industrial HPP effects on local cow and goat milks in terms of its physico-chemical and microbiological properties. The changes in milk properties were measured at storage temperature of 8 °C following commercial handing practices for 22 days and compared with pasteurized milk.

2. Materials and Methods

2.1. Milk Sampling and Treatment

Total of 10 liters of fresh cow milk was collected from the university farm, Ladang 16, Universiti Putra Malaysia while fresh goat milk was collected from a private farm, Janggut Ternak Farm, Dengkil, Selangor, Malaysia. The milk was maintained below 10 °C during transportation to the HPP facility (Kara F&B Productions, Sdn. Bhd.) within 1 h for bottling into HPP grade plastic bottles of 250 mL after a thorough stirring. Three bottles of milk sample, each for cow and goat was used directly for analysis as control or fresh milk samples.

The two HPP pressures applied were 450 MPa and 600 MPa for two durations, 5 min or 7 min at 15 ± 2 °C. The three HPP treatments combinations were 450 MPa/7 min, 600 MPa/5 min, and 600 MPa/7 min giving a total of 27 milk samples, i.e., 3 HPP treatments × 3 storage durations of 0, 11-, and 22-days × 3 analysis which include the physico-chemical and microbiological properties. Both cow and goat milk samples were loaded into the pressure chamber of HPP unit with 55 L volume (Hiperbaric 55, Hiperbaric, Spain) at the same time for the same treatment.

Pasteurization of cow and goat milk were done at 72 °C for at least 15 sec (HTST–72 °C/15 s). Milk was poured into beakers which were pre-heated with boiling water to ensure sterile condition and heated in the water bath (BS-21, Jeio Tech, South Korea) with temperature monitored using clean thermocouples with regular stirring. After pasteurization, the milk was cooled rapidly to room temperature in an ice cooler box filled with ice cubes. Then, the milk was bottled into 9 plastic bottles (3 storage days × 3 analysis) each for cow and goat milk. All treated milk samples were kept in a chiller (RV710, Hitachi, Japan) at 8 ± 2 °C for storage until analysis.

2.2. Physico-Chemical Analysis

The pH values of both cow and goat milk samples were determined using a portable pH meter (Mi805, Milwaukee, Hungary). The pH meter was calibrated to the pH of the buffer solution at 25 °C. Titratable acidity was determined following AOAC Official Method 947.05. The specific gravity was determined following the method of AOAC 925.22 using a pycnometer. Total protein was determined using Pyne's method which is known as formaldehyde titration [23]. This method was proven to be having the same accuracy as compared to Kjeldahl method, as the differences were insignificant between protein contents measured by using both methods [24]. The total protein was calculated using factor of 1.74. The total solid of both cow and goat milk was determined using the oven drying method (UM500, Memmert, Germany) in accordance to AOAC Official Method 16.032 where total solid in percentage was calculated as follow:

$$\text{Total solid } (\%) = \frac{\text{Weight of dish + dried sample} - \text{Weight of empty dish}}{\text{Weight of sample}} \times 100 \quad (1)$$

The total fat percentage was determined following the AOAC Official Method 905.02, which is also known as the Roese-Gottlieb method. The total fat was calculated as follows:

$$\text{Total Fat } (\%) = \frac{\text{Weight of flask containing fat} - \text{weight of flask after fat washing}}{\text{Weight of sample}} \times 100 \quad (2)$$

The milk solid-non-fat (SNF) was determined by subtracting the total solid (%) with the total fat content (%).

2.3. Microbiological Analysis

Microbiological analysis was conducted in an aseptic environment to prevent cross contamination. The total plate count (TPC) and total yeast and mold count in milk were determined following methods in FDA Bacteriological Analytical Manual (BAM) Chapter 3 and 18, respectively. The total coliform count was determined following AOAC Official Method 991.14. The mesophilic, thermophilic aerobic spore count and psychrotrophic bacteria count were determined following Compendium of Methods for the Microbiological Examination of Foods (CMMEF) Chapter 23, 26, and 13, respectively. *Bacillus cereus*, *Staphylococcus aureus*, and *Listeria monocytogenes* were determined following FDA BAM Chapter 14, 12, and 10, respectively. *Escherichia coli* and *Clostridium perfringens* were determined by AOAC Official Method 991.14 and 976.30, respectively. *Salmonella* spp. was determined using enzyme-linked immunosorbent assay (ELISA) method.

2.4. Experimental Design and Statistical Analysis

All measurements were done in duplicates with samples prepared from the same batch of milk. Minitab software (Version 16, Minitab Statistical Software, United States) was used to perform analysis of variance (ANOVA). Tukey's test was performed to compare the differences among the mean values at a confidence level of 0.05.

3. Results and Discussion

3.1. Effects of HPP on Physico-Chemical Properties

Physico-chemical properties of cow and goat milk before and after the treatment are presented in Table 1. The pH of pasteurized cow milk decreased significantly ($p < 0.05$) from 6.59 ± 0.01 to 6.52 ± 0.00. In contrast, the pH of pasteurized goat milk increased significantly ($p < 0.05$) from 6.34 ± 0.00 to 6.38 ± 0.01. Ma and Barbano [25] reported that pH of cow skim milk decreases linearly with increasing temperature during treatment. However, José et al. [26] reported that pasteurization increased pH of bovine milk because of the lower whey protein associated with the casein micelles. Tamime [10] explained that the increase of pH is due to the reduced casein micelles where heat denatured its whey protein. For HPP-treated milk, pH for cow milk only changed slightly but statistically insignificant ($p > 0.05$). Nonetheless, pH of goat milk has increased significantly ($p < 0.05$). This increment is mainly due to the alteration of the distribution of minerals by HPP, which includes calcium, phosphate, and other ionized minerals such as calcium [27]. Other studies which have observed increase in the milk pH after HPP include Liepa et al. [28] who found that pressure of 500–600 MPa for 15 min increased the pH of cow milk; while Zobrist et al. [29] reported that pressure of 100–600 MPa for 30 min also caused an average 0.08 units of pH increment in cow milk.

Table 1. Physico-chemical properties of cow and goat milk before and after treatment.

Analysis	Before Treatment	450 MPa/7 min	600 MPa/5 min	600 MPa/7 min	Pasteurized
			Cow Milk		
pH	6.59 ± 0.01 [a]	6.56 ± 0.01 [a]	6.58 ± 0.00 [a]	6.61 ± 0.02 [a]	6.52 ± 0.00 [b]
Titratable acidity	0.16 ± 0.01 [a]	0.15 ± 0.00 [a]	0.16 ± 0.01 [a]	0.17 ± 0.01 [a]	0.16 ± 0.01 [a]
Specific gravity	1.03 ± 0.00 [a]	1.03 ± 0.00 [a]	1.02 ± 0.00 [a]	1.03 ± 0.00 [a]	1.03 ± 0.00 [a]
Total protein	3.13 ± 0.00 [a]	2.87 ± 0.12 [a]	3.04 ± 0.12 [a]	3.04 ± 0.12 [a]	2.96 ± 0.25 [a]
Total fat	3.33 ± 0.08 [a]	3.29 ± 0.06 [a]	3.24 ± 0.09 [a]	3.30 ± 0.14 [a]	3.32 ± 0.08 [a]
Total solid	11.65 ± 0.01 [a]	11.61 ± 0.21 [a]	11.55 ± 0.19 [a]	11.64 ± 0.11 [a]	11.88 ± 0.06 [a]
Solid-non-fat	8.32 ± 0.09 [a]	8.32 ± 0.15 [a]	8.31 ± 0.28 [a]	8.34 ± 0.03 [a]	8.56 ± 0.14 [a]
			Goat Milk		
pH	6.34 ± 0.00 [c]	6.38 ± 0.00 [b]	6.42 ± 0.01 [a]	6.38 ± 0.01 [b]	6.38 ± 0.01 [b]
Titratable acidity	0.22 ± 0.01 [a]	0.22 ± 0.01 [a]	0.22 ± 0.01 [a]	0.21 ± 0.01 [a]	0.23 ± 0.01 [a]
Specific gravity	1.03 ± 0.00 [a]	1.03 ± 0.00 [a]	1.03 ± 0.00 [a]	1.03 ± 0.00 [a]	1.03 ± 0.00 [a]
Total protein	3.57 ± 0.12 [a]	3.39 ± 0.12 [a]	3.57 ± 0.12 [a]	3.31 ± 0.00 [a]	3.48 ± 0.00 [a]
Total fat	4.51 ± 0.05 [a]	4.43 ± 0.49 [a]	4.44 ± 0.41 [a]	4.42 ± 0.45 [a]	4.34 ± 0.61 [a]
Total solid	13.88 ± 0.06 [a]	13.73 ± 0.14 [a]	13.69 ± 0.00 [a]	13.70 ± 0.13 [a]	13.08 ± 1.42 [a]
Solid-non-fat	9.37 ± 0.10 [a]	9.30 ± 0.35 [a]	9.35 ± 0.41 [a]	9.28 ± 0.58 [a]	8.74 ± 2.03 [a]

Note: Mean ± standard deviation in same row with different superscripts letters are significantly different at $p < 0.05$.

In general, the results show that the titratable acidity, specific gravity, total protein, total fat, total solid, and solid-non-fat in both cow and goat milk were not influenced by both HPP and pasteurization. The values changed slightly after HPP treatment but they were statistically insignificant when compared with untreated samples ($p > 0.05$).

3.2. Effects of HPP on Microbiological Properties

Table 2 summarizes the microbiological properties of cow and goat milk before and after the treatment. According to the Malaysian Standard for pasteurized milk (MS 410:1995), the total plate count (TPC) should not be more than 10^5 CFU/mL. In this study, the initial TPC of raw cow and goat milk was 1.5×10^5 CFU/mL and 6.9×10^6 CFU/mL, respectively. Both HPP and pasteurization produced desirable results of decreased TPC in both milk samples. The pressure regime of 600 MPa/5 min led to the highest TPC reduction in both milk, i.e., 99.98% for the cow milk and 99.99% for the goat milk. The lowest TPC reduction was found in the pressure regime of 450 MPa/7 min (96%) for the cow milk and for goat milk, it was from the pasteurization process (99.98%). This is in line with the findings from Alexandros et al. [14] who found that HPP (600 MPa/3 min) led to a more pronounced decrease of TPC than that of pasteurization.

Table 2. Microbiological properties of cow and goat milk before and after treatment.

Bacteria Count	Before Treatment	450 MPa/7 min	600 MPa/5 min	600 MPa/7 min	Pasteurized
		Cow Milk			
Total plate count (CFU/mL)	1.5×10^5	6×10^3	2.6×10^1	2.5×10^2	1.5×10^2
Psychrotrophic bacteria count (CFU/mL)	9.2×10^4	8.3×10^2	1	<1	5.4×10^1
Mesophilic aerobic spore count (CFU/mL)	7.0×10^1	<1	<1	<1	2.0×10^1
Total coliform count (CFU/mL)	3.7×10^1	<1	<1	<1	<1
Total yeast and mold count (CFU/mL)	6.0×10^1	<1	<1	<1	<1
Bacillus cereus (CFU/mL)	<1	2.0×10^1	6	4	8
Clostridium perfringens (CFU/mL)	<1	<1	<1	<1	<1
Staphylococcus aureus (CFU/mL)	<1	<1	<1	<1	<1
Total *E. coli* count (CFU/mL)	<1	<1	<1	<1	<1
Thermophilic aerobic spore count (CFU/mL)	<1	<1	<1	<1	<1
Listeria monocytogenes per 25 g	Absent	Absent	Absent	Absent	Absent
Salmonella spp. per 25 g	Absent	Absent	Absent	Absent	Absent
		Goat Milk			
Total plate count (CFU/mL)	6.9×10^6	6.4×10^2	6	4.0×10^1	8.0×10^2
Psychrotrophic bacteria count (CFU/mL)	8.0×10^5	<1	<1	<1	3.6×10
Mesophilic aerobic spore count (CFU/mL)	<1	<1	<1	<1	<1
Total coliform count (CFU/mL)	2.1×10^5	<1	<1	<1	2.6×10^1
Total yeast and mold count (CFU/mL)	8.0×10^1	<1	<1	<1	<1
Bacillus cereus (CFU/mL)	<1	<1	<1	<1	<1
Clostridium perfringens (CFU/mL)	<1	<1	<1	<1	<1
Staphylococcus aureus (CFU/mL)	<1	<1	<1	<1	<1
Total *E. coli* count (CFU/mL)	<1	<1	<1	<1	<1
Thermophilic aerobic spore count (CFU/mL)	<1	<1	<1	<1	<1
Listeria monocytogenes per 25 g	Absent	Absent	Absent	Absent	Absent
Salmonella spp. per 25 g	Absent	Absent	Absent	Absent	Absent

Psychrotrophic bacteria are undesirable for prolonged storage of milk because they are able to grow below 7 °C even though their optimum temperature ranges from 20 °C to 30 °C. They also produce heat resistant enzymes, including proteolytic and lipolytic enzymes at low temperatures, which are able to hydrolyze milk fat and protein leading to the formation of off-flavors [30]. According to Malaysians Food Act 1983 and Food Regulations 1985, psychrotrophic bacteria count in milk should not exceed 10^5 CFU/mL. In this research, the initial psychrotrophic bacteria count in raw cow and goat milk were 9.2×10^4 CFU/mL and 8.0×10^5 CFU/mL, respectively. Both HPP and pasteurization have successfully decreased psychrotrophic bacteria count of both milk samples. The lowest psychrotrophic bacteria count in cow milk was recorded in the pressure regime of 600 MPa/7 min where bacteria was undetectable (<1 CFU/mL) while the highest count was found in the pressure regime of 450 MPa/7 min (8.3×10^2 CFU/mL). For goat milk HPP treatment at all pressure regimes have reduced the bacteria count to undetectable level (<1 CFU/mL). The highest psychrotrophic bacteria count recorded was in the pasteurized sample which is 3.6×10 CFU/mL.

Mesophilic spores are known as the most prevalent spore-former found in bulk milk processing tank. In spore form, these organisms are capable of surviving environmental stresses including low pH, exposure to sanitizers, high pressure and temperature [31]. Mesophilic spores were undetected in all goat milk samples both before and after treatments. For cow milk, the initial mesophilic aerobic spore count was 70 CFU/mL. All pressure regimes in HPP successfully inactivated the spores. However, pasteurization only managed to reduce it to 20 CFU/mL. McGuiggan et al. [32] explained that these aerobic spores have the ability to survive in high temperature as well as pasteurization.

Coliforms are Gram-negative and non-spore forming bacteria that ferment lactose with the production of acid at 35 °C within 48 h [33]. According to MS 410:1995, the total coliform should be absent in 0.1 mL of sample. In this study, the total coliform count for untreated cow milk was 37 CFU/mL, but after treatment, the coliform became undetected in all HPP-treated and pasteurized samples. While all pressure regimes of HPP were able to reduce the total coliform count to undetectable level in goat milk, pasteurization was capable to reduce the counts from 2.1×10^5 CFU/mL to 26 CFU/mL (99.98% reduction).

Yeast and mold are a large and diverse group which include several hundred species. Although yeast and mold are relatively HPP sensitive [12], mascospores of heat-resistant molds such as

Byssochlamys, *Neosartorya*, and *Talaromyces* are generally considered to be extremely HPP resistant. In this research, the total yeast and mold count of untreated cow and goat milk were 60 CFU/mL and 80 CFU/mL, respectively. Both HPP and pasteurization were effective in inactivating yeast and mold in cow and goat milk fully. Similar result was reported by Evrendilek [34], where most vegetative yeast and mold were inactivated within 5–10 min by 300–400 MPa at 25 °C.

Aerobic spore-forming bacteria of the genus *Bacillus* are commonly present in raw milk. Their spores survive pasteurization and subsequently germinate, outgrow, and multiply [35]. In this study although *B. cereus* was not found in raw cow milk, it was present in HPP-treated and pasteurized milk with the highest count of 20 CFU/mL in milk subjected to 450 MPa for 7 min. The unexpected presence of *B. cereus* in treated cow milk may be due to several factors. McClements et al. [35] reported that *B. cereus* spores were more resistant to pressure than vegetative cells. The contamination, germination, and outgrowth of *Bacillus* spores in milk can occur during simple transfer during milking and bottling when hygiene conditions are not fully observed. Post-pasteurization *Bacillus cereus* adherence to stainless steel contact surfaces in the form of biofilm formation has been reported to occur on milk contact surfaces [36]. Nutrient germinant such as amino acids and unsaturated fatty acids, levels of some indigenous antibacterial factors such as lysozyme, lactoferrin, and lactoperoxidase, somatic cells and metal ions have been suggested as factors influencing the germination of *Bacillus* species [32,37]. *B. cereus* was not found in both raw and treated goat milk.

C. perfringens, *S. aureus*, *E. coli*, *L. monocytogenes*, *Salmonella* spp., and thermophilic spores were undetected in both cow and goat milk. The absence of these bacteria and spores is due to good milking hygiene, breeds, climate and geographic location. In India, Chaturvedi and Shukla [38] reported that 51.61% of raw milk samples were found to be highly contaminated with *Clostridium* species due to unhygienic milking practice. Mohamed et al. [39] observed no germs of *S. aureus* and spore of *Clostridium* spp. in all cow and goat milk samples at Djibouti. However, in the same research, the *E. coli* count of 2.58 log CFU/mL was significantly higher than standard value of 2 log CFU/mL in their milk samples. Chye et al. [40] reported that only 1.9% and 1.4% of cow milk samples in Malaysia were contaminated by *L. monocytogenes* and *Salmonella* spp., respectively. Higher incidence of *Listeria* was obtained from eastern region while *Salmonella* spp. was found more prevalent in central region of Peninsular Malaysia. Hence, the presence of these bacteria is highly dependent on the location and hygiene.

3.3. Changes in Physico-Chemical Properties in Milk during Storage

The pH for both HPP-treated and pasteurized cow and goat milk decreased significantly ($p < 0.05$) during a 22-day storage period (Figure 1). Similar results were obtained by Siddique et al. [41] who reported pH declination at the end of 90 days storage of UHT milk. Brodziak et al. [42] mentioned about statistically significant pH reduction of open drinking cow milk after 7 days of storage. The decreasing trend in pH during milk storage is due to the acidity increase in milk caused by lactic acid and fatty acid [43]. It is obvious that pH of goat milk was always lower than cow milk which may be due to its higher acidity content as shown in Figure 2. This is in agreement with Mahmood and Usman [44] who also found lower pH in goat milk (6.48) than cow milk (6.59).

Generally, the titratable acidity of both cow and goat milk increased during the 22-day storage (Figure 2). The increment found in all HPP-treated cow milk was statistically insignificant ($p > 0.05$) as compared to significant increment ($p < 0.05$) for pasteurized cow milk with highest acidity of 0.23% at day 11. For goat milk, all HPP and pasteurized samples titratable acidity was found to have increased significantly, from 0.25 to 0.26% and 0.30%, respectively at the end of 22-day storage from the range of 0.21 to 0.23% at day 0. It is thus obvious that both pasteurized cow and goat milk deteriorated faster than HPP-treated milk. Brodziak et al. [42] also reported increase of titratable acidity in open drinking cow milk after 7-day storage in pasteurized milk, followed by micro-filtered milk and UHT milk. The increase in titratable acidity is mainly due to increasing lactic acid concentration after degradation of lactose caused by predominant spoilage lactic acid bacteria (LAB) under low oxygen, temperature, and acidic condition [41,45].

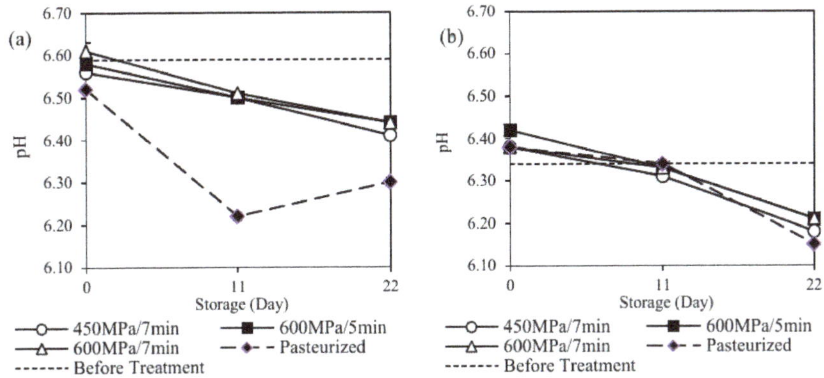

Figure 1. Changes in pH of (**a**) cow and (**b**) goat milk during storage period of 22 days. Solid lines represent HPP-treated milk (450 MPa/7 min, 600 MPa/5 min, 600 MPa/7 min), dashed lines represent pasteurized milk, and dotted lines represent fresh milk before treatment.

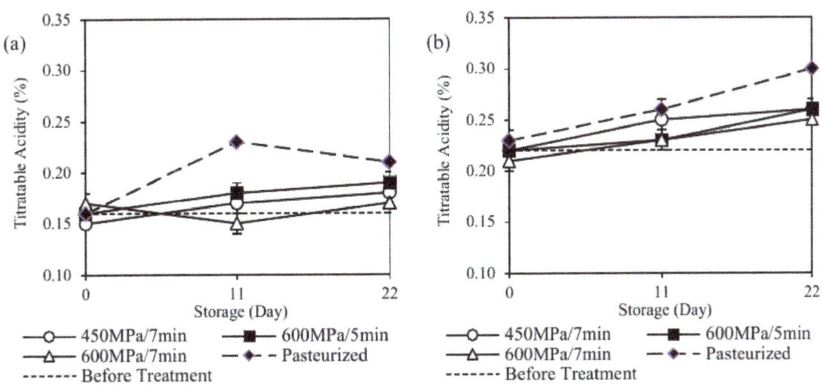

Figure 2. Changes in titratable acidity of (**a**) cow and (**b**) goat milk during storage period of 22 days. Solid lines represent HPP-treated milk (450 MPa/7 min, 600 MPa/5 min, 600 MPa/7 min), dashed lines represent pasteurized milk, and dotted lines represent fresh milk before treatment.

The specific gravity of both cow and goat milk increased gradually when approaching day 22 of storage (Figure 3). All HPP-treated and pasteurized samples showed a statistically significant increase ($p < 0.05$) in specific gravity except for goat milk in the pressure regime of 450 MPa/7 min which shows significant decrease ($p < 0.05$). The increase in specific gravity agrees with Aldubhany et al. [46]'s work who reported increment in specific gravity in cow milk samples after 6 months of storage at various storage temperatures. The specific gravity of milk could be influenced by the proportion of its constituents such as fat and lactose which degrade during storage [45]. Ahmad et al. [47] proved that milk specific gravity decreased due to mastitis. The authors explained that presence of mastitis contributes to the decrease in lactose and increase in chloride contents as specific gravity is positively correlated with lactose and negatively correlated with chloride.

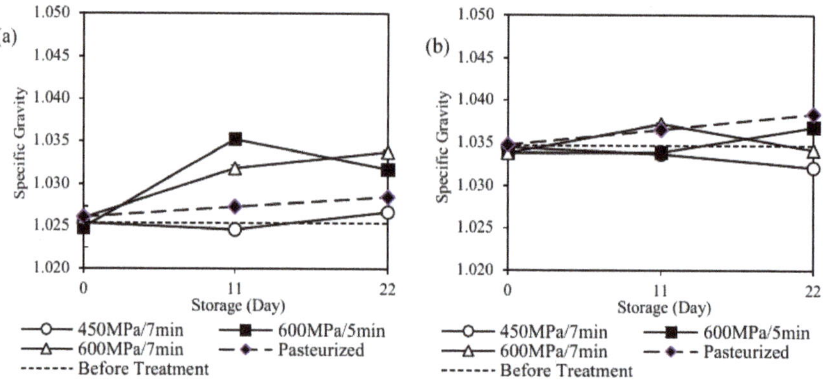

Figure 3. Changes in specific gravity of (**a**) cow and (**b**) goat milk during storage period of 22 days. Solid lines represent HPP-treated milk (450 MPa/7 min, 600 MPa/5 min, 600 MPa/7 min), dashed lines represent pasteurized milk and dotted lines represent fresh milk before treatment.

Figure 4 shows that the total protein content in both cow and goat milk samples from both HPP and pasteurization treatment did not change significantly ($p > 0.05$) after 22 days. This is supported by Brodziak et al. [42] who found that storage duration of milk had no statistically significant effect on its protein content. Figure 5, however shows that the total fat content of both cow and goat milk samples from pasteurization treatment have reduced significantly ($p < 0.05$) starting from day 11 for cow milk and day 22 for goat milk while no significant difference ($p > 0.05$) was found for HPP samples. The total fat for both pasteurized cow and goat milk have reduced from 3.32% to 2.94% while for goat milk from 4.34% to 3.11%. HPP has been proven to have no damage on the milk fat globule membrane where lipolysis incidence could be prevented, thus the milk fat content could be well preserved [34,48]. Zajác et al. [43] has reported the significant decrease in fat content of raw cow milk after 24 h of storage at a temperature of 4 °C to be due to the lipolysis of milk fat initiated by both indigenous milk lipases and also microbial lipases which usually grew in number during storage. Significant increase of free fatty acids as a consequence of lipolysis in UHT cow milk after three weeks of storage was also confirmed by Janstová et al. [49]. These analyses have also shown that goat milk has relatively higher total protein and fat contents than cow milk.

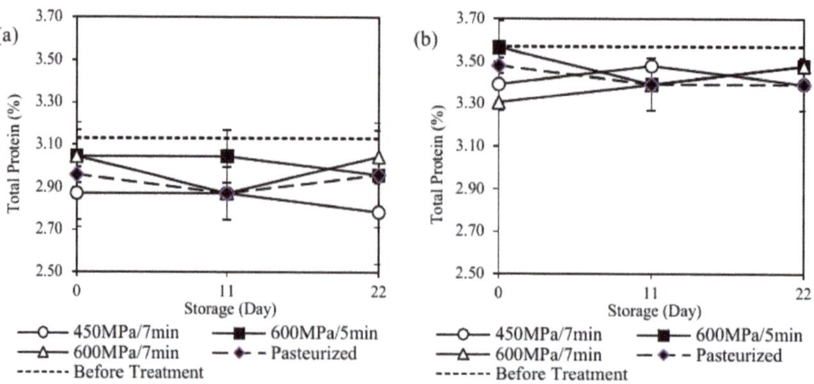

Figure 4. Changes in total protein of (**a**) cow and (**b**) goat milk during storage period of 22 days. Solid lines represent HPP-treated milk (450 MPa/7 min, 600 MPa/5 min, 600 MPa/7 min), dashed lines represent pasteurized milk and dotted lines represent fresh milk before treatment.

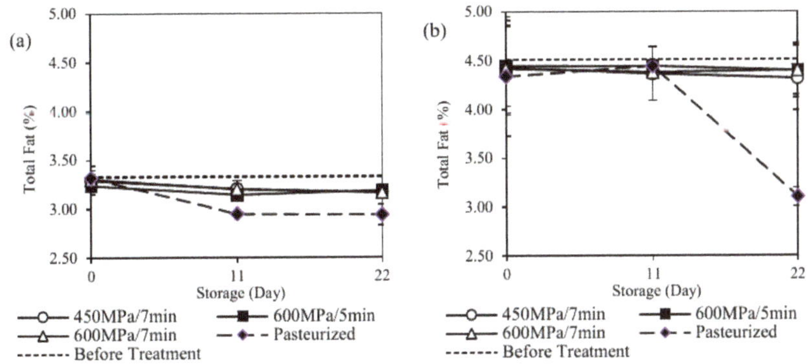

Figure 5. Changes in total fat of (**a**) cow and (**b**) goat milk during storage period of 22 days. Solid lines represent HPP-treated milk (450 MPa/7 min, 600 MPa/5 min, 600 MPa/7 min), dashed lines represent pasteurized milk, and dotted lines represent fresh milk before treatment.

Figures 6 and 7 show a mimicking trend of total solid and solid non-fat contents of cow and goat milk over the fat content trend, i.e., no significant difference during 22 storage for HPP treated samples and significant decrease ($p < 0.05$) for pasteurized samples. The significant decrease in fat content could affect milk composition in terms of total solid and non-fat contents [43]. Omer and Eltinay [50] have reported similar results where total solid and non-fat content of camel milk reduced significantly after two weeks of storage at 7 °C.

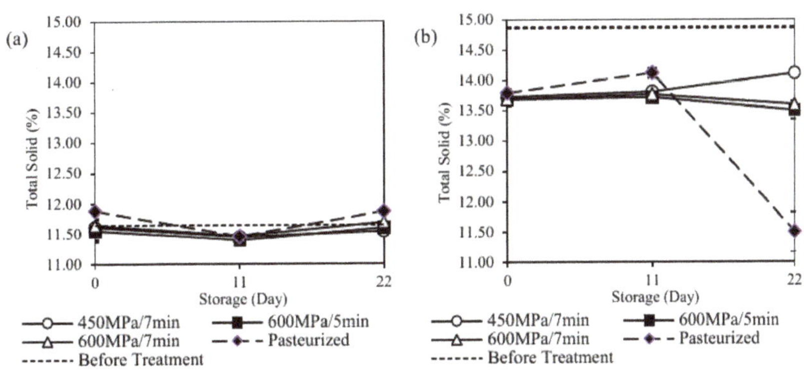

Figure 6. Changes in total solid of (**a**) cow and (**b**) goat milk during storage period of 22 days. Solid lines represent HPP-treated milk (450 MPa/7 min, 600 MPa/5 min, 600 MPa/7 min), dashed lines represent pasteurized milk, and dotted lines represent fresh milk before treatment.

3.4. Changes in Microbiological Properties in Milk during Storage

Table 3 presents the count of two microbiological measurements that turned positive during 22 days of storage. The psychrotrophic bacteria population in both cow and goat milk grew progressively throughout the storage period. At the end of 22 days, the best treatment was HPP at 600 MPa/7 min giving the smallest colony count in cow milk, followed by 600 MPa/5, pasteurization and 450 MPa/7 min. All these counts, however, are still far below the limit of 10^5 CFU/mL set by the Malaysians Food Act 1983 and Food Regulations 1985. In the case for goat milk, the largest colony count of 9.0×10^8 CFU/mL was observed in the pasteurized sample. It exceeded the permissible limit thus considered spoilt at 22 days of storage. The results are in line with Doll et al. [51] who found that 90% of pasteurized cow milk samples stored at 8 °C contained 10^6 to 10^7 CFU/mL of psychrotrophs at the end of 24 days storage.

Pressure regime of 600 MPa/7 min presented the best result as the psychrotrophic bacteria in goat milk remained undetectable at day 22. Garcia-Risco et al. [52] reported that the bovine milk sample treated at 400 MPa for 3 min has a psychrotrophic bacteria count of 10^6 at day 30 of storage at 7 °C.

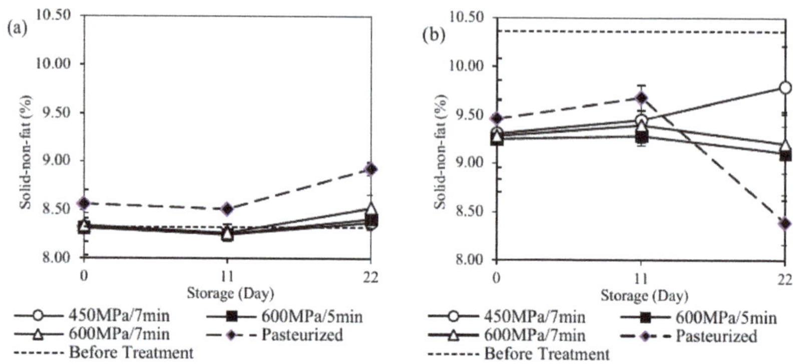

Figure 7. Changes in solid-non-fat of (**a**) cow and (**b**) goat milk during storage period of 22 days. Solid lines represent HPP-treated milk (450 MPa/7 min, 600 MPa/5 min, 600 MPa/7 min), dashed lines represent pasteurized milk, and dotted lines represent fresh milk before treatment.

Table 3. Psychrotrophic bacteria count and total plate count (CFU/mL) in milk for 22 days of storage.

Treatment	Cow Milk			Goat Milk		
	Day 0	Day 11	Day 22	Day 0	Day 11	Day 22
			Psychrotrophic Bacteria Count			
450 MPa/7 min	8.3×10^2	9.0×10^3	3.0×10^3	<1	1.0×10^4	6.8×10^1
600 MPa/5 min	1	1.2×10^4	4.6×10^2	<1	1.8×10^2	2.2×10^1
600 MPa/7 min	<1	6.4×10^4	1.9×10^2	<1	3	<1
Pasteurized	5.4×10^1	1.2×10^4	2.0×10^3	3.6×10	9.0×10^5	9.0×10^8
			Total Plate Count			
450 MPa/7 min	6.0×10^3	2.5×10^3	8.0×10^3	6.4×10^2	2.0×10^3	7.6×10^2
600 MPa/5 min	2.6×10^1	6.0×10^3	3.7×10^2	6	8.8×10^2	3.6×10^4
600 MPa/7 min	2.5×10^2	7.5×10^4	2.8×10^1	4.0×10^1	9.4×10^1	8.8×10^1
Pasteurized	1.5×10^2	1.1×10^4	3.0×10^3	8.0×10^2	6.0×10^5	3.5×10^8

For TPC, progressive increase was observed during storage period of 22 days. The lowest TPC observed in cow milk at day 22 was in the pressure regime of 600 MPa/7 min, followed by 600 MPa/5 min, then pasteurization and lastly 450 MPa/7 min. Despite the increase, the TPC was still below the limit set by MS 410:1995, which states that the TPC in pasteurized milk should not have TPC above 10^5 CFU/mL. For goat milk at day 22, the lowest TPC of 8.8×10^1 CFU/mL was milk treated at 600 MPa/7 min. Pasteurized milk had the highest TPC of 3.5×10^8 CFU/mL, which exceeded the permissible limit. These results are in agreement with Alexandros et al. [14] who found that TPC for HPP milk was always lower compared to pasteurized where TPC in HPP milk reached 10^7 CFU/mL after 28 days compared to pasteurized milk which took only 14 days to reach $>10^7$ CFU/mL. From both the psychrotrophic bacteria population and the TPC microbial count, HPP pressure regime of 600 MPa/7 min is the best for shelf life extension of both milk, renders them the shelf life of more than 22 days.

Other monitored microbes which include *C. perfringens*, *S. aureus*, *E. coli*, *L. monocytogenes*, *Salmonella* spp., and thermophilic aerobic spore remained undetected throughout the 22 days of storage. Yeast and mold that has been inactivated by HPP or pasteurization also remained undetected throughout the storage period. Although the *Bacillus cereus* was found to be present in low number in cow milk at Day 0 after treatment (Table 2), they were not present in all samples on storage Day 11 and

22. The same situation goes to coliform and mesophilic aerobic spores, which were also present in low number in some pasteurized milk samples at Day 0 after treatment; they were undetected on storage Day 11 and 22. This inconsistent result is most probably due to the randomized sampling and testing of milk from different bottles for each analysis on respective experimental day. The storage results from Table 3 have actually reaffirmed the unexplained presence of coliform and mesophilic aerobic spores after HPP and pasteurization treatments (Table 2) suggesting either it was experimental randomization error or that microbes were inactivated during storage at low storage temperature which may have prevented microbial growth [52] or reduced the microbial count to undetectable levels.

4. Conclusions

The results recognized that HPP treatment of 7 min at 450 MPa is sufficient to render both cow and goat milk a shelf life of 22 days at 8 °C, although higher pressure of 600 MPa can give a better safety margin with best treatment results in the order of 600 MPa/7 min, 600 MPa/5 min, and lastly 450 MPa/7 min. The HPP treatment has shown better capability of preserving milk than pasteurization where pasteurized goat milk was considered spoilt during storage since both TPC and psychrotrophic bacteria count exceeded the permissible limit. For storage studies, the HPP-treated milk, particularly the cow milk displayed better physico-chemical properties stability compared to the pasteurized. The HPP-treated milk did not show any significant changes in total fat, total solid, and SNF but pasteurized cow and goat milk demonstrated significant changes in pH, titratable acidity, specific gravity, total fat, total solid, and SNF. In terms of microbiological properties, although both HPP and pasteurization were effective in reducing the numbers of TPC, psychrotrophic bacteria, mesophilic spores, coliform, yeast, and mold to far beyond the permissible limit, pasteurization however did not fully inactivate the mesophilic spores and coliform while HPP did. Pressure of 600 MPa for holding time of both 5 and 7 min have resulted in very low TPC and psychrotrophic bacteria count which is favorable to prolong milk shelf life. The HPP treatment is suggested as a viable alternative treatment for milk if cost is not a consideration, as cost of HPP treatment is much higher.

Author Contributions: S.F.T. designed the experiments, collected and analyzed the data, and wrote the manuscript. N.L.C. conceptualized the study, supervised the research, edited and revised the manuscript. T.P.T. supervised the experiments and S.K.C. provided technical know-how in running the industrial HPP unit. All authors have read and agreed to the published version of the manuscript.

Funding: This research received no external funding.

Conflicts of Interest: The authors declare no conflict of interest.

References

1. Weaver, C.; Bettoni, R.W.; McMahon, D.; Spence, L. Milk and dairy products as part of the diet. In *Milk and Dairy Products in Human Nutrition*; Food and Agriculture Organization of the United Nations: Rome, Italy, 2013; pp. 103–183. ISBN 9789251078631.
2. Clark, S.; Garcia, M.B.M. Introduction. A 100-Year Review: Advances in goat milk research. *J. Dairy Sci.* **2017**, *100*, 100026–100044. [CrossRef] [PubMed]
3. Ulusoy, B.H. Nutritional and health aspects of goat milk consumption. *Akad. Gida* **2015**, *13*, 56–60.
4. Woldemariam, H.W.; Asres, A.M. Microbial and physicochemical qualities of pasteurized milk. *J. Food Process. Technol.* **2017**, *08*, 1–5.
5. Barraquio, V.L. Which milk is fresh. *Int. J. Dairy Process. Res.* **2014**, *1*, 1–6. [CrossRef]
6. Holsinger, V.H.; Rajkowski, K.T.; Stabel, J.R. Milk pasteurization and safety: A brief history and update. *Rev. Sci. Technol.* **1997**, *16*, 441–451. [CrossRef]
7. Ziarno, M.; Molska, I.; Gronczynska, M. Quality and shelf life of milk pasteurized at 84 °C/19-22 sec and free from post pasteurization contamination. *Med. Weter.* **2005**, *61*, 1165–1168.
8. Nagla, B.A.H.; Mohamed, O.M.A.; Abdel, A.M.N. Microbial quality of heat-treated milk during storage. *Pak. J. Nutr.* **2009**, *8*, 1845–1848.

9. Incecco, P.D.; Rosi, V.; Cabassi, G.; Hogenboom, J.A.; Pellegrino, L. Microfiltration and ultra-high-pressure homogenization for extending the shelf-storage stability of UHT milk. *Food Res. Int.* **2018**, *107*, 477–485. [CrossRef]
10. Tamime, A.Y. *Milk Processing and Quality Management*; Wiley-Blackwell: West Sussex, UK, 2014; ISBN 9781444301649.
11. Bogahawaththa, D.; Buckow, R.; Chandrapala, J.; Vasiljevic, T. Comparison between thermal pasteurization and high pressure processing of bovine skim milk in relation to denaturation and immunogenicity of native milk proteins. *Innov. Food Sci. Emerg. Technol.* **2018**, *47*, 301–308. [CrossRef]
12. Considine, K.M.; Kelly, A.L.; Fitzgerald, G.F.; Hill, C.; Sleator, R.D. High-pressure processing - effects on microbial food safety and food quality. *Fed. Eur. Microbiol. Soc.* **2008**, 1–9. [CrossRef]
13. Wang, C.; Huang, H.; Hsu, C.; Yang, B.B. Recent Advances in food processing using high hydrostatic pressure recent advances in food processing using high hydrostatic pressure technology. *Food Sci. Nutr.* **2015**, *56*, 527–540.
14. Alexandros, C.S.; Elena, S.I.; Mark, L.; Joan, T.; Liam, M.; Nicolae, C.; Anastasios, K.; Brijesh, K.T. Effect of high pressure processing on the safety, shelf life and quality of raw milk. *Innov. Food Sci. Emerg. Technol.* **2019**, *52*, 325–333.
15. Chawla, R.; Patil, G.R.; Singh, A.K. High hydrostatic pressure technology in dairy processing: A review. *J. Food Sci. Technol.* **2011**, *48*, 260–268. [CrossRef] [PubMed]
16. Elamin, W.M.; Endan, J.B.; Yusof, Y.A.; Shamsudin, R.; Ahmedov, A. High pressure processing technology and equipment evolution: A Review. *J. Eng. Sci. Technol. Rev.* **2015**, *8*, 75–83. [CrossRef]
17. Pereda, J.; Ferragut, V.; Quevedo, J.M.; Guamis, B.; Trujillo, A.J. Effects of ultra-high pressure homogenization on microbial and physicochemical shelf life of milk. *J. Dairy Sci.* **2007**, *90*, 1081–1093. [CrossRef]
18. Mussa, D.M.; Ramaswamy, H.S. Ultra high pressure pasteurization of milk: Kinetics of microbial destruction and changes in physico-chemical characteristics. *Food. Sci. Technol.* **1997**, *30*, 551–557. [CrossRef]
19. Rademacher, B.; Kessler, H.G. High pressure inactivation of microorganisms and enzymes in milk and milk products. In *High Pressure Bio-Science and Biotechnology*; Elsevier: Amsterdam, The Netherlands, 1997; pp. 291–293.
20. Sharabi, S.; Okun, Z.; Shpigelman, A. Changes in the shelf life stability of ribo fl avin, vitamin C and antioxidant properties of milk after (ultra) high pressure homogenization: Direct and indirect e ff ects. *Innov. Food Sci. Emerg. Technol.* **2018**, *47*, 161–169. [CrossRef]
21. Amador-espejo, G.G.; Gallardo-chacon, J.J.; Nykänen, H.; Juan, B.; Trujillo, A.J. Effect of ultra high-pressure homogenization on hydro- and liposoluble milk vitamins. *FRIN* **2015**, *77*, 49–54. [CrossRef]
22. Yang, B.; Shi, Y.; Xia, X.; Xi, M.; Wang, X.; Ji, B.; Meng, J. Inactivation of foodborne pathogens in raw milk using high hydrostatic pressure. *Food Control* **2012**, *28*, 273–278. [CrossRef]
23. Pyne, G.T. The determination of milk-proteins by formaldehyde titration. *Biochem. J.* **1932**, *26*, 1006. [CrossRef]
24. Gomaa, M.S.; Abdel-Aziz, M.E.; Hafez, E.H.; Salama, S.I. New method for determination of protein content of goat's milk and compared with different methods. *J. Food Dairy Sci. Mansoura Univ.* **2014**, *5*, 95–101. [CrossRef]
25. Ma, Y.; Barbano, D.M. Milk pH as a Function of CO_2 concentration, temperature, and pressure in a heat exchanger. *J. Dairy Sci.* **2003**, *86*, 3822–3830. [CrossRef]
26. José, M.P.; Adriano, G.; Bruna, W.M.; Daniel, N.L.; Claucia, F.V. Effects of pasteurization and ultra-high temperature processes on proximate composition and fatty acid profile in bovine milk. *Am. J. Food Technol.* **2015**, *10*, 265–272.
27. Chopde, S.S.; Deshmukh, M.A.; Kalyankar, S.D.; Changade, S.P. Applications of high pressure technology for milk processing. *Res. J. Anim. Husb. Dairy Sci.* **2014**, *5*, 143–147. [CrossRef]
28. Liepa, M.; Zagorska, J.; Galoburda, R. Effect of high pressure processing on milk coagulation properties. *Res. Rural Dev.* **2017**, *1*, 223–229.
29. Zobrist, M.R.; Huppertz, T.; Uniacke, T.; Fox, P.F.; Kelly, A.L. High-pressure-induced changes in the rennet coagulation properties of bovine milk. *Int. Dairy J.* **2005**, *15*, 655–662. [CrossRef]
30. Samaržija, D.; Zamberlin, Š.; Pogačić, T. Psychrotrophic bacteria and their negative effects on milk and dairy products quality psychrotrophic bacteria and milk and dairy products quality. *Mijekarstvo* **2012**, *62*, 75–95.

31. Kent, D.J.; Chauhan, K.; Boor, K.J.; Wiedmann, M.; Martin, N.H. Spore test parameters matter: Mesophilic and thermophilic spore counts detected in raw milk and dairy powders differ significantly by test method. *J. Dairy Sci.* **2016**, *99*, 5180–5191. [CrossRef]
32. McGuiggan, J.T.M.; McCleery, D.R.; Hannan, A.; Gilmour, A. Aerobic spore-forming bacteria in bulk raw milk: Factors influencing the numbers of psychrotrophic, mesophilic and thermophilic Bacillus spores. *Int. J. Dairy Technol.* **2002**, *55*, 100–107. [CrossRef]
33. Wanjala, W.N.; Nduko, J.M.; Mwende, M.C. Coliforms Contamination and hygienic status of milk chain in emerging economies. *J. Food Qual. Hazards Conrol* **2018**, *5*, 3–10. [CrossRef]
34. Evrendilek, G. Non-thermal processing of milk and milk products for microbial safety. In *Dairy Microbiology and Biochemistry: Recent Developments*; CRC Press: Boca Raton, FL, USA, 2017; pp. 322–355.
35. McClements, J.M.; Patterson, M.F.; Linton, M. The effect of growth stage and growth temperature on high hydrostatic pressure inactivation of some psychrotrophic bacteria in milk. *J. Food Prot.* **2001**, *64*, 514–522. [CrossRef] [PubMed]
36. Chmielewski, R.A.N.; Frank, J.F. Biofilm formation and control in food processing facilities. *Compr. Rev. Food Sci. F* **2006**, *2*, 22–32. [CrossRef]
37. Christiansson, A.; Bertilsson, J.; Svensson, B. *Bacillus cereus* spores in raw milk: Factors affecting the contamination of milk during the grazing period. *J. Dairy Sci.* **1999**, *82*, 305–314. [CrossRef]
38. Chaturvedi, A.; Shukla, S. Occurance of Clostridium species in different dairy products and its associated health risk. *Int. J. Recent Sci. Res.* **2015**, *6*, 2827–2829.
39. Mohamed, A.F.; Somda, M.K.; Fourreh, A.E.; Okieh, A.A.; Said, C.N.; Merito, A.; Yagi, S. Evaluation of microbiological quality of raw milk from farmers and dairy producers in six districts of Djibouti. *J. Food Microbiol. Saf. Hyg.* **2017**, *02*.
40. Chye, F.Y.; Abdullah, A.; Ayob, M.K. Bacteriological quality and safety of raw milk in Malaysia. *Food Microbiol.* **2004**, *21*, 535–541. [CrossRef]
41. Siddique, F.; Riffat, S.; Arshad, M.; Farooq, U.; Iftikhar, K.; Kharal, S. Effect of storage temperature on the physiochemical properties of UHT milk. *Int. J. Food Allied Sci.* **2016**, *2*, 52. [CrossRef]
42. Brodziak, A.; Krol, J.; Litwinczuk, Z.; Zaborska, A.; Czernecki, T. Effect of storage time under home refrigeration conditions on the quality of opened drinking milk. *Mljekarstvo* **2017**, *67*, 283–296. [CrossRef]
43. Zajác, P.; Čapla, J.; Vietoris, V.; Zubrická, S.; Čurlej, J. Effects of storage on the major constituents of raw milk. *Potravin. Sci. J. Food Ind.* **2015**, *9*, 375–381. [CrossRef]
44. Mahmood, A.; Usman, S. A comparative study on the physicochemical parameters of milk samples collected from buffalo, cow, goat and sheep of Gujrat, Pakistan. *Pak. J. Nutr.* **2010**, *9*, 1192–1197. [CrossRef]
45. Rawat, S. Food Spoilage: Microorganisms and their prevention. *Pelagia Res. Libr. Asian J. Plant Sci. Res.* **2015**, *5*, 47–56.
46. Aldubhany, T.A.W.; Gouda, E.; Khattab, A.; Dabour, N. Effects of storage on some physico-chemical characteristics of uht milk stored at different temperature. *Alexandria Sci. Exch. J.* **2014**, *35*, 107–114.
47. Ahmad, T.; Bilal, M.Q.; Ullah, S.; Muhammad, G. Effect of severity of mastitis on pH and specific gravity of buffalo milk. *Pak. J. Agri. Sci.* **2005**, *42*, 64–67.
48. Huppertz, T.; Kelly, A.L.; Fox, P.F. Effects of high pressure on constituents and properties of milk. *Int. Dairy J.* **2002**, *12*, 561–572. [CrossRef]
49. Janstová, B.; Lukasová, J.; Dracková, M.; Vorlová, L. Influence of *Bacillus* spp. enzymes on ultra high temperature-treated milk proteins. *Acta Vet. Brno* **2004**, *73*, 393–400.
50. Omer, R.H.; Eltinay, A.H. Changes in chemical composition of camel raw milk during storage. *Pak. J. Nutr.* **2009**, *8*, 607–610.
51. Doll, E.V.; Scherer, S.; Wenning, M. Spoilage of Microfiltered and pasteurized extended shelf life milk is mainly induced by psychrotolerant spore-forming bacteria that often originate from recontamination. *Front. Microbiol.* **2017**, *8*, 1–13. [CrossRef] [PubMed]
52. Garcia-Risco, M.R.; Cortes, E.; Carrascosa, A.V.; Lopez-Fandino, R. Microbiological and chemical changes in high-pressure- treated milk during refrigerated storage. *J. Food Prot.* **1998**, *61*, 735–737. [CrossRef] [PubMed]

processes

MDPI

Article

Classification, Force Deformation Characteristics and Cooking Kinetics of Common Beans

Ebenezer M. Kwofie [1,2], Ogan I. Mba [1] and Michael Ngadi [1,*]

1 Department of Bioresource Engineering, McGill University, 21111 Lakeshore Road, Sainte-Anne-de-Bellevue, QC H9X 3V9, Canada; ebenezer.kwofie@mcgill.ca (E.M.K.); ogan.mba@mcgill.ca (O.I.M.)

2 Department of Biological and Agricultural Engineering, University of Arkansas, 4183 Bell Engineering Center, Fayetteville, AR 72701, USA

* Correspondence: michael.ngadi@mcgill.ca; Tel.: +1-514-398-7779; Fax: +1-514-398-8387

Received: 4 September 2020; Accepted: 29 September 2020; Published: 1 October 2020

Abstract: Post-harvest characteristics of common beans influences its classification, which significantly affects processing time and energy requirements. In this work, ten bean cultivars were classified as either easy-to-cook (ETC) or hard-to-cook (HTC) based on a traditional subjective finger pressing test and a scientific objective hardness test. The hardness study used seed coat rigidity to explain the structural deformation observed during cooking. The result shows that the average hardness of raw dry ETC and HTC beans was 102.4 and 170.8 N, respectively. The maximum seed coat resistance is observed within the first 30 min of cooking regardless of the classification. The results show that a modified three-parameter non-linear regression model could accurately predict the rate of bean softening (R^2 = 0.994–0.999 and RMSE = 3.3–14.7%). The influence of bean softeners such as potassium carbonate (K_2CO_3) and sodium chloride (NaCl) to reduce cooking time was also investigated. The results showed that the addition of K_2CO_3 to the cooking water significantly reduced the cooking time by up to 50% for ETC and 57% for HTC.

Keywords: common beans; beans classification; hard-to-cook; bean softening

1. Introduction

Common beans are an essential food and export commodity in many regions of the world. They are unique sources of plant protein and energy in the diets of more than 300 million people in parts of East Africa and Latin America. Beans are low in fats and rich in proteins, complex carbohydrates, and micronutrients [1]. Despite the nutritional benefits, the presence of flatulence oligosaccharides, antinutrients, and the hard-to-cook (HTC) behavior reduces its potential global consumption [2]. This HTC phenomenon necessitates cooking beans for long hours and requires very high energy demand. There is evidence that the long cooking time is partly influenced by the properties of the beans, such as seed size, variety, and storage conditions, as well as pre-cooking treatments and cooking methods [3–5].

Cooking time is, therefore, an important criterion for evaluating bean cooking quality [6]. Several authors have proposed different theories, including the middle lamella-cation-phytate-phytase theory [7] and the lignification theory [8,9], to explain the hard-to-cook behavior of beans. Several techniques have been deployed to soften beans, reduce the cooking time, and increase consumption. These include soaking [10], ultrasound-assisted hydration [11], salts addition [12], blanching [13], autoclaving, radiation, and extrusion [14–16]. Of all these techniques, the most common approach for household bean processing has been soaking. The typical rural community approach has been to soak common beans overnight or to add table salt or potash to the water during the cooking of the beans. There is evidence that depending on the salt added, the cooking time may reduce significantly or take a much longer time. For instance, Njoroge, Kinyanjui [17] found that cooking stored dried beans in deionized and Na_2CO_3 took 257 and 193 min, respectively. However, they reported that thermally treated beans with $CaCl_2$ did not cook at all within an experimental time of 300 min.

Although beans generally begin to lose their hardness with the onset of cooking and continue until the desired softness is achieved, the force deformation during cooking and impact of the physical parameters with respect to cooking time seems to be missing in the bean cooking narrative. Understanding and optimizing the softening phenomenon through the kinetics of cell wall rigidity and overall structural deformation at different stages during cooking is important. The objectives of this work were, therefore, to (1) classify ten common bean cultivars as either ETC or HTC based on the typical household subjective mechanism and a scientific objective penetration tests and by examining the impact of physical parameters in the classification; (2) examine the bean structure deformation of ETC and HTC during cooking; (3) study the influence of bean softening techniques during cooking of selected ETC and HTC bean cultivars; (4) model the softening behavior of ETC and HTC common beans samples during cooking.

2. Materials and Methods

2.1. Materials

Ten cultivars of common beans cultivated in the Chitipa district of Malawi were used for this work. The bean cultivars included Boma, Mandondo, Sugar, White, Satana, Msiska, Masusu, Kabulangeti, and Magungulu. Before the experiments, the beans were thoroughly cleaned and sorted. All cultivars were from the same farming season and within the same geographical area. The bean samples were part of the lot from the March–April 2018 harvest season. Cooking grade sodium chloride (NaCl) and ACS grade potassium carbonate (K_2CO_3) were supplied by Fisher Scientific (Fair Lawn, NJ, USA).

2.2. Common Beans Classification

2.2.1. Cooking of Beans

One hundred grams of dry unsoaked beans from each of the ten cultivars were selected for the cooking test. The beans were cooked in 800 mL distilled water in a water bath at 96 °C until fully cooked. At 30 min intervals, randomly selected ten-gram seeds were cooled in an ice bath for 1 min prior to the cookability test. Bean cultivars that were fully cooked within 120 min were classified as ETC. Above 120 min, the beans were considered HTC. The 120 min benchmark mimicked the length of cooking practiced by the bean-growing communities in Malawi.

2.2.2. Subjective Cookability Evaluation

The local subjective cookability test to assess softness is to press the cooked beans between the thumb and the index finger. Like other researchers, this "pinching test" was adopted to determine the beans' cookability (softening) after 120 min cooking [18,19]. The beans were considered cooked if the cotyledons disintegrated upon pressing.

2.2.3. Objective Cookability Evaluation

An objective cookability test was determined using a TA-HD Plus texture analyzer (Stable Micro Systems Ltd., Surrey, UK). The hardness of the samples was measured by a penetration test and recording the peak of maximum force (hardness) as further described in Section 2.4. For each variety, a predetermined average compression force (3.75 ± 0.88 N) for fully cooked beans was used to differentiate the uncooked and cooked beans. The cookability data generated were used to generate cooking curves for each bean variety.

2.3. Bean Texture Measurement

Following the method described by Gowen, Abu-Ghannam [13], changes in bean hardness (in Newtons) and seed coat rigidity of raw, soaked, and cooked beans were determined as a function of time and temperature using the TA-HD Plus texture analyzer (Stable Micro Systems Ltd., Surrey, UK). At return-to-start (RTS) mode, a 2 mm cylindrical stainless-steel probe (P2) and 50 kg load cell were used to measure the compression force. The selected probe can impact the beans' tegument, which helps differentiate similar samples [20]. Bean samples were compressed axially to 75% of their original height, applying a crosshead speed of 1.0 mm/s and a pre-test and post-test speed of 1 mm/s [21]. Due to the significant variation of individual bean hardness, ten-gram bean seeds were chosen to represent each treatment. For consistency, the orientation of the seeds on the texture analyzer platform was kept uniform. The seed coat rigidity was determined as the distance the probe traveled before penetrating the cotyledon.

2.4. Bean Softening Experiments

2.4.1. Soaking

The procedure described by Shafaei, Masoumi [22] was adopted for the soaking of representative bean cultivars. Ten grams of randomly selected seeds of three cultivars representing ETC (Masusu), intermediate (Magungulu), and HTC (Msiska) were used. The seeds were soaked in 50 mL of distilled water at room temperature for 10 h. Preliminary experiments showed that the time for equilibrium water absorption in all cultivars was 8–10 h. The soaking experiments were conducted in triplicates. The water absorption capacity was evaluated using Equation (1) [22,23].

$$W_a = \frac{W_f - W_i}{W_i} \times 100 \quad (1)$$

where W_a is the water absorption (d.b.%), W_f is the final weight of seeds after soaking (g), and W_i is the initial weight of seeds prior to soaking (g).

2.4.2. Cooking in Salt

For each cooking test, one hundred gram dry unsoaked seeds of Masusu, Magungulu, and Msiska) were separately cooked in 800 mL of 5% table salt (NaCl) or potash based (K_2CO_3) salt solution. NaCl and K_2CO_3 are commonly used for cooking and softening, respectively, in the local communities. The control bean samples were cooked in distilled water only. All the cooking was done in a water bath at 96 °C until the beans were fully cooked. At 30 min intervals, ten-gram seeds were randomly selected and cooled in an ice bath for 1 min prior to the texture evaluation. The beans were considered cooked when the predetermined cooked bean softness for a cultivar was reached.

2.5. Statistical Analysis

The parameters of the proximate composition analysis were stated as means ± SD (standard deviation, n = 3). The results were statistically analyzed by analysis of variance (ANOVA), and the means were compared using Tukey-Kramer HSD (α = 0.05) on JMP software.

2.6. Kinetics of Bean Softening

The rate of bean softening was determined using a 3-parameter non-linear exponential regression model shown in Equation (2). The 3-parameter model is a modified version of those used by [2,13,24].

$$F = F_e + F_d \exp^{-K_F t} \quad (2)$$

where F and F_e represent the force (N) at a given cooking time (min) and the equilibrium force during cooking, respectively. F_d is the differential softening, representing the total degree of softening over the cooking period (i.e., the difference between the initial and final force), and F_d represents the rate constant of bean softening.

For simplicity, the experimental data were refitted to a simplified 2-parameter non-linear exponential regression model as shown in Equation (3). The force at any given time is defined in terms of only the initial force prior to cooking, K_F, and the cooking time, t.

$$F = F_o \exp^{-K_F t} \quad (3)$$

3. Results and Discussion

3.1. Bean Classification

The result of the subjective finger-pressing test is shown in Figure 1. The results showed that without any softening technique employed, cooking of common beans will take 2–4 h, depending on the cultivar. Using the locally accepted cooking time classifier of two hours (designated as "subjective bean classification line" in Figure 1), the bean varieties were classified as either easy-to-cook beans (ETC) or as hard-to-cook (HTC). The plot shows that Masusu, Satana, and Sugar beans could be classified as ETC. These results confirmed what the communities had known for years. These varieties sell quickly in the communities because they are cooked in a relatively shorter time and required less fuel. Other cultivars, including Maine, Magungulu, White, Kabulangeti, Mandondo, Boma, and Msiska beans, classified as hard-to-cook (HTC), take up to 4 h to be fully cooked.

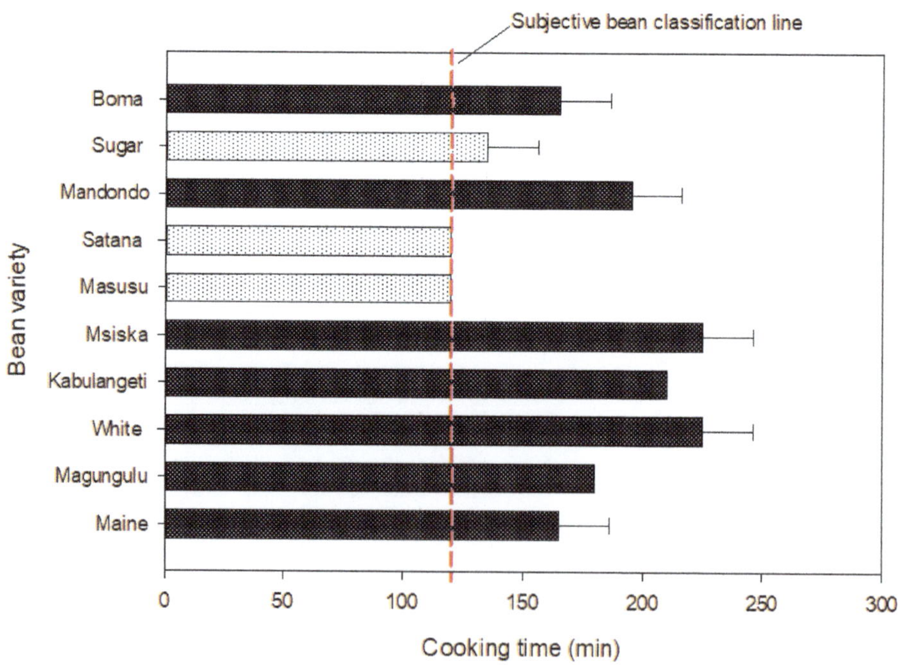

Figure 1. Subjective hardness test of the bean cultivars. (The red line represents the subjective bean classification line. Bean cultivars below the line are easy-to-cook (ETC), while those above the line are hard-to-cook (HTC).

The cooking profile of the common beans by the objective compression test showed that the maximum force showed similar categorization as the pinching test classifying Masusu, Satana, and Sugar beans as ETC (Figure 2a) and the other cultivars as HTC (Figure 2b). The average hardness of raw dry ETC and HTC beans was 102.4 and 170.8 N, respectively. Reduction in hardness was observed in the first 30 min of cooking for all the cultivars. In the ETC cultivars, the average hardness reduction was 77%, while 87% was recorded for the HTC cultivars. Nevertheless, the ETC reached their cooking texture much faster. Among the HTC cultivars, White and Msiska cultivars took the longest time to cook. This may not be unexpected as their average hardness in the dry unsoaked state was 36–47% higher than all the ETC beans. When compared to the other HTC cultivars, the average hardness was 16–36%. This can be attributed to the inherent seed characteristics such as cell wall structure, seed composition, and compactness of cells in the seeds [25,26].

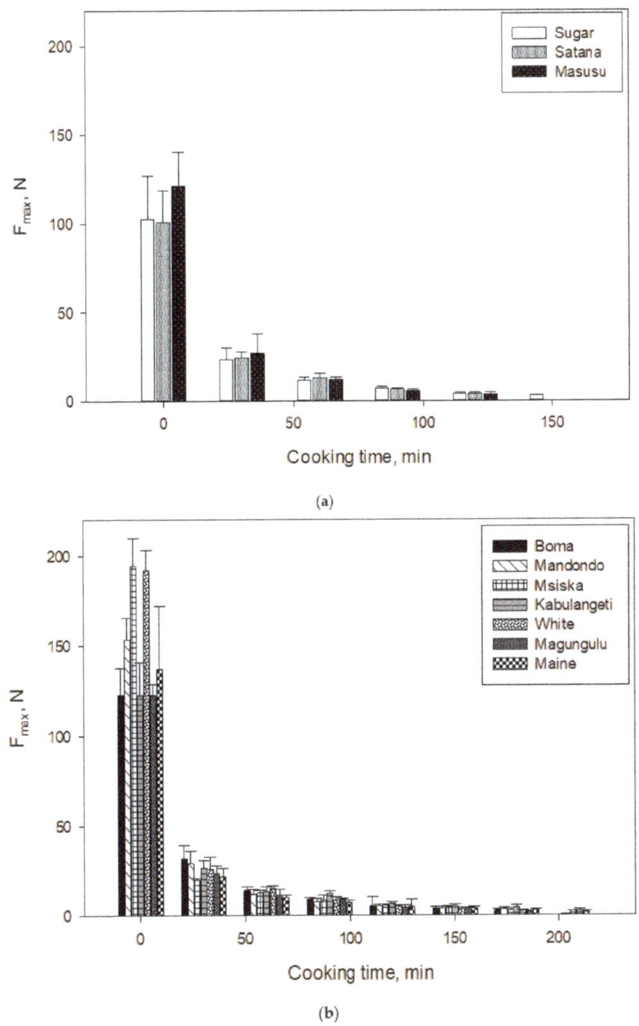

Figure 2. Objective hardness test of the bean cultivars (**a**) ETC cultivars (**b**) HTC cultivars.

Based on the physical parameters, it is expected that Boma and Magungulu cultivars with the higher equivalent diameter and seed volume would have a higher rate of water absorption and cook faster compared to the other beans. However, these physical properties had no significant impact on their softening characteristics. This may be due to the high temperature of cooking water. High temperature tends to diminish the influence of chemical composition, pore structure, and physical properties on water absorption and by extension textural deformation [27]. This result implies that in communities where the cost of energy (especially firewood) for domestic thermal applications is high, preference would be given to ETC rather than the HTC cultivars.

3.2. Bean Structure Deformation during Cooking

Uncooked beans are generally hard and strong. The bean-cooking process leads to some structural deformation that significantly reduces the hardness and produces a softened edible product. The typical force deformation curves for dry and cooked beans are shown in Figure 3. The sequence of deformation occurs in stages. Stage A represents when the probe just touches the bean. Stage B represents the force required for compressing the seed coat and outside layers of the cotyledon. The graphs show that in the uncooked beans (Figure 3a), AB and BC have the same slope. However, in the cooked sample (Figure 3b), BC had a steeper gradient than AB. This can be attributed to some level of plasticity attained by the seed coat during high-temperature cooking with water. A similar effect was observed for red kidney beans immersed in boiling distilled water at 100 °C for 1.5 min [28]. At stage C, the probe penetrates the seed coat and further compresses the bean. A rapid reduction in force was observed in the dry bean (Figure 3a) compared to the cooked beans. Between D and E, a constant force was observed as the probe moved through the bean. The higher force required to penetrate the dry bean resulted in breakage and fragmentation. As the probe traveled through the empty space, a near-zero force was recorded (stage DE). Similarly, DE was relatively constant in the cooked beans as the probe traveled after penetrating the seed coat (Figure 3b).

The force deformation curve changes significantly during cooking regardless of the cultivar. The variation in deformation curves of ETC and HTC bean cultivars is shown in Figures 3c and 3d, respectively. The plots show that the maximum force to penetrate the bean decreased with increasing cooking time. The maximum distance the probe traveled prior to penetration decreased with increasing cooking time. This observation suggested that the seed coat rigidity or fracturability may considerably determine the softening behavior of beans. The figures also show that the force after penetration of the seed coat leveled with cooking time. This may be attributed to the fact that starch gelatinization, protein denaturation, and other phase changes due to cooking become uniform with time.

The evaluation of the seed coat rigidity/fracturability (Figure 4) shows that the dry seed coat is less rigid. The rigidity of the seed coat increased as it absorbed moisture and presented increased resistance to the probe during compression. This rigidity declined as the cooking time increased. The bean seed coat is composed of a thin cuticle above a layer of palisade cells. These palisade cells are prismatic and thick-walled contiguous cells [29]. The thickness of the palisade cells and the lignin and cellulose content of the seed coat are essential determiners of the cooking quality [29]. Seed coat thickness, seed volume, and protein content are known structural features responsible for water absorption [30]. It has been reported that beans with thinner seed coats absorb water more rapidly during the initial soaking of beans [31]. At higher temperatures such as those used in cooking, absorption could be higher, leading to higher initial resistance of the probe through the seed coat.

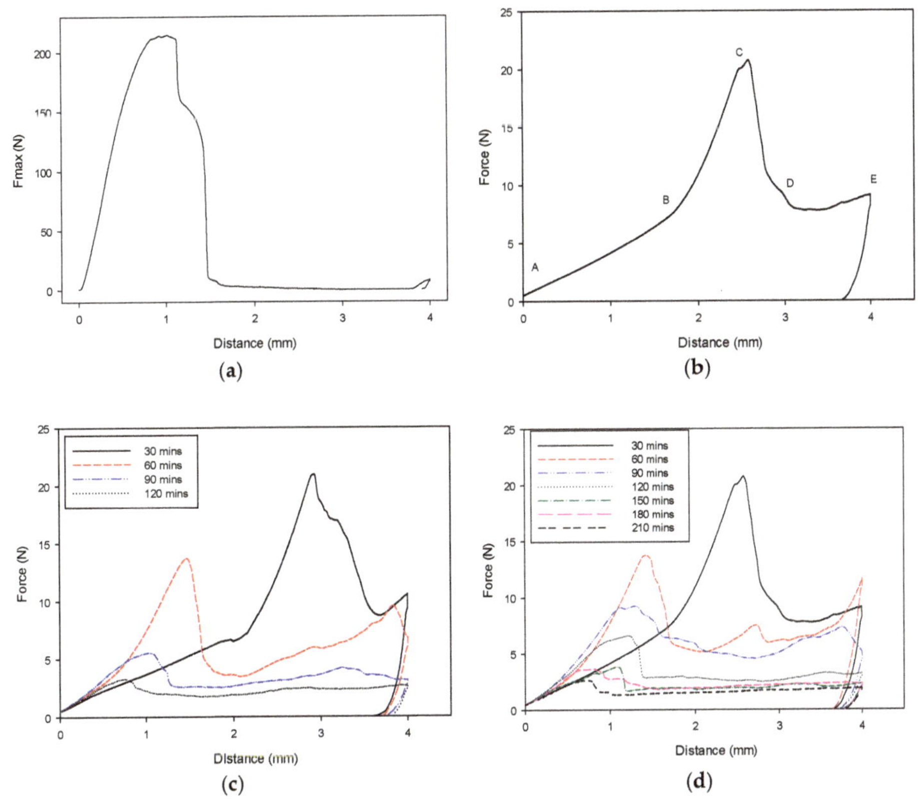

Figure 3. Variation of force deformation in common beans: (**a**) typical uncooked beans (**b**) typical cooked beans (**c**) cooked ETC cultivar (Masusu), and (**d**) cooked HTC cultivar (Msiska).

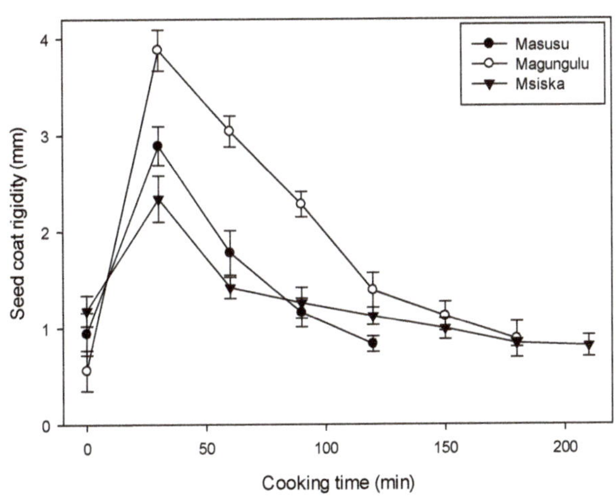

Figure 4. Variation of seed coat rigidity of common beans during cooking (ETC – Masusu, HTC -Msiska, Intermediate – Magungulu)

3.3. Modeling Bean Texture as a Function of Time

The experimental data were fitted to the three-parameter model (Equation (2)). Using non-linear regression analysis, the model constants were determined and presented in Table 1. The results showed that the model accurately predicted softening characteristics of the cultivars considered in this study with R^2 of 0.994–0.999 and RMSE of 3.3–14.7%. Apart from Msiska and White beans, whose equilibrium maximum forces were 0.5 and 3.5%, respectively, above the experimentally determined cooking threshold, the final force determining fully cooked beans were accurately predicted. However, these variations were not statistically significant ($p < 0.05$). The rate constants revealed that Masusu and Sugar bean cultivars, already classified as ETC, had a relatively higher softening rate. The softening rates of Msiska, Boma, and Mandondo cultivars were relatively slower. It can be deduced that the model is useful as a classification tool to differentiate ETC and HTC cultivars and provide meaningful insight on bean softness.

Table 1. Model parameters of the 3-parameter and simplified 2-parameter models.

	3-Parameter Model					2-Parameter Model			
Bean Cultivar	F_e	F_d	K_F	R^2	RMSE (%)	F_0	K_F	R^2	RMSE (%)
Boma	4.39	103.8	0.0428	0.999	12.5	107.8	0.0367	0.991	14.5
Sugar	4.33	104.0	0.0574	0.998	3.3	108.0	0.0463	0.990	8.9
Mandondo	3.66	104.7	0.0459	0.998	10.4	107.9	0.0404	0.991	18.4
Satana	4.22	104.1	0.0538	0.997	4.5	107.1	0.0445	0.990	8.5
Masusu	3.61	104.7	0.0584	0.998	3.9	108.7	0.0427	0.994	4.8
Msiska	4.56	102.9	0.0424	0.994	14.7	107.9	0.0498	0.978	29.7
Kabulangeti	4.12	102.1	0.0514	0.994	9.2	107.5	0.0405	0.975	22.4
White	4.39	103.6	0.0486	0.994	10.4	107.6	0.0409	0.984	14.7
Magungulu	4.43	103.4	0.0549	0.996	12.8	107.9	0.0455	0.986	16.1
Maine	4.05	103.6	0.0537	0.999	11.8	108.1	0.0490	0.988	10.2

F_e = equilibrium force during cooking; F_d = differential softening representing the total degree of softening over the cooking period; K_F = rate constant of bean softening; F_o = initial force prior to cooking. (Tukey–Kramer HSD ($\alpha = 0.05$)).

The model parameters of the simplified two-parameter model (Table 1) showed that the model could accurately predict the softening characteristics of common beans with R^2 of 0.975–0.994 and RMSE of less than 30%. When compared with the three-parameter model, the rate of softening is 14.2–15.1% lower. Considering the standard deviation of texture measurement for each cultivar, these variations in the rate of softening were significantly different ($p < 0.05$).

Cooking kinetics of cereal and legumes have largely relied on Peleg's hydrodynamic model [32]. The hydrodynamic models used in cooking kinetics assume a diffusion process; however, there is evidence that liquid transport during cooking cannot solely be described by the diffusion mechanism [33]. The proposed two-parameter model could provide the benefit of predicting the softening characteristics using the initial bean hardness, which can easily be measured using hardness testing equipment including simple handheld texture analyzers.

3.4. Softening Techniques and Their Impact during Cooking

The effects of the different softening techniques on the cookability of the bean samples are presented in Figure 5.

3.4.1. Effect of Soaking on the Texture of Common Beans

The comparative cookability profile of soaked and unsoaked beans is shown in Figure 5a. The maximum penetration force representing the hardness was significantly lower for soaked beans compared to unsoaked beans. Cooking time was reduced both for ETC varieties like Masusu and HTC varieties like Msiska. The reduction was 42, 50, and 50% for Msiska, Magungulu, and Masusu, respectively. This result seems to suggest that the rate of cooking time reduction due to soaking does not depend on the classification. This reduction may be attributed to increased cell wall hydration as enough water is absorbed, causing the softening of the cell wall and enhancing the breakdown of the middle lamella of the cotyledon [34].

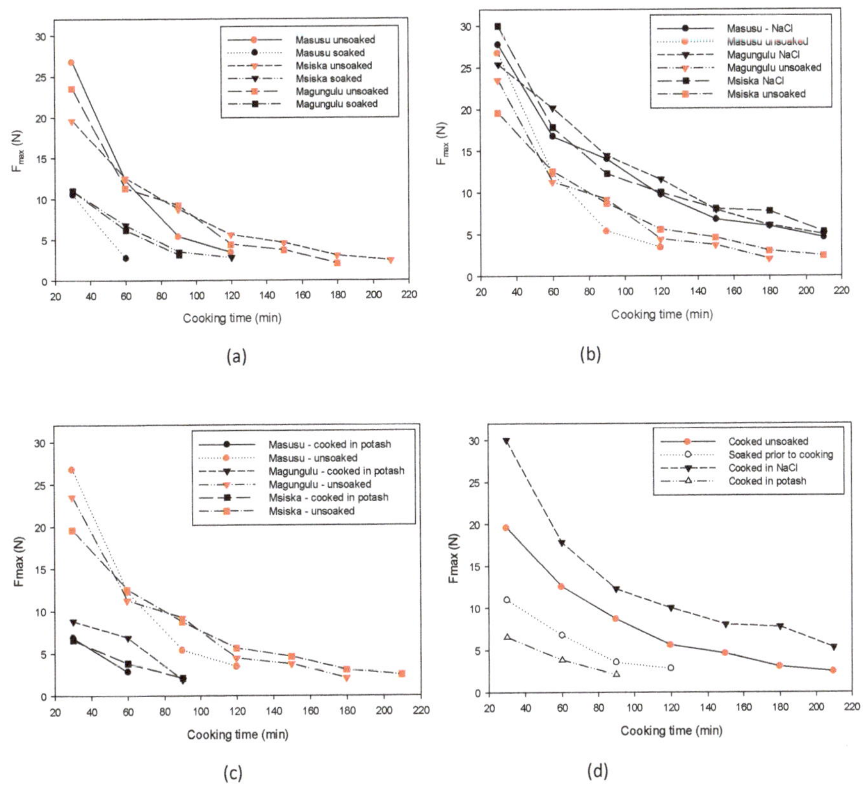

Figure 5. Effect of softening techniques on common beans cookability profile: (**a**) soaking before cooking, (**b**) cooking in NaCl solution, (**c**) cooking in potash solution, and (**d**) comparison of softening techniques for HTC cultivar (Msiska).

3.4.2. Effect of Salt Addition on Texture Profile during Cooking

Salt addition has been a common technique for softening. The results show that the addition of NaCl had no softening impact (Figure 5b). It had no influence on the HTC cultivars and even increased the hardness of the ETC after 60 min. During the cooking of whole (unhulled) beans, early addition of common salt NaCl before the seed coat has absorbed enough cooking water appears to make the seed coat impervious, toughen the beans middle lamella, and impede further water absorption. These lead to the failure of underlying cells to achieve the needed hydration to separate upon cooking and interfere with the solubility of the pectic substances during cooking. This phenomenon has been mostly associated with the ability of Ca^{2+} and Mg^{2+} to form water-resistant pectin complexes with the pectin-rich intercellular middle lamella [4,35]. Na^{+} is recognized to share similar chemical and biochemical reaction properties with Ca^{2+} and Mg^{2+}. This finding is consistent with the report of Kinyanjui, Njoroge [18], especially with respect to the HTC cultivars. On the other hand, the K_2CO_3 salt had a pronounced effect on the hardness of both the ETC and HTC cultivars (Figure 5c). The addition of K_2CO_3 salt reduced the cooking time of ETC and HTC beans by up to 50 and 57%, respectively. As shown in Figure 5d, the addition of potassium salt resulted in the highest level of bean softening.

The impact of increased potassium load on the functional and nutrient quality of the product must be balanced with the desired degree of bean softening.

4. Conclusions

This study correlated the classification of ten beans cultivar as easy-to-cook (ETC) or hard-to-cook (HTC) based on a subjective pinching test with an objective penetration test. The penetration test showed that three cultivars, namely Sugar, Satana, and Masusu, stood out as ETC cultivars, while the other seven cultivars could be described as HTC cultivars. Force deformation curves and seed coat rigidity/fracturability tests were used to monitor the textural changes during cooking. The average hardness of raw dry beans was found to be 102.4 and 170.8 N for ETC and HTC, respectively. Seed coat rigidity plays important role in bean softening during cooking. It was found that the seed coat resistance increased sharply during the initial 30 min and gradually declined until the bean was fully cooked. The results also showed that the three-parameter non-linear regression model and its simplified version can be used to adequately and satisfactorily model the softening characteristics of common bean during cooking. Potassium-based aqueous solutions and pre-soaking can considerably reduce cooking time up to 57%.

Author Contributions: E.M.K. performed the experiments with O.I.M. This research was led by M.N., who was also instrumental in the initiation, conceptualization, and planning of this work. All authors have read and agreed to the published version of the manuscript.

Funding: This research was funded by the International Fund for Agricultural Development (IFAD) for providing financial assistance through IFAD project grant 2000000974.

Conflicts of Interest: The authors declare no conflict of interest.

References

1. Anton, A.A.; Ross, K.A.; Beta, T.; Fulcher, R.G.; Arntfield, S.D. Effect of pre-dehulling treatments on some nutritional and physical properties of navy and pinto beans (Phaseolus vulgaris L.). *LWT Food Sci. Technol.* **2008**, *41*, 771–778. [CrossRef]
2. Yildirim, A.; Öner, M.D.; Bayram, M. Effect of soaking and ultrasound treatments on texture of chickpea. *J. Food Sci. Technol.* **2013**, *50*, 455–465. [CrossRef]
3. Uebersax, M.A.; Ruengsakulrach, S.; Occea, L.G. Strategies and procedures for processing dry beans. *J. Food Tech.* **1991**, *44*, 7.
4. Nyakuni, G.A.; Kikafunda, J.K.; Muyonga, J.H.; Kyamuhangire, W.M.; Nakimbugwe, D.; Ugen, M. Chemical and nutritional changes associated with the development of the hard-to-cook defect in common beans. *Int. J. Food Sci. Nutr.* **2008**, *59*, 652–659. [CrossRef] [PubMed]
5. Shiga, T.M.; Lajolo, F.M.; Filisetti, T.M.C.C. Changes in the cell wall polysaccharides during storage and hardening of beans. *Food Chem.* **2004**, *84*, 53–64. [CrossRef]
6. Moscoso, W.; Bourne, M.C.; Hood, L.F. Relationships Between the Hard-to-Cook Phenomenon in Red Kidney Beans and Water Absorption, Puncture Force, Pectin, Phytic Acid, and Minerals. *Int. J. Food Sci.* **1984**, *49*, 1577–1583. [CrossRef]
7. Galiotou-Panayotou, M.; Kyriakidis, N.B.; Margaris, I. Phytase–phytate–pectin hypothesis and quality of legumes cooked in calcium solutions. *J. Sci. Food Agric.* **2008**, *88*, 355–361. [CrossRef]
8. Hincks, M.J.; Stanley, D.W. Lignification: Evidence for a role in hard-to-cook beans. *J. Food Biochem.* **1987**, *11*, 18. [CrossRef]
9. Srisuma, N.; Hammerschmidt, R.; Uebersax, M.A.; Ruengsakulrach, S.; Bennink, M.R.; Hosfield, G.L. Storage Induced Changes of Phenolic Acids and the Development of Hard-To-Cook in Dry Beans (Phaseolus vulgaris var. Seafarer). *J. Food Sci.* **1989**, *54*, 311–314. [CrossRef]
10. Pujolà, M.; Farreras, A.; Casañas, F. Protein and starch content of raw, soaked and cooked beans (Phaseolus vulgaris L.). *Food Chem.* **2007**, *102*, 1034–1041. [CrossRef]

11. Ghafoor, M.; Misra, N.N.; Mahadevan, K.; Tiwari, B.K. Ultrasound assisted hydration of navy beans (Phaseolus vulgaris). *Ultrason. Sonochem.* **2014**, *21*, 409–414. [CrossRef] [PubMed]
12. de León, L.F.; Elías, L.G.; Bressani, R. Effect of salt solutions on the cooking time, nutritional and sensory characteristics of common beans (Phaseolus vulgaris). *Food Res. Int.* **1992**, *25*, 131–136. [CrossRef]
13. Gowen, A.; Abu-Ghannam, N.; Frias, J.; Oliveira, J. Influence of pre-blanching on the water absorption kinetics of soybeans. *J. Food Eng.* **2007**, *78*, 965–971. [CrossRef]
14. Ai, Y.; Cichy, K.A.; Harte, J.B.; Kelly, J.D.; Ng, P.K. Effects of extrusion cooking on the chemical composition and functional properties of dry common bean powders. *Food Chem.* **2016**, *211*, 538–545. [CrossRef] [PubMed]
15. Alonso, R.; Aguirre, A.; Marzo, F. Effects of extrusion and traditional processing methods on antinutrients and in vitro digestibility of protein and starch in faba and kidney beans. *Food Chem.* **2000**, *68*, 159–165. [CrossRef]
16. Alonso, R.; Orúe, E.; Marzo, F. Effects of extrusion and conventional processing methods on protein and antinutritional factor contents in pea seeds. *Food Chem.* **1998**, *63*, 505–512. [CrossRef]
17. Njoroge, D.M.; Kinyanjui, P.K.; Chigwedere, C.M.; Christiaens, S.; Makokha, A.O.; Sila, D.N.; Hendrickx, M.E. Mechanistic insight into common bean pectic polysaccharide changes during storage, soaking and thermal treatment in relation to the hard-to-cook defect. *Food Res. Int.* **2016**, *81*, 39–49. [CrossRef]
18. Kinyanjui, P.K.; Njoroge, D.M.; Makokha, A.O.; Christiaens, S.; Ndaka, D.S.; Hendrickx, M. Hydration properties and texture fingerprints of easy- and hard-to-cook bean varieties. *Food Sci. Nutr.* **2015**, *3*, 9–47. [CrossRef]
19. Vindiola, O.L.; Seib, O.A.; Hoseney, R.C. Accelerated development of hard-to-cook state in beans. *Cereal Foods World* **1986**, *31*, 14.
20. dos Santos Siqueira, B.; Vianello, R.P.; Fernandes, K.F.; Bassinello, P.Z. Hardness of carioca beans (Phaseolus vulgaris L.) as affected by cooking methods. *LWT Food Sci. Technol.* **2013**, *54*, 13–17. [CrossRef]
21. Revilla, I.; Vivar-Quintana, A.M. Effect of canning process on texture of Faba beans (Vicia Faba). *Food Chem.* **2008**, *106*, 310–314. [CrossRef]
22. Shafaei, S.M.; Masoumi, A.A.; Roshan, H. Analysis of water absorption of bean and chickpea during soaking using Peleg model. *J. Saudi Soc. Agric. Sci.* **2016**, *15*, 135–144. [CrossRef]
23. McWatters, K.H.; Chinnan, M.S.; Phillips, R.D.; Beuchat, L.R.; Reid, L.B.; Mensa-Wilmot, Y.M. Functional, Nutritional, Mycological, and Akara-making Properties of Stored Cowpea Meal. *J. Food Sci.* **2002**, *67*, 2229–2234. [CrossRef]
24. Cunningham, S.E.; Mcminn, W.A.; Magee, T.R.; Richardson, P.S. Effect of processing conditions on the water absorption and texture kinetics of potato. *J. Food Eng.* **2008**, *84*, 214–223. [CrossRef]
25. Khazaei, J.; Mohammadi, N. Effect of temperature on hydration kinetics of sesame seeds (Sesamum indicum L.). *J. Food Eng.* **2009**, *91*, 542–552. [CrossRef]
26. Miano, A.C.; Augusto, P.E.D. From the sigmoidal to the downward concave shape behavior during the hydration of grains: Effect of the initial moisture content on Adzuki beans (Vigna angularis). *Food Bioprod. Process.* **2015**, *96*, 43–51. [CrossRef]
27. Joshi, M.; Adhikari, B.; Panozzo, J.; Aldred, P. Water uptake and its impact on the texture of lentils (Lens culinaris). *J. Food Eng.* **2010**, *100*, 61–69. [CrossRef]
28. Abu-Ghannam, N. Interpretation of the force deformation curves of soaked red kidney beans (Phaseolus vulgaris L.). *Int. J. Food Sci. Technol.* **1998**, *33*, 509–515. [CrossRef]
29. Stanley, D.W.; Aguilera, J.M. A review of textural defects in cooked reconstituted legumes—The influence od structure and composition. *J. Food Biochem.* **1985**, *9*, 277–323. [CrossRef]
30. Sefa-Dedeh, S.; Stanley, D.W. The relationship of microstructure of cowpeaa to water absorption and dehulling properties. *Cereal Chem.* **1979**, *56*, 7.
31. Sefa-Dedeh, S.; Stanley, D.W. Textural implications of the microstructure of legumes. *Food Technol.* **1979**, *33*, 7.
32. Peleg, M. An Empirical Model for the Description of Moisture Sorption Curves. *J. Food Sci.* **1988**, *53*, 1216–1217. [CrossRef]
33. Sharanagat, V.S.; Kansal, V.; Kumar, K. Modeling the effect of temperature on the hydration kinetic whole moong grain. *J. Saudi Soc. Agric. Sci.* **2016**. [CrossRef]

34. Sefa-Dedeh, S.; Stanley, D.W.; Voisey, P.W. Effects of soaking time and cooking conditions on texture and microstructure of cowpeas (Vigna unguiculata). *J. Food Sci.* **1978**, *43*, 7. [CrossRef]
35. Bressani, R. Grain quality of common beans. *Food Rev. Int.* **1993**, *9*, 80. [CrossRef]

processes

MDPI

Article

Application of Computational Intelligence in Describing the Drying Kinetics of Persimmon Fruit (*Diospyros kaki*) During Vacuum and Hot Air Drying Process

Alfadhl Yahya Khaled [1,*], Abraham Kabutey [1], Kemal Çağatay Selvi [1], Čestmír Mizera [1], Petr Hrabe [2] and David Herák [1]

[1] Department of Mechanical Engineering, Faculty of Engineering, Czech University of Life Sciences Prague, Kamýcká 129, 165 00 Prague, Czech Republic; kabutey@tf.czu.cz (A.K.); kcselvi74@gmail.com (K.Ç.S.); mizera@tf.czu.cz (Č.M.); herak@tf.czu.cz (D.H.)

[2] Department of Material Science and Manufacturing Technology, Faculty of Engineering, Czech University of Life Sciences Prague, Kamýcká 129, 165 00 Prague, Czech Republic; hrabe@tf.czu.cz

* Correspondence: f_yahya87@hotmail.com; Tel.: +420-7763-17207

Received: 23 March 2020; Accepted: 28 April 2020; Published: 7 May 2020

Abstract: This study examines the potential of applying computational intelligence modelling to describe the drying kinetics of persimmon fruit slices during vacuum drying (VD) and hot-air-drying (HAD) under different drying temperatures of 50 °C, 60 °C and 70 °C and samples thicknesses of 5 mm and 8 mm. Kinetic models were developed using selected thin layer models and computational intelligence methods including multi-layer feed-forward artificial neural network (ANN), support vector machine (SVM) and k-nearest neighbors (kNN). The statistical indicators of the coefficient of determination (R^2) and root mean square error (RMSE) were used to evaluate the suitability of the models. The effective moisture diffusivity and activation energy varied between 1.417×10^{-9} m^2/s and 1.925×10^{-8} m^2/s and 34.1560 kJ/mol to 64.2895 kJ/mol, respectively. The thin-layer models illustrated that page and logarithmic model can adequately describe the drying kinetics of persimmon sliced samples with R^2 values (>0.9900) and lowest RMSE (<0.0200). The ANN, SVM and kNN models showed R^2 and RMSE values of 0.9994, 1.0000, 0.9327, 0.0124, 0.0004 and 0.1271, respectively. The validation results indicated good agreement between the predicted values obtained from the computational intelligence methods and the experimental moisture ratio data. Based on the study results, computational intelligence methods can reliably be used to describe the drying kinetics of persimmon fruit.

Keywords: persimmon fruit; drying methods; computational intelligence methods; artificial neural network model; support vector machine model; k-nearest neighbors

1. Introduction

Persimmon (*Diospyros kaki*) is an edible fruit with a sweet taste and rich in vitamin A, C, calcium, condensed tannins, carotenoid, phenolic compounds and iron [1–3]. Besides the nutritional value of persimmon, it has many health benefits such as therapeutic effect on cardiovascular system disease, helps effectively to reduce cholesterol and blood pressure, strengthens the immune system, prevents cancer and remedy for digestion [4]. Persimmon has high moisture content resulting in susceptibility to spoilage even at refrigerator temperatures. Thus, it has to be preserved by proper drying processes to increase the shelf life [5,6].

Drying is considered one of the commonly used postharvest preservation techniques. The drying process of harvested agricultural products assists in reducing spoilage, increasing shelf-life and reducing the bulk weight of products during transportation. Drying of agricultural products causes the enzymatic reactions to be inactivated as a result of heat and mass transfer leading to a reduction of the moisture content inside the product [7]. It also helps the extraction of bioactive compounds from food products. Drying methods such as hot-air drying (HAD), freeze-drying (FD), vacuum drying (VD), microwave drying (MWD) and infrared drying (IRD) have been used in drying agricultural crops [7–10]. Amongst these drying methods, the HAD and VD are the most common commercially used drying methods as they provide more uniform dried product, naturally harmless and nontoxic [11]. For the VD method, the use of low temperatures in the absence of oxygen can preserve heat-sensitive and easily oxidizable foods [12]. Consequently, discoloration and decomposition of the flavor and some nutritional substances can be prevented [10]. However, using HAD and VD without appropriate operating parameters can negatively affect the essential properties of food products such as nutritional and phytochemical properties. Hence, the determination of the optimum operating parameters, drying conditions using suitable drying models are indispensable for achieving quality along with minimum product cost and maximum yield [7,13,14].

Previous studies have explored several mathematical thin-layer drying models (empirical, semi-theoretical and theoretical) to describe the drying kinetics of fruits and vegetables for enhancing the overall performance of the drying process [5,15–17]. These models are derived from the physical laws that govern the process such as mass, reaction kinetics and thermodynamics. The models accuracy of a thermal process are limited to other factors like physical properties, which may vary during the thermal process. Besides that, these models can achieve satisfying regression results in comparison with experimental data in specific conditions. Nevertheless, mathematical thin-layer models are empirical in nature, do not give the physical interpretation of the drying process and they are product dependent [18].

Computational intelligence tools such as artificial neural networks (ANN), support vector machine (SVM) and k-nearest neighbors (kNN) are considered as complex tools for complex systems and dynamic modelling [19]. The application of ANN, SVM and kNN offer many advantages compared to conventional modeling techniques due to the learning ability, increased flexibility, online non-destructive measurements, reduced assumptions, suitability to the non-linear process and tolerance of incomplete data [7,13,14]. For example, ANN is inspired by the biological neural system as a useful statistical tool for nonparametric regression [19]. SVM is highly recognized in terms of its superior performance of the regression data due to its excellent performance capability when working with multi-dimensional data [20]. The kNN is an easy-to-implement algorithm that can be used to solve prediction problems [21].

Computational intelligence modelling methods can be applied as a potential alternative to mathematical thin-layer models in the drying of fruits and vegetables due to the stability and precision in case of uncertainties in the input parameters. ANN and SVM have been successfully applied in modelling and optimizing the drying processes of fruits and vegetables such as pomelo [22], ginkgo biloba seeds [13], mushroom [23], tomato [24], wood [25], celeriac slices [26], pumpkin [11], pepper [27], eggplant [28] and tea leaves [29]. The application of computational intelligence modelling methods in many areas such as cybersecurity and health-care has proven their unique ability to learn such complex data interactions [30,31]. Therefore, including these models in the drying techniques can enhance the prediction and optimization of the drying kinetics of agricultural crops. They provide a significant advantage to overcome the large errors in predictions, by automatically adapting the models until the minimum error rate. In view of this, ANN, SVM and kNN techniques can be applied to predicting and optimizing the drying kinetics of persimmon fruit under VD and HAD. It can help in the evaluation of drying parameters in real conditions, the optimization of processing conditions and the increase in the overall drying efficiency. However, this knowledge is limited in the literature regarding the application of ANN, SVM and kNN techniques in modelling the drying kinetics of persimmon fruit under VD and HAD. Application of these models can optimize energy and quality for persimmon fruit and further minimize the required time and production cost.

The objectives of this study are to investigate the drying characteristics of persimmon fruit at different temperatures using VD and HAD, to evaluate the feasibility of applying ANN, SVM and kNN modelling as a non-destructive technique in describing the drying behavior of persimmon fruit under different drying conditions and to compare the results to mathematical thin-layer models.

2. Materials and Methods

2.1. Samples Preparation

Persimmon fruit bought from a market in Prague, Czech Republic, was used for the experiment. A total of 30 persimmon fruits were selected based on similar physical appearances (shape, color and size). Before the drying experiments, the persimmon fruits were stored in a refrigerator at 5 °C until further processing. Prior to each experiment, samples were peeled and washed under running tap water. Afterwards they were sliced into two thickness levels of 5 mm and 8 mm with a diameter of 56 mm using a Sencor slicer (SFS 4050SS, Prague, Czech Republic). The initial moisture content of the fresh samples was determined to be 3.98 kg/kg (dry basis) based on the ASABE standard by drying 25 g of selected samples at 70 °C for 24 h using the conventional oven [32].

2.2. Drying Experiments

2.2.1. Vacuum Drying (VD) Technique

The VD technique was carried out using a laboratory-scale drying unit (l 450 Gold brunn, Poland). The vacuum was regulated at a 50 mbar ultimate pressure and 2 L/s pump speed by a vacuum pump (VE 135, RoHS, Shanghai, China). However, it is worth mentioning that the pressure was monitored through the vacuum gauge which was unstable during the drying process. This problem was solved manually by fixing the pressure at approximately 50 mbar. This meant that when the pressure increased above or decreased below 50 mbar, the pump was opened for the adjustment. The VD operates by heating the samples with a conduction heat from a heater plate in the container. The vacuum pump reduces the pressure around the sample to be dried and further ensures less atmospheric pressure. This decreases the boiling point of the water inside the product and thereby increases the rate of evaporation. The sliced samples were dried at three temperatures (50 °C, 60 °C and 70 °C). Prior to the experiments, the VD was set-up to the required temperature for 30 min to enable the dryer temperature to reach equilibrium with the surrounding air temperature. The weight of the persimmon sample was measured at 1 h interval using a digital scale (HR-250AZ, A&D Company Limited) weighing balance of 252 g 0.1 mg^{-1} precision. All the drying experiments were carried-out in triplicates and the average values were used for further analyses.

2.2.2. Hot-Air Drying (HAD) Technique

A laboratory-scale convective HAD (UF 110, Memmert, Germany) was used. Similar to VD, three temperatures (50 °C, 60 °C and 70 °C) at a constant air velocity of 1.10 m/s was attained until constant weight between two successive readings. The air velocity was measured using a Thermo-Anemometer (Model 451104, EXTECH Instruments, Tainan, Taiwan; with the accuracy of 62% velocity). Before starting the experiments, the HAD was set-up to the required temperature for a half-hour to enable the dryer temperature to reach equilibrium with the surrounding air temperature. Similar to VD, the weight of the persimmon sample was measured 1 h interval using a digital scale weighing balance; whereby the samples were removed from the dryer and measured and then returned to the dryer. The experiments were conducted in triplicate and the average values used in further analyses.

2.3. Drying Kinetics

The variation in moisture content during VD and HAD techniques was expressed in the form of moisture ratio (dimensionless) as described in Equation (1).

$$\mathrm{MR} = \frac{(M_t - M_e)}{(M_o - M_e)} \tag{1}$$

where M_t, M_e and M_o are the moisture content of the samples at time t, equilibrium moisture content and initial moisture content, respectively. According to Aghbashlo et al. [33], *Me* values did not change because they were relatively low compared to M_t and M_o values, resulting in negligible error during simplification, thus, in this study, the moisture ratio was expressed as shown in Equation (2):

$$\mathrm{MR} = \frac{\mathrm{M_t}}{\mathrm{M_o}} \tag{2}$$

2.4. Effective Moisture Diffusivity

To effectively assess the behavior of the VD and HAD methods of samples, it is important to understand the mechanisms of moisture movement within the samples during drying. Fick's diffusion equation as a dimensional approach was applied due to its simplicity to describe the mass transfer of drying samples. The effective moisture diffusivity of samples for VD and HAD methods was estimated using Crank's solution [33] of Fick's diffusion equation as described in Equation (3) [16].

$$\frac{\partial M_t}{\partial \mathrm{t}} = \nabla .\left(D_{eff}\, \nabla\, M_t\right) \tag{3}$$

Assuming constant diffusion and uniform initial moisture distribution, the Crank's solution for cylindrical shaped sample is shown in Equation (4).

$$\mathrm{MR} = \frac{8}{\pi^2} \sum_{\mathrm{n}=1}^{\infty} \frac{1}{(2\mathrm{n}+1)^2} \exp\left(-\frac{(2\mathrm{n}+1)^2 D_{eff}\, \mathrm{t}}{\mathrm{r}^2} \right) \tag{4}$$

where D_{eff} is the effective moisture diffusivity (m^2/s), n is the positive integer, r is the radius of the sample (m) and t is the drying time (s). For the sake of mathematical simplicity, Equation (4) was restricted to the first term, resulting in Equation (5).

$$\mathrm{MR} = \frac{8}{\pi^2} \exp\left(-\frac{\pi^2 D_{eff}\, \mathrm{t}}{\mathrm{r}^2} \right) \tag{5}$$

2.5. Activation Energy

In general, activation energy is the minimum energy needed in order for drying to occur. The activation energies for VD and HAD methods were calculated from the relationship between effective moisture diffusivity and the average temperature of the samples based on the Arrhenius equation as shown in Equation (6) [34].

$$D_{eff} = D_o \exp\left(-\frac{E_a}{\mathrm{R}(\mathrm{T} + 273.15)} \right) \tag{6}$$

where D_o is the pre-exponential factor, E_a is the activation energy (kJ/mol), R is universal gas constant (8.3143×10^{-3} kJ/mol) and T is the average temperature of the sample (K). The values of E_a for VD and HAD methods for different persimmon thickness levels were measured from the resulting slope values by plotting the fitting curve between *ln*D and 1/(T + 273.15) (Equation (7)).

$$\mathrm{Slope} = -\frac{E_a}{\mathrm{R}} \tag{7}$$

2.6. Mathematical Thin-Layer Modelling

The experimental drying data measured were fitted to ten selected mathematical thin-layer drying models. The selected mathematical models are listed in Table 1 namely Newton, Page, Modified page, Logarithmic, Two-term, Two-term exponential, Henderson and Pabis, Modified Henderson and Pabis, Midilli et al. and Hii et al. The coefficients of the mathematical models were determined based on non-linear least squares regression analysis using Sigma plot software (Version12.0, Systat Software Inc., California, USA). The application of these models gives a better prediction with fewer assumptions [35].

Table 1. Mathematical thin-layer drying models.

Model No.	Model Name	Model Expression	Reference
1.	Newton model	$MR = \exp(-kt)$	[36]
2.	Page model	$MR = \exp(-kt^n)$	[15]
3.	Modified page	$MR = \exp[-(kt)^n]$	[37]
4.	Logarithmic model	$MR = a \exp(-kt) + c$	[38]
5.	Two-term model	$MR = a \exp(-k_1 t) + b \exp(-k_2 t)$	[39]
6.	Two-term exponential model	$MR = a \exp(-k_0 t) + (1-a) \exp(-k_1 a t)$	[40]
7.	Henderson and Pabis model	$MR = a \exp(-kt)$	[41]
8.	Modified Henderson and Pabis model	$MR = a \exp(-kt) + b \exp(-gt) + c \exp(-ht)$	[42]
9.	Midilli et al. model	$MR = a \exp(-kt) + bt$	[43]
10.	Hii et al. model	$MR = a \exp(-k_1 t^n) + b \exp(-k_2 t^n)$	[11]

2.7. Computational Intelligence Methods

2.7.1. Artificial Neural Network

The structure of a neural network is in the form of interconnected layers [44,45]. Haykin [46] divided an ANN into three groups of structures based on their connection namely, single layer feed-forward network, the multi-layer feed-forward network and the recurrent network. Among these structures, the multi-layer feed-forward network is widely applied in modelling of agricultural and food systems. The feed-forward neural network has an input layer (n), an output layer (m) and one or more hidden layers (h). The number of neurons in the input and output layers is representative of the number of independent variables (input) and dependent variables (output) respectively. Each of the nodes has connected weight to all the nodes in the next layer calculated to give the sum of the nodes (x) representing an activation input value function of the node. The value of x is computed first followed by computing the activation function of the node until the output nodes activation function is acquired. Hidden layer with different nodes is used to process the information received by the input nodes through activation function. In this study, a multilayer feed-forward network structure was used with three input parameters (temperature, thickness and drying time), 1–3 hidden layers and one output parameter (moisture ratio) as shown in Figure 1. A back-propagation algorithm was applied in training of the model because it is stable when a small learning rate is used and sigmoid function was used in all cases as illustrated in Equation (8) [47].

$$f(x) = \frac{1}{1 + e^{-x}} \tag{8}$$

This algorithm passed through four steps; initialization, activation, weight training and iteration to train data. In this case, the error minimization can be obtained by several procedures; gradient descent, conjugate gradient and Levenberge-Marquardt. Several methods for speeding up back-propagation algorithm have been used like using a variable learning rate and adding a momentum term. In this process, the parameters of learning rate, momentum and number of epoch were set as 0.3, 0.2 and 500, respectively. It is worth noting that the training time in this study depended on the number of iterations which was chosen to be 200. This is because at 200 iterations, the minimum error rate was reached. The datasets were prepared by randomly dividing the data into training (70%) and testing (30%) [48,49]. The chosen hidden layer architectures were (3), (6), (9), (3, 3), (6, 6), (9, 9), (3, 3, 3), (6, 6, 6) and (9, 9, 9) matrix, where for example, (3, 3) and (3, 3, 3), represent the 2 and 3 hidden layers with 6 and 9 neurons each (Figure 1). The number of the hidden layers and neurons are affected directly by the simplicity of the ANN topology, while few numbers of hidden layers and neurons are recommended for simplicity. The software (Weka 3.6, Hamilton, New Zealand) was used to analyze the ANN model. Overall, ANN has the ability to performing the neural fitting and prediction and in this case, could be used for future predictions without the need for the neural network tools for the specific drying conditions.

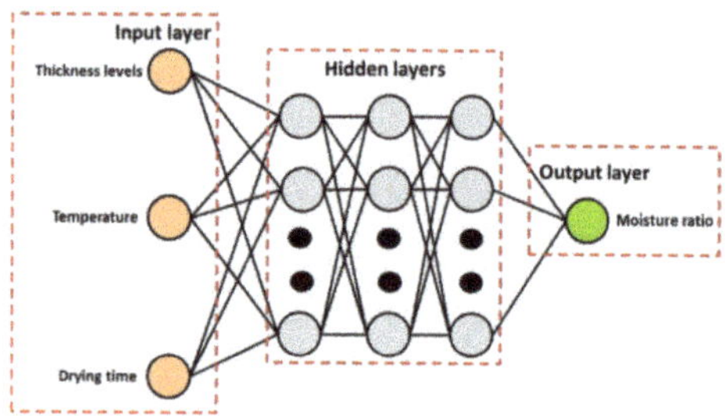

Figure 1. Artificial Neural networks topology with three hidden layers and different number of neurons.

2.7.2. Support Vector Machine

Support Vector Machine (SVM) is a supervised learning model which combines theoretical solutions with numerical algorithms used for classification and regression methods [50]. In 1999, Vapnik [51] developed the SVM algorithm and other important feature information and patterns. SVM as a regression method is considered an effective approach due to its capability of capturing non-linear relationships in the feature space.

The SVM used for the moisture ratio values was determined by the SMOreg sequence in the Waikato Environment for Knowledge Analysis (WEKA) software, whereby, the SMOreg implements SVM for regression. In this regard, the parameters of SVM learned by setting the RegOptimizer with ReqSMOImproved as a learning algorithm [52]. Figure 2 shows the workflow of using the SVM according to ReqSMOImproved learning algorithm.

The three input variables used for the SVM model were temperature, thickness and drying time with the output as the moisture ratio. Two filter types were applied, namely normalize and standardize in order to determine how/if the data need to be transformed. Additionally, three different kernel models: polynomial, Pearson universal and Radial Basis Function (RBF) were used to construct the predictive model of the calculated moisture ratio values. The data required to be optimized in the three kernels' parameters for maximum performance were obtained using the Grid Search technique. Similar to ANN, the number of iterations used was 200 due to the minimum error rate limit. The three kernels' parameters were computed mathematically as described in Equations (9)–(11):

$$f(x,\ y) = \frac{((x \times y) + 1)^d}{\sqrt{((x \times y) + 1)^d((y \times y) + 1)^d}} \tag{9}$$

$$f(x,\ y) = \frac{1}{\left[1 + \left(\frac{2\times\sqrt{\|x-y\|_2\sqrt{2^{(1/\omega)}-1}}}{\sigma}\right)^2\right]^{\omega}} \tag{10}$$

$$f(x,\ y) = e^{-'Y\|(x-y)\|^2} \tag{11}$$

where x represents the feature vector, y is referred to as the label for each x, d is the degree of polynomial, ω and σ are Pearson width parameters and 'Y is the kernel dimension.

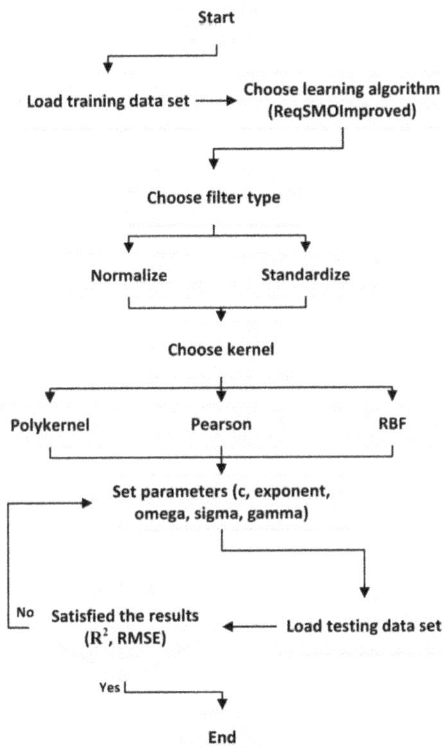

Figure 2. The workflow of support vector machine (SVM).

2.7.3. k-Nearest Neighbors

The kNN is a simple algorithm that predicts the test samples category according to the k training samples, which are the nearest neighbors to the test sample and classifies it to the category that has the largest category probability (Zhang and Zhou, 2005). kNN can select the appropriate value of k based on cross-validation and also kNN can perform distance weighting. In this study, the numbers of neighbors were used to be 3, 5, 7, 9 and 11, in order to select the best k-value in kNN based on the highest results of statistical indicators. The parameters of debug are meanSquared, crossValidate, distanceWeighting, nearestNeighbourSearchAlgorithm, windowSize and doNotCheckCapabilities did not change in Weka software. The value of the batchSize parameter in the kNN classifier was set to be 150 since the highest accuracy was observed at this value. The number of iterations was set to be 200 due to the minimum error rate. The number of decimal places used for the output of numbers in the model was 2 (numDecimalPlaces = 2).

2.8. Color Measurements

Color is one of the most important quality evaluation attributes for fruits and vegetables during drying. For the color measurements, first, the images of fresh (reference) and dried samples obtained from VD and HAD methods were captured using a smartphone camera (oppo F7, Dongguan, China). The smartphone camera is equipped with a 16 mega-pixel charged-coupled device (CCD). Samples were put in a glass plate located on the white paper as a background during image capture for more focus and the distance between sample and smartphone camera was set-up to be 18 cm vertically. Then the images were transferred to ImageJ software for determining the color parameters: lightness (L*), redness/greenness (a*) and yellowness/blueness (b*) (http://rsb.info.nih.gov/ij/). The total color difference (ΔE) was estimated based on Equation (12). For each sample, three replications were performed.

$$E = \sqrt{(L^* - L_o^*)^2 + (a^* - a_o^*)^2 + (b^* - b_o^*)^2} \tag{12}$$

where L_o^*, a_o^* and b_o^* indicate the reference values of fresh sample.

2.9. Statistical Analysis for Mean Comparison

Statistical analysis was performed using the Statistical Analysis System software (SAS version 9.2, Institute, Inc., Cary, NC, USA). ANOVA at 5% level of significance and 95% confidence interval was performed using the Duncan test to compare the mean significant differences between quality attributes (L*, a*, b* and ΔE) for different sample thickness levels of 5 mm and 8 mm, drying time intervals between (0 and 600 minutes) and drying techniques (VD and HAD). The results were presented as mean ± standard error values. The fit accuracy of experimental data to the mathematical thin-layer and computational intelligence (ANN, SVM and kNN) models was determined by the statistical indicators: coefficient of determination (R^2) and root mean square error (RMSE) which are described mathematically in Equations (13) and (14):

$$R^2 = 1 - \frac{\sum_{i=1}^{N}\left(V_{pred} - V_{exp}\right)^2}{\sum_{i=1}^{N}\left(V_{pred} - V_m\right)^2} \tag{13}$$

$$RMSE = \sqrt{\frac{\sum_{i=1}^{N}\left(V_{pred} - V_{exp}\right)^2}{N}} \tag{14}$$

where V_{pred} is the predicted value, V_{exp} is the actual observation from experimental data, V_m is the mean of the actual observation and N is number of observations. From the values of R^2 and RMSE, the higher the value of R^2 and the lower the RMSE value, the better the goodness of fit.

3. Results and Discussion

3.1. Drying Process Behavior

The variations of moisture ratio with time for VD and HAD techniques at different temperatures (50 °C, 60 °C and 70 °C) and samples thicknesses (5 mm and 8 mm) are presented in Figure 3. From the plot, the moisture ratio of the sliced samples for all techniques decreased with an increase in drying time. The drying rates for VD and HAD methods occurred in the falling rate period. Based on Figure 3a, it is clear that the drying time thus reduces as the drying temperature increases. The moisture ratio values of 0.18 and 0.28 were determined at drying time of 170 min and temperatures of 50 ° C and 60 °C. At a drying time of 360 min at 70 °C was found the moisture ratio of 0.22. As shown in Figure 3b, as the drying time increased, for example, the moisture ratio values at 300 min at 50 °C and 70 °C were 0.44 and 0.12 respectively. Similar results were observed for HAD at samples thickness levels of 5 mm and 8 mm as illustrated in Figure 3c,d. For example, the final time found from HAD of 5 mm sample thickness at 50 °C, 60 °C and 70 °C was 420 min, 300 min and 240 min respectively. The results indicate that the moisture transfer rate from the inner layers to its surface thus increases as the drying air temperature increases. The rate of moisture evaporation at the surface of dried material to the atmosphere also increases as the temperature increase leading to a higher drying rate. The results are in agreement with other researchers on the drying behavior of various varieties of persimmon [1,2,5].

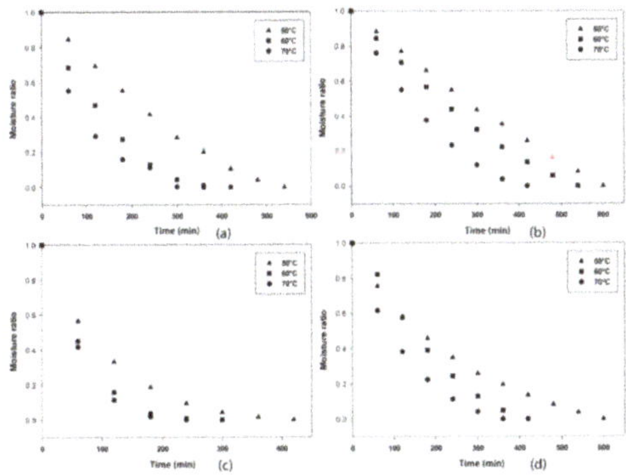

Figure 3. Drying characteristics of persimmon fruit sliced samples; (**a**) vacuum drying (VD) of sample thickness of 5 mm; (**b**) VD of 8 mm; (**c**) hot-air drying (HAD) of 5 mm; (**d**) HAD of 8 mm.

3.2. Results of Effective Moisture Diffusivity

The values of the effective moisture diffusivity (D_{eff}) are presented in Table 2. The Deff values were varied between the range of 1.417×10^{-9} m^2/s and 1.925×10^{-8} m^2/s. As it can be seen from Table 2, the D_{eff} increased significantly with increasing drying temperature from 50 °C to 70 °C. The HAD with a 5 mm thickness of persimmon samples illustrated the highest values of D_{eff} as compared to VD. The result is attributed to the increase in water molecules activity at higher temperatures leading to higher moisture diffusivity [53]. Moreover, the D_{eff} values of samples thickness of 5 mm were higher than samples of 8 mm thickness for VD and HAD drying techniques (Table 2). This is because of the increased heat arising from the increase in drying temperature of the product resulting to the water molecules activities compared to samples of bigger thickness. The values of D_{eff} obtained in this study were within the general range of 10^{-6} to 10^{-12} m^2/s for drying of food materials [5,54]. The values of D_{eff} are agreement with published works for strawberry drying (2.40–12.1 $\times 10^{-9}$ m^2/s), apple drying (2.27–4.97 $\times 10^{-10}$ m^2/s), persimmon slices (2.60–5.40 $\times 10^{-10}$ m^2/s) and pumpkin drying (1.19–4.27 $\times 10^{-9}$ m^2/s) [17,55–57].

Table 2. Values for effective moisture diffusivity, D_{eff}.

Thickness (mm)	Drying Method	D_{eff} (m^2/s)	R^2	RMSE
5	VD-50	3.316×10^{-9}	0.9647	0.5564
	VD-60	8.742×10^{-9}	0.9897	0.4563
	VD-70	1.330×10^{-8}	0.8094	0.3829
8	VD-50	1.417×10^{-9}	0.9670	0.6056
	VD-60	2.688×10^{-9}	0.9710	0.5708
	VD-70	2.959×10^{-9}	0.9786	0.5273
5	HAD-50	9.221×10^{-9}	0.9900	0.3632
	HAD-60	1.712×10^{-8}	0.9956	0.3280
	HAD-70	1.925×10^{-8}	0.9991	0.3461
8	HAD-50	2.905×10^{-9}	0.9314	0.4644
	HAD-60	4.834×10^{-9}	0.9811	0.5641
	HAD-70	7.890×10^{-9}	0.9888	0.4395

3.3. Results of Activation Energy

The activation energy (Ea) of persimmon fruit samples was calculated from the values of effective moisture diffusivity, D_{eff}. The relationship between Ea and D_{eff} was described by an Arrhenius-type equation (Equation (6)). The values of activation energy were obtained by plotting

$ln(D_{eff})$ versus 1/(T + 273.15) for VD and HAD methods. The activation energy is equal to the slope times the universal gas constant (R) as shown in Figure 4. The activation energy values for the VD and HAD for samples thicknesses (5 mm and 8 mm) were further estimated as given in Table 3. The values of *Ea* for the VD method were 64.2895 kJ/mol and 34.1785 kJ/mol for 5 mm and 8 mm respectively. While for the HAD method, the *Ea* was 34.1560 kJ/mol at 5 mm thickness, and that of 8 mm was 46.0715 kJ/mol at temperatures between 50 °C and 70 °C. The results found in this study are similar to those reported for persimmon sliced samples with activation energy between 30.64 and 43.26 kJ/mol for blanched and control respectively [5]. The *Ea* values of persimmon samples dried using VD and HAD methods were consistent with those in the literature for different fruits and vegetables, for example,30.46–35.57 kJ/mol in strawberry, 25.26–72.47 kJ/mol in yams and 22.66–30.92 kJ/mol in apples [17,58,59]. The values of *Ea* were within the acceptable range from 12.7 to 110 kJ/mol for various food materials [60].

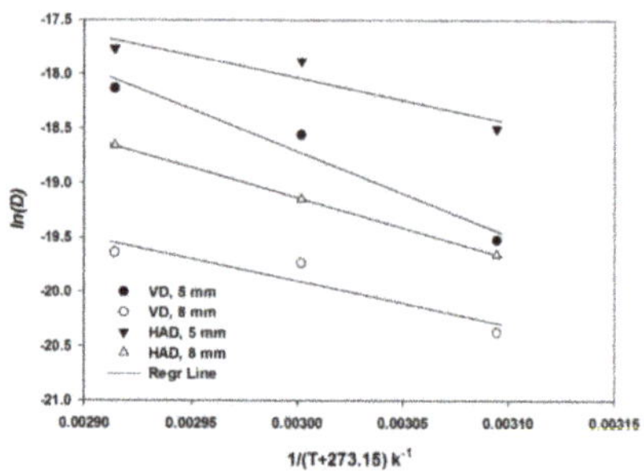

Figure 4. Arrhenius-type relationship of effective moisture diffusivity versus temperature for VD and HAD methods at different samples thickness levels.

Table 3. Activation energy of persimmon fruit samples for VD and HAD at different thickness levels.

Drying Method	R^2	E_a (kJ/mol)
VD-5 mm	0.9577	64.2895
VD-8 mm	0.8583	34.1785
HAD-5 mm	0.8774	34.1560
HAD-8 mm	0.9999	46.0715

3.4. Comparison of Mathematical Thin-Layer Models

The mathematical thin-layer models were used to describe the drying kinetics of persimmon fruit samples for VD and HAD methods. Tables 4–7 show the selected mathematical models that fitted the experimental moisture content data in relation to samples thickness levels of 5 mm and 8 mm. Although all the selected ten models adequately fitted the experimental data, the logarithmic model sufficiently described the drying kinetics of persimmon samples with R^2 values (> 0.9900) and lowest RMSE values (< 0.0200) for VD technique at all drying temperatures as given in Table 4. Similar results were obtained for VD at 8 mm thickness indicating that the logarithmic model showed the best model to fit the experimental data with the highest R^2 of > 0.0990 and the lowest RMSE of < 0.0100 at all the three temperatures (50 °C, 60 °C and 70 °C). For HAD technique, the Page, Logarithmic, Midilli et.al and Hii et al. models significantly described the drying kinetics of persimmon samples of thicknesses (5 mm and 8 mm) with R^2 of more than 0.9990 and lowest RMSE of less than 0.0010 (Tables 6 and 7). For example, the validation of the logarithmic model by comparing the predicted moisture data and those obtained from the experiments is shown in Figure 5. The moisture ratio data predicted using the logarithmic model lied closely along a straight regression line for different drying conditions indicating the suitability of the model for describing the VD and HAD behaviors of persimmon fruit samples. Onwude et al. [61] also reported the adequacy of page, logarithmic, Midilli et al. and Hii et al. models for predicting the drying kinetics of sweet potato. Similarly, Younis et al. [62] indicated the appropriateness of page, logarithmic, Midilli et al. and Hii et al. models for describing the drying performance of garlic slices.

Table 4. Statistical evaluation of the mathematical drying models for persimmon samples of 5 mm thickness for VD.

Drying Temperature (°C)	Model No.	Model Parameters	R^2	RMSE
50	1	$k = 0.2491$	0.9577	0.0682
	2	$k = 0.1224$, n = 1.4734	0.9940	0.2559
	3	$k = 0.2259$, n = 1.1031	0.9577	0.0762
	4	$a = 1.5532$, $k = 0.1229$, $c = -0.5369$	0.9980	0.0146
	5	$a = 0.5606$, $k_1 = 0.2663$, $b = 0.5108$, $k_2 = 0.2663$	0.9650	0.0620
	6	a = 0.6224, $k_0 = 0.2491$, $k_1 = 0.4003$	0.9577	0.0815
	7	$a = 1.0713$, $k = 0.2663$	0.9650	0.0693
	8	$a = 0.3715$, $k = 0.2663$, $b = 0.3610$, $g = 0.2663$, $c = 0.3388$, $h = 0.2663$	0.9650	0.0620
	9	$a = 1.0177$, $k = 0.1639$, $b = -0.0287$	0.9976	0.0194
	10	$a = 0.4916$, $k_1 = 0.1068$, $b = 0.4821$, $k_2 = 0.1068$, n = 1.5424	0.9948	0.0339
60	1	$k = 0.4503$	0.9821	0.0457
	2	$k = 0.3292$, n = 1.3069	0.9960	0.0217
	3	$k = 0.4097$, n = 1.0990	0.9821	0.0527
	4	$a = 1.1435$, $k = 0.3445$, $c = -0.1331$	0.9964	0.0206
	5	$a = 0.5347$, $k_1 = 0.4632$, $b = 0.4988$, $k_2 = 0.4632$	0.9836	0.0438
	6	$a = 0.6046$, $k_0 = 0.4503$, $k_1 = 0.7448$	0.9821	0.0577
	7	$a = 1.0335$, $k = 0.4632$	0.9836	0.0505
	8	$a = 0.3531$, $k = 0.4632$, $b = 0.3503$, $g = 0.4632$, $c = 0.3300$, $h = 0.4632$	0.9836	0.0438
	9	$a = 1.0125$, $k = 0.3867$, $b = -0.0143$	0.9955	0.0289
	10	$a = 0.4936$, $k_1 = 0.3194$, $b = 0.4955$, $k_2 = 0.3194$, n = 1.3254	0.9961	0.0348
70	1	$k = 0.6145$	0.9957	0.0220
	2	$k = 0.5871$, n = 1.0608	0.9963	0.0205
	3	$k = 0.5954$, n = 1.0320	0.9957	0.0260
	4	$a = 1.0325$, $k = 0.5613$, $c = -0.0345$	0.9974	0.0173
	5	$a = 0.5045$, $k_1 = 0.6171$, $b = 0.5003$, $k_2 = 0.6171$	0.9958	0.0219
	6	$a = 0.6832$, $k_0 = 0.6145$, $k_1 = 0.8993$	0.9957	0.0290
	7	$a = 1.0049$, $k = 0.6171$	0.9958	0.0259
	8	$a = 0.3329$, $k = 0.6171$, $b = 0.3408$, $g = 0.6171$, $c = 0.3312$, $h = 0.6171$	0.9958	0.0219
	9	$a = 0.9985$, $k = 0.5811$, $b = -0.0054$	0.9975	0.0223
	10	$a = 0.4869$, $k_1 = 0.5862$, $b = 0.5122$, $k_2 = 0.5862$, $n = 1.0617$	0.9963	0.0383

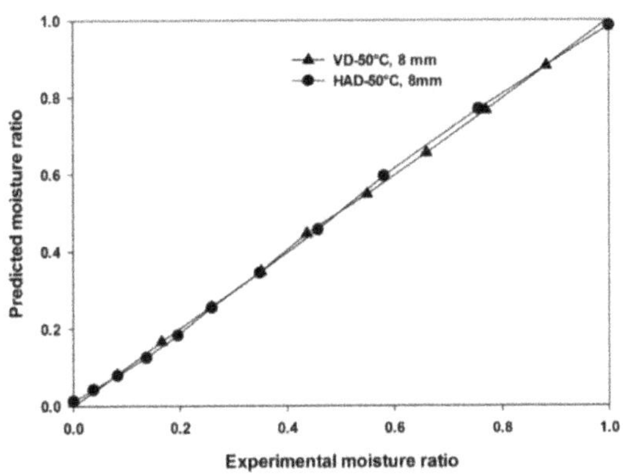

Figure 5. Predicted versus experimental moisture ratio data for logarithmic model at 50 °C and 8 mm samples thickness for VD and HAD methods.

Table 5. Statistical evaluation of the mathematical drying models for persimmon samples of 8 mm thickness for VD.

Drying Temperature (°C)	Model No.	Model Parameters	R^2	RMSE
50	1	$k= 0.1854$	0.9420	0.0765
	2	$k = 0.0754$, $n = 1.5266$	0.9878	0.0350
	3	$k = 0.1702$, $n = 1.0897$	0.9420	0.0846
	4	$a = 2.8487, k = 0.0435, c = -1.8449$	0.9998	0.0040
	5	$a = 0.5626, k_1 = 0.2010, b = 0.5161, k_2 = 0.2010$	0.9522	0.0694
	6	$a = 0.5761, k_0 = 0.1854, k_1 = 0.3219$	0.9420	0.0897
	7	$a = 1.0787, k = 0.2010$	0.9522	0.0768
	8	$a = 0.3734, k = 0.2010, b = 0.3626, g = 0.2010, c = 0.3427, h = 0.2010$	0.9522	0.0694
	9	$a = 1.0046, k = 0.0795, b = -0.0456$	0.9998	0.0050
	10	$a = 0.4864, k_1 = 0.0580, b = 0.4751, k_2 = 0.0580, n = 1.6487$	0.9896	0.0439
60	1	$k = 0.2355$	0.9579	0.0667
	2	$k = 0.1183, n = 1.4480$	0.9921	0.0288
	3	$k = 0.2152, n = 1.0947$	0.9579	0.0746
	4	$a = 1.6897, k = 0.1027, c = -0.6819$	0.9994	0.0076
	5	$a = 0.5557, k_1 = 0.2512, b = 0.5109, k_2 = 0.2512$	0.9646	0.0611
	6	$a = 0.6343, k_0 = 0.2355, k_1 = 0.3713$	0.9579	0.0797
	7	$a = 1.0665, k = 0.2512$	0.9646	0.0683
	8	$a = 0.3682, k = 0.2512, b = 0.3594, g = 0.2512, c = 0.3389, h = 0.2512$	0.9646	0.0611
	9	$a = 1.0092, k = 0.1438, b = -0.0323$	0.9992	0.0107
	10	$a = 0.4885, k_1 = 0.1009, b = 0.4817, k_2 = 0.1009, n = 1.5277$	0.9931	0.0382
70	1	$k = 0.3601$	0.9736	0.0548
	2	$k = 0.2336, n = 1.3657$	0.9955	0.0228
	3	$k = 0.3303, n = 1.0903$	0.9736	0.0633
	4	$a = 1.3017, k = 0.2238, c = -0.2918$	0.9987	0.0120
	5	$a = 0.5419, k_1 = 0.3756, b = 0.5052, k_2 = 0.3756$	0.9768	0.0513
	6	$a = 0.6891, k_0 = 0.3601, k_1 = 0.5225$	0.9736	0.0693
	7	$a = 1.0470, k = 0.3756$	0.9768	0.0593
	8	$a = 0.3581, k = 0.3756, b = 0.3542, g = 0.3756, c = 0.3347, h = 0.3756$	0.9768	0.0513
	9	$a = 1.0117, k = 0.2714, b = -0.0249$	0.9983	0.0177
	10	$a = 0.4921, k1 = 0.2225, b = 0.4942, k_2 = 0.2225, n = 1.3937$	0.9957	0.0363

Table 6. Statistical evaluation of the mathematical drying models for persimmon samples of 5 mm thickness for HAD.

Drying Temperature (°C)	Model No.	Model Parameters	R^2	RMSE
50	1	$k = 0.5737$	0.9986	0.0120
	2	$k = 0.5514, n = 1.0481$	0.9990	0.0103
	3	$k = 0.5582, n = 1.0277$	0.9986	0.0139
	4	$a = 1.0228, k = 0.5349, c = -0.0255$	0.9998	0.0047
	5	$a = 0.5027, k_1 = 0.5753, b = 0.5005, k_2 = 0.5753$	0.9987	0.0120
	6	$a = 0.6842, k_0 = 0.5737, k_1 = 0.8384$	0.9986	0.0152
	7	$a = 1.0032, k = 0.5753$	0.9987	0.0138
	8	$a = 0.3315, k = 0.5753, b = 0.3405, g = 0.5753, c = 0.3313, h = 0.5753$	0.9987	0.0120
	9	$a= 0.9980, k = 0.5505, b= -0.0035$	0.9998	0.0059
	10	$a = 0.4894, k_1 = 0.5493, b = 0.5083, k_2 = 0.5493, n= 1.0503$	0.9990	0.0167
60	1	$k = 0.8838$	0.9957	0.0235
	2	$k = 0.7945, n = 1.2519$	0.9998	0.0047
	3	$k = 0.8213, n = 1.0762$	0.9957	0.0288
	4	$a = 1.0409, k = 0.8064, c = -0.0359$	0.9982	0.0152
	5	$a = 0.5142, k_1 = 0.8898, b = 0.4950, k_2 = 0.8898$	0.9958	0.0232
	6	$a = 0.6039, k_0 = 0.8838, k_1 = 1.4636$	0.9957	0.0333
	7	$a = 1.0092, k = 0.8898$	0.9958	0.0284
	8	$a = 0.3401, k= 0.8898, b = 0.3409, g = 0.8898, c = 0.3282, h = 0.8898$	0.9958	0.0232
	9	$a = 1.0058, k = 0.8451, b = -0.0064$	0.9978	0.0240
	10	$a = 0.4975, k_1 = 0.7941, b = 0.5022, k_2 = 0.7941, n = 1.2522$	0.9998	0.0117
70	1	$k = 0.9781$	0.9943	0.0284
	2	$k = 0.8745, n = 1.3348$	0.9999	0.0031
	3	$k = 0.8897, n = 1.0994$	0.9943	0.0367
	4	$a = 1.0564, k = 0.8581, c = -0.0522$	0.9979	0.0171
	5	$a = 0.5158, k_1 = 0.9840, b = 0.4928, k_2 = 0.9840$	0.9944	0.0281
	6	$a = 0.5740, k_0 = 0.9781, k_1 = 1.7039$	0.9943	0.0449
	7	$a = 1.0086, k = 0.9840$	0.9944	0.0363
	8	$a = 0.3409, k = 0.9840, b = 0.3408, g = 0.9840, c = 0.3270, h = 0.9840$	0.9944	0.0281
	9	$a = 1.0047, k = 0.9118, b = -0.0110$	0.9974	0.0301
	10	$a = 0.4973, k_1 = 0.8743, b = 0.5026, k_2 = 0.8743, n = 1.3349$	0.9999	-

Table 7. Statistical evaluation of the mathematical drying models for persimmon samples of 8 mm thickness for HAD.

Drying Temperature (°C)	Model No.	Model Parameters	R^2	RMSE
50	1	$k = 0.2797$	0.9925	0.0261
	2	$k = 0.2448$, n = 1.0909	0.9946	0.0225
	3	$k = 0.2702$, n = 1.0350	0.9925	0.0292
	4	$a = 1.0943$, $k = 0.2204$, $c = -0.1083$	0.9989	0.0099
	5	$a = 0.5092$, $k_1 = 0.2828$, $b = 0.5022$, $k_2 = 0.2828$	0.9927	0.0261
	6	$a = 0.6403$, $k_0 = 0.2797$, $k_1 = 0.4368$	0.9925	0.0310
	7	$a = 1.0114$, $k = 0.2828$	0.9927	0.0288
	8	$a = 0.3350$, $k= 0.2828$, $b = 0.3444$, $g = 0.2828$, $c = 0.3320$, $h = 0.2828$	0.9927	0.0261
	9	$a = 0.9876$, $k = 0.2407$, $b = -0.0080$	0.9992	0.0100
	10	$a = 0.4845$, $k_1 = 0.2343$, $b = 0.5018$, $k_2 = 0.2343$, n = 1.1112	0.9948	0.0299
60	1	$k = 0.3411$	0.9622	0.0670
	2	$k = 0.1887$, n = 1.4915	0.9979	0.0157
	3	$k = 0.3134$, n = 1.0882	0.9622	0.0774
	4	$a = 1.3969$, $k = 0.1992$, $c = -0.3693$	0.9956	0.0229
	5	$a = 0.5562$, $k_1 = 0.3624$, $b = 0.5121$, $k_2 = 0.3624$	0.9689	0.0608
	6	$a = 0.6258$, $k_0 = 0.3411$, $k_1 = 0.5450$	0.9622	0.0847
	7	$a = 1.0684$, $k = 0.3624$	0.9689	0.0702
	8	$a = 0.3669$, $k = 0.3624$, $b = 0.3610$, $g = 0.3624$, $c = 0.3405$, $h= 0.3624$	0.9689	0.0608
	9	$a = 1.0289$, $k= 0.2489$, $b = -0.0294$	0.9949	0.0313
	10	$a = 0.4986$, $k_1 = 0.1849$, $b = 0.4961$, $k_2 = 0.1849$, n = 1.5034	0.9980	0.0254
70	1	$k = 0.5139$	0.9942	0.0255
	2	$k = 0.4559$, n = 1.1338	0.9972	0.0177
	3	$k = 0.4882$, n = 1.0527	0.9942	0.0302
	4	$a = 1.0830$, $k = 0.4234$, $c = -0.0857$	0.9998	0.0044
	5	$a = 0.5125$, $k_1 = 0.5195$, $b = 0.4998$, $k_2= 0.5195$	0.9944	0.0251
	6	$a = 0.6353$, $k_0 = 0.5139$, $k_1 = 0.8089$	0.9942	0.0338
	7	$a = 1.0123$, $k = 0.5195$	0.9944	0.0296
	8	$a = 0.3385$, $k = 0.5195$, $b = 0.3431$, $g = 0.5195$, $c = 0.3307$, $h = 0.5195$	0.9944	0.0251
	9	$a = 0.9984$, $k = 0.4570$, $b = -0.0112$	0.9999	0.0053
	10	$a = 0.4915$, $k_1 = 0.4516$, $b = 0.5040$, $k_2= 0.4516$, n= 1.1393	0.9972	0.0329

3.5. Results of Artificial Neural Network

The parameters namely time, temperature and samples thickness were used to predict moisture ratio using ANN model for VD and HAD techniques. Tables 8 and 9 show the statistical results with respect to training and validation of the multilayer feed-forward network structure of samples drying data for VD and HAD. The training data set were used to assess the optimum number of neurons and hidden layers for multilayer neural network modelling for determining the best predictive power. For VD, it was found that the architecture with 1 and 2 hidden layers with 9 and 12 (6, 6 neurons), obtained the best results for training set ($R^2 = 0.9991$) and test set ($R^2 = 0.9881$) as compared to those of 1 hidden layer (3 and 6 neurons), 2 hidden layers (6 and 18 neurons) and 3 hidden layers (9, 18 and 27 neurons), respectively (Table 8). Moreover, the networks were found to be susceptible to the number of neurons in their hidden layers. Thus, smaller neurons led to under fitting, while too many neurons contributed to overfitting. In addition, Figure 6 showed the predicted and experimental moisture ratio values for the optimal ANN for training and testing sets.

Table 9 presented the statistical results of drying kinetics of persimmon samples for the ANN model using HAD. Similar to VD, the results showed that the architecture with 1 and 2 hidden layers with 9 and 6 (3, 3) neurons, obtained the best results as compared to those of 1 hidden layer (3 and 6 neurons), 2 hidden layers (12 and 18 neurons) and 3 hidden layers (9, 18 and 27 neurons), respectively. The highest result of R^2 values found for training and testing set were 0.9994 and 0.9979, respectively. Therefore, to compare the results obtained from the thin-layer mathematical models and ANN results, it can be seen that the ANN model is very close to the highest model in the theoretical mathematical models with R^2 and RMSE values of 0.9994 and 0.0124, respectively as compared to those of the logarithmic model ($R^2 = 0.9998$ and RMSE = 0.0044). Additionally, the predicted and experimental moisture ratio values for the optimal ANN for training and testing data sets are illustrated in Figure 7.

Zenoozian et al. [63] demonstrated that 1 and 2 hidden layers and 30 neurons for ANN adequately predicted the moisture changes of pumpkin during osmotic dehydration. ANN with 1 and 2 hidden layers have also been successful in predicting the drying behavior of other fruits and vegetables such as pepper, apple slices, mushroom during microwave-vacuum drying [36,64,65].

Table 8. Statistical results of drying kinetics of persimmon fruit samples for the artificial neural network (ANN) model using VD.

No. Hidden Layer	No. Neurons	Training		Testing	
		R^2	RMSE	R^2	RMSE
1	3	0.9979	0.0249	0.9791	0.0696
1	6	0.9990	0.0245	0.9794	0.0692
1	9	0.9991	0.0213	0.9812	0.0671
2	3, 3	0.9981	0.0209	0.9820	0.0658
2	6, 6	0.9978	0.0237	0.9881	0.0572
2	9, 9	0.9976	0.0312	0.9848	0.0800
3	3, 3, 3	0.9980	0.0275	0.9803	0.0704
3	6, 6, 6	0.9982	0.0264	0.9854	0.0576
3	9, 9, 9	0.9987	0.0269	0.9866	0.0661

Table 9. Statistical results of drying kinetics of persimmon samples for the ANN model using HAD.

No. Hidden Layer	No. Neurons	Training		Testing	
		R^2	RMSE	R^2	RMSE
1	3	0.9952	0.0655	0.9885	0.0557
1	6	0.9962	0.0296	0.9882	0.0878
1	9	0.9992	0.0183	0.9979	0.0351
2	3, 3	0.9994	0.0124	0.9983	0.0281
2	6, 6	0.9975	0.0529	0.9943	0.0605
2	9, 9	0.9989	0.0297	0.9979	0.0459
3	3, 3, 3	0.9518	0.1348	0.8025	0.3419
3	6, 6, 6	0.9969	0.0344	0.9947	0.0649
3	9, 9, 9	0.9983	0.0233	0.9952	0.0431

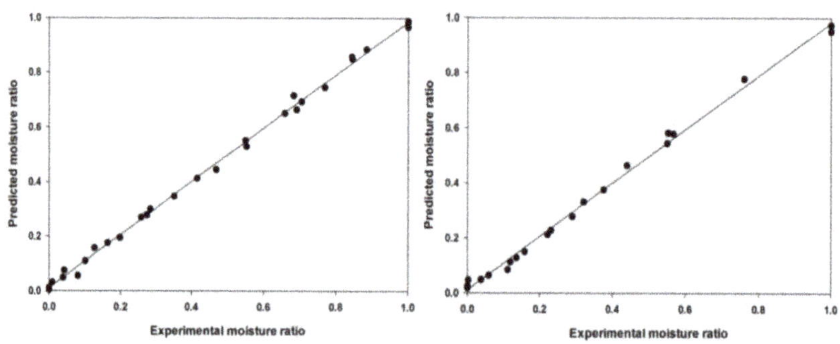

Figure 6. Predicted and experimental moisture ratio using VD from training and testing data set of ANN model.

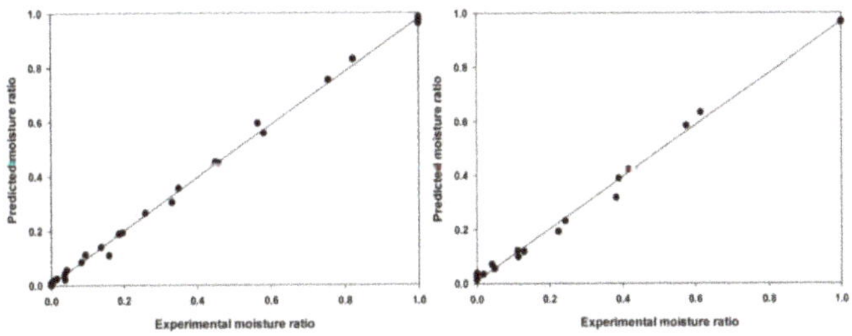

Figure 7. Predicted and experimental moisture ratio using HAD from training and testing data set of ANN.

3.6. Results of Support Vector Machine

The statistical results related to the training and validation of the SVM of persimmon drying experimental data using VD and HAD are given in Tables 10 and 11. Similar to the ANN model, the training data set was used to evaluate the best filter and kernel type modeling for determining the best predictive power. For VD, the training data set at standardize filter with Pearson universal kernel type found the best result of R^2 and RMSE values of 1.0000 and 0.0004 as compared to those of normalize with polynomial, Pearson universal kernel and RBF kernel and also standardize with polynomial and RBF kernel, respectively. For the testing data set, standardize filter with polynomial kernel type obtained the best results for R^2 and RMSE values of 0.9996 and 0.1213 (Table 10).

Table 10. Statistical results of drying kinetics of persimmon samples for SVM model using VD method.

Filter Type	Kernel Type	Training		Testing	
		R^2	RMSE	R^2	RMSE
Normalize	Polynomial kernel	0.9672	0.0904	0.9234	0.1339
Normalize	Pearson universal kernel	1.0000	0.0014	0.9564	0.1041
Normalize	RBF kernel	0.9998	0.0067	0.9575	0.1000
Standardize	Polynomial kernel	0.9672	0.0903	0.9996	0.1213
Standardize	Pearson universal kernel	1.0000	0.0004	0.9258	0.2174
Standardize	RBF kernel	0.9999	0.0040	0.9577	0.1042

Table 11. Statistical results of drying kinetics of persimmon samples for SVM model using HAD method.

Filter Type	Kernel Type	Training		Testing	
		R^2	RMSE	R^2	RMSE
Normalize	Polynomial kernel	0.8712	0.1697	0.8797	0.1612
Normalize	Pearson universal kernel	0.9996	0.0103	0.9680	0.0862
Normalize	RBF kernel	0.9980	0.0233	0.9674	0.0871
Standardize	Polynomial kernel	0.9339	0.1337	0.8797	0.1612
Standardize	Pearson universal kernel	1.0000	0.0005	0.8933	0.2308
Standardize	RBF kernel	1.0000	0.0004	0.9690	0.0912

However, for HAD, the standardize filter type with RBF kernel obtained the best results for training and testing data set of R^2 and RMSE values of 1.0000, 0.0004, 0.9690 and 0.0912 respectively (Table 11). From the results, it is clear that the SVM model showed the highest results as compared to the theoretical mathematical models and also ANN (Tables 7 and 9). Few studies used SVM as a model in drying techniques. Das and Akpinar [50] applied SVM to investigate pear drying performance by different ways of convective heat transfer. The authors applied normalization and standardization filter to the target attribute with three kernel models (polynomial kernel, Pearson universal kernel and RBF kernel). They found that the polynomial kernel showed the lowest RMSE of value 0.3351.

This indicates that the applied algorithm for SVM prediction for the moisture ratio during the drying process for both VD and HAD techniques can be promised.

3.7. Results of k-Nearest Neighbors

Tables 12 and 13 summarized the R^2 and RMSE of kNN tool developed with training and testing data set for VD and HAD methods. For VD method, the prediction model developed using a k-value of 3, produced the highest results for training and testing data set of R^2 values of 0.9327 and 0.8782 and lowest RMSE values of 0.1271 and 0.1829 respectively. Meanwhile, k-values of 11 and 9 indicated the lowest R^2 values of 0.8355 and 0.5638 and the highest RMSE values of 0.2214 and 0.6399 for training and testing data set, respectively (Table 12). For HAD technique, the results of the training set showed R^2 values between 0.6347 and 0.8638 and RMSE values between 0.2145 and 0.2907. The k-value of 7 showed the highest result compared to other k-values (3, 5, 9 and 11). The prediction model developed using the testing data set indicated R^2 values of ranging from 0.4877 to 0.7873 and RMSE values from 0.2385 to 0.3099. The k-value of 5 showed the highest values (Table 13).

Table 12. Statistical results of drying kinetics of persimmon samples for k nearest neighbors (kNN) model using VD method.

k	Training		Testing	
	R^2	RMSE	R^2	RMSE
3	0.9327	0.1271	0.8782	0.1829
5	0.9209	0.1548	0.7881	0.2320
7	0.8969	0.1730	0.6123	0.2799
9	0.8383	0.2059	0.5638	0.2877
11	0.8355	0.2214	0.6399	0.6399

Table 13. Statistical results of drying kinetics of persimmon for kNN model using HAD method.

k	Training		Testing	
	R^2	RMSE	R^2	RMSE
3	0.7923	0.2143	0.6819	0.2510
5	0.8135	0.2253	0.7873	0.2385
7	0.8638	0.2320	0.7160	0.2673
9	0.6347	0.2860	0.4877	0.3099
11	0.7101	0.2907	0.7484	0.2955

3.8. Comparison between Computational Intelligence and Mathematical Thin-Layer Models

The highest results obtained from the computational intelligence models (ANN, SVM and kNN) and the top three mathematical thin-layer models (page, logarithmic and Midilli et al.) for moisture ratio prediction are summarized in Table 14. The R^2 value of 0.9991 and RMSE value of 0.0213 at 1 hidden layer with 9 neurons by applying ANN model showed the best results for VD method. For the HAD method applying ANN model, the R^2 value of 0.9994 and RMSE value of 0.0124 were found. For kNN, the results of R^2 and RMSE were 0.9327, 0.1271, 0.8638 and 0.2320 for VD and HAD respectively. On the other hand, the mathematical thin-layers modelling found the range of R^2 values from 0.9963 to 0.9999 and RMSE from 0.0205 to 0.0031 for both VD and HAD methods.

Table 14. Statistical results of drying kinetics of persimmon samples for computational intelligence and mathematical thin-layer models using VD and HAD.

Model		VD		HAD	
		R^2	RMSE	R^2	RMSE
Computational intelligence	ANN	0.9991	0.0213	0.9994	0.0124
	SVM	1.0000	0.0004	1.0000	0.0004
	kNN	0.9327	0.1271	0.8638	0.2320
Mathematical model	Page	0.9963	0.0205	0.9999	0.0031
	Logarithmic	0.9998	0.0040	0.9998	0.0047
	Midilli et al.	0.9998	0.0050	0.9999	0.0053

It can be seen that the prediction of moisture ratio with the model developed using SVM gave the highest R^2 and the lowest RMSE values of 1.0000 and 0.0004 compared to the models developed using other computational intelligence methods (ANN and kNN) and mathematical thin-layers models; page, logarithmic and Midilli et al. (Table 14). The results indicate that SVM gave the best results which are in agreement with the findings of prior studies [19,66] that used SVM as a data prediction method to improve outcomes. In summary, the results from this study demonstrate the application of computational intelligence methods for the drying processes of persimmon samples at different conditions. Thus, applying computational intelligence methods in drying generally improves the drying performance by controlling the drying input parameters for optimizing energy, quality and production cost.

3.9. Results of Color Measurements

The color properties of persimmon samples of 5 mm and 8 mm thickness for VD and HAD methods were determined based on the change in color parameters of L *, a * and b * and the total color change ΔE as given in Table 15. It can be seen clearly from Table 4 that the lightness (L *) of persimmon samples dried using VD and HAD methods decreased significantly ($P \leq 0.05$) compared to the fresh samples. However, there was no significant difference between the lightness of samples using HAD-60 for 5 mm and HAD-60 and HAD-70 for 8mm, with those of the fresh samples. The VD method significantly ($P \leq 0.05$) reduced the lightness of samples compared to HAD. The VD-70 for 8 mm showed the lowest lightness value of 33.479 ± 6.888. The redness/greenness (a *) of all dried persimmon samples reduced significantly ($P \leq 0.05$) compared to fresh samples, except VD-50, which resulted in higher a* value than the fresh. Generally, the values of redness/greenness (a *) found using HAD were lower than the values found using VD. The lowest value of redness/greenness (a *) was found to be 6.032 ± 1.740 using HAD-60. Similar results were found for the yellowness/blueness (b *), where all dried persimmon samples were reduced significantly ($P \leq 0.05$) compared to fresh samples, except HAD-60, which resulted in higher b* than the fresh. In addition, there was no significant difference between the b* fresh samples with those of dried samples using VD-50, VD-60, VD-70 for 5 mm; VD-60 for 8 mm; HAD-50, HAD-60, HAD-70 for 5 mm and HAD-50, HAD-60, HAD-70 for 8 mm. Dried samples using VD-50 and VD-70 methods for 8 mm showed different groups of values of 45.250 ± 2.542 and 38.733 ± 6.948 respectively. The values of total color change using VD method showed higher values compared to the HAD method for the persimmon samples. The highest value was found using VD-70 for 8 mm of 38.733 ± 6.948. Generally, the change of color properties of the fresh samples under different drying methods is due to the increased sample temperature resulting from increased enzymatic and non-enzymatic chemical reactions of the product [67,68].

Table 15. Drying methods and color parameters of persimmon fruit samples.

Thickness (mm)	Drying Method	Color Properties for Different Sample Thickness			
		L *	a *	b *	ΔE
5	Fresh	61.425 ± 2.533 ab	26.282 ± 0.747 b	65.698 ± 2.126 ab	-
	VD-50	52.560 ± 3.680 bc	20.69 ± 1.95bc d	56.040 ± 3.410 b	17.790 ± 4.100 bcd
	VD-60	56.161 ± 3.476 bc	16.280 ± 3.598 de	61.073 ± 2.834 ab	15.859 ± 1.822 bcd
	VD-70	59.646 ± 0.413 bc	12.804 ± 0.764 e	62.265 ± 0.076 ab	14.568 ± 0.669 cd
8	VD-50	39.637 ± 2.554 d	31.815 ± 1.901 a	45.250 ± 2.542 c	27.069 ± 3.852 ab
	VD-60	50.341 ± 2.610 c	15.736 ± 0.868 de	55.656 ± 2.020 b	16.624 ± 1.715 bcd
	VD-70	33.479 ± 6.888 d	22.842 ± 0.784 bc	38.733 ± 6.948 c	35.875 ± 9.638 a
5	HAD-50	56.930 ± 1.152 bc	23.438 ± 0.258 b	61.762 ± 1.088 ab	9.528 ± 1.568 d
	HAD-60	60.250 ± 2.229 ab	11.684 ± 0.259 e	64.831 ± 2.231 ab	14.928 ± 0.680 bcd
	HAD-70	59.846 ± 1.149 bc	17.259 ± 1.164 cde	64.491 ± 0.750 ab	10.140 ± 0.405 d
8	HAD-50	56.301 ± 1.562 bc	20.571 ± 2.890 bcd	60.907 ± 0.971 ab	8.817 ± 0.789 d
	HAD-60	60.310 ± 1.478 ab	6.032 ± 1.740 f	69.432 ± 0.556 a	24.618 ± 2.301 abc
	HAD-70	60.781 ± 1.973 ab	11.541 ± 0.798 e	64.015 ± 1.325 ab	15.959 ± 0.876 bcd

Different letters at the same column indicates statistical difference for Duncan test, $P < 0.05$.

4. Conclusions

This study investigated the potential of using computational intelligence as a modelling tool for predicting the drying process of persimmon fruit samples. The performance of vacuum drying (VD) and hot-air-drying (HAD) techniques for drying persimmon fruit samples was evaluated. The results showed that VD and HAD had a significant effect on the drying kinetics, moisture diffusivity, activation energy and color properties of persimmon fruit samples. An increase in drying temperature and samples thickness influenced the drying kinetics and moisture diffusivity of samples. The effective moisture diffusivity and activation energy varied between 1.417×10^{-9} m^2/s and 1.925×10^{-8} m^2/s and 34.1560 kJ/mol to 64.2895 kJ/mol respectively. Persimmon fruit samples using HAD showed significant color attributes compared to VD method.

The mathematical thin-layer modelling results showed that page and logarithmic models can adequately ($R^2 = 0.9999$) describe the drying kinetics of persimmon fruit samples. The highest R^2 values of 0.9994, 1.0000 and 0.9327 were observed for ANN (2 hidden layers with (3, 3) neurons), SVM (standardize filter and RBF kernel) and kNN (k-value of 3) models, respectively. SVM tool as a computational intelligence method produced higher results compared to mathematical thin-layer. Therefore, SVM models are able to describe a wider range of experimental data whereas the application of theoretical models is limited to specific experimental conditions in most cases. Thus, computational intelligence models may be considered as a suitable alternative modelling method for describing the drying behavior of persimmon sliced samples. On the other hand, computational intelligence methods can be successfully applied to industrial drying processes and operations as well as online monitoring and control. However, further study is required to ascertain the suitability of ANN, SVM and kNN for predicting the nutritional composition of fruits and vegetables during drying.

Author Contributions: Conceptualization, A.Y.K. and D.H.; methodology, A.Y.K.; software, A.Y.K.; validation and investigation, A.K., K.Ç.S.; formal analysis, A.Y.K.; resources, D.H.; writing—original draft preparation, A.Y.K.; writing—review and editing, A.K., K.Ç.S., Č.M. and P.H.; supervision and project administration, D.H. All authors have read and agreed to the published version of the manuscript.

Funding: The research was funded by EU, Managing Authority of the Czech Operational Programme Research, Development and Education through the project "supporting the development of international mobility of research staff at CULS Prague", Grant Number: CZ.02.2.69/0.0/0.0/16_027/0008366. The APC was funded through the project "supporting the development of international mobility of research staff at CULS Prague".

Conflicts of Interest: The authors declare no conflict of interest.

References

1. Bozkir, H.; Rayman, A.; Serdar, E.; Metin, G.; Baysal, T. Ultrasonics—Sonochemistry influence of ultrasound and osmotic dehydration pretreatments on drying and quality properties of persimmon fruit. *Ultrason.-Sonochem.* **2019**, *54*, 135–141. [CrossRef] [PubMed]
2. Çalışkan, G.; Nur Dirim, E. Freeze drying kinetics of persimmon puree. *GIDA* **2015**, *40*, 9–14.
3. Heras, R.M.-L.; Landines, E.F.; Heredia, A.; Castelló, M.L.; Andrés, A. Influence of drying process and particle size of persimmon fibre on its physicochemical, antioxidant, hydration and emulsifying properties. *J. Food Sci. Technol.* **2017**, *54*, 2902–2912. [CrossRef]
4. Bolek, S.; Obuz, E. Quality characteristics of Trabzon persimmon dried at several temperatures and pretreated by different methods. *Turk. J. Agric. For.* **2014**, *38*, 242–249. [CrossRef]
5. Doymaz, I. Evaluation of some thin-layer drying models of persimmon slices (Diospyros kaki L). *Energy Convers. Manag.* **2012**, *56*, 199–205. [CrossRef]
6. Senthilkumar, T.; Jayas, D.S.; White, N.D.; Fields, P.; Gräfenhan, T. Detection of fungal infection and Ochratoxin A contamination in stored barley using near-infrared hyperspectral imaging. *Biosyst. Eng.* **2016**, *147*, 162–173. [CrossRef]
7. Rodríguez, J.; Clemente, G.; Sanjuan, N.; Bon, J. Modelling drying kinetics of thyme (Thymus vulgaris L.): Theoretical and empirical models, and neural networks. *Food Sci. Technol. Int.* **2013**, *20*, 13–22. [CrossRef]
8. Onwude, D.; Hashim, N.; Abdan, K.; Janius, R.; Chen, G. The potential of computer vision, optical backscattering parameters and artificial neural network modelling in monitoring the shrinkage of sweet potato (Ipomoea batatas L.) during drying. *J. Sci. Food Agric.* **2018**, *98*, 1310–1324. [CrossRef]
9. Si, X.; Wu, X.; Yi, J.Y.; Li, Z.; Chen, Q.; Bi, J.; Zhou, L. Comparison of different drying methods on the physical properties, bioactive compounds and antioxidant activity of raspberry powders. *J. Sci. Food Agric.* **2015**, *96*, 2055–2062. [CrossRef]
10. Tekin, Z.H.; Baslar, M. The effect of ultrasound-assisted vacuum drying on the drying rate and quality of red peppers. *J. Therm. Anal. Calorim.* **2018**, *132*, 1131–1143. [CrossRef]
11. Onwude, D.I.; Hashim, N.; Janius, R.B.; Nawi, N.; Abdan, K. Evaluation of a suitable thin layer model for drying of pumpkin under forced air convection. *Int. Food Res. J.* **2016**, *23*, 1173.
12. Karasu, S.; Akcicek, A.; Kayacan, S. Effects of different drying methods on drying kinetics, microstructure, color, and the rehydration ratio of minced meat. *Foods* **2019**, *8*, 216.
13. Bai, J.; Xiao, H.; Ma, H.; Zhou, C. Artificial neural network modeling of drying kinetics and color changes of ginkgo biloba seeds during microwave drying process. *J. Food Qual.* **2018**, *2018*, 1–8. [CrossRef]
14. Behroozi, N.; Tavakoli, T.; Ghassemian, H. Applied machine vision and artificial neural network for modeling and controlling of the grape drying process. *Comput. Electron. Agric.* **2013**, *98*, 205–213. [CrossRef]
15. Akoy, E.O.M. Experimental characterization and modeling of thin-layer drying of mango slices. *Int. Food Res. J.* **2014**, *21*, 7–11.
16. Erbay, Z.; Icier, F. A review of thin layer drying of foods: Theory, a review of thin layer drying of foods: Theory, modeling, and experimental results. *Food Sci. Nutr.* **2010**, *50*, 441–464.
17. Lee, G.; Hsieh, F. Thin-layer drying kinetics of strawberry fruit leather. *Trans. ASABE* **2008**, *51*, 1699–1705. [CrossRef]
18. Liu, Z.; Bai, J.; Yang, W.; Wang, J.; Deng, L. Effect of high-humidity hot air impingement blanching (HHAIB) and drying parameters on drying characteristics and quality of broccoli florets. *Dry. Technol.* **2019**, *37*, 1251–1264. [CrossRef]
19. Khaled, A.Y.; Aziz, S.A.; Bejo, S.K.; Nawi, N.M.; Abu Seman, I.A. Spectral features selection and classification of oil palm leaves infected by Basal stem rot (BSR) disease using dielectric spectroscopy. *Comput. Electron. Agric.* **2018**, *144*, 297–309. [CrossRef]
20. Zhang, R.; Ma, J. Feature selection for hyperspectral data based on recursive support vector machines. *Int. J. Remote Sens.* **2009**, *30*, 3669–3677. [CrossRef]
21. Xu, G.; Shen, C.; Liu, M.; Zhang, F.; Shen, W. A user behavior prediction model based on parallel neural network and k-nearest neighbor algorithms. *Cluster Comput.* **2017**, *20*, 1703–1715. [CrossRef]
22. Kirbas, I.; Tuncer, A.D.; Şirin, C.; Usta, H. Modeling and developing a smart interface for various drying methods of pomelo fruit (Citrus maxima) peel using machine learning approaches. *Comput. Electron. Agric.* **2019**, *165*, 104928. [CrossRef]

23. Omari, A.; Behroozi-Khazaei, N.; Sharifian, F. Drying kinetic and artificial neural network modeling of mushroom drying process in microwave-hot air dryer. *Food Process Eng.* **2018**, *41*, 1–10. [CrossRef]
24. Movagharnejad, K.; Nikzad, M. Modeling of tomato drying using artificial neural network. *Comput. Electron. Agric.* **2007**, *59*, 78–85. [CrossRef]
25. Wen, S.; Deng, M.; Inoue, A. Moisture content prediction of wood drying process using VM-based model. *Int. J. Inov. Comput. Infomation Control* **2012**, *8*, 4083–4093.
26. Beigi, M.; Ahmadi, I. Artificial neural networks modeling of kinetic curves of celeriac (Apium graveolens L) in vacuum drying. *Food Sci. Technol.* **2018**, *2061*, 1–6. [CrossRef]
27. Jafari, S.M.; Ghanbari, V.; Ganje, M.; Dehnad, D. Modeling the drying kinetics of green bell pepper in a heat pump assisted fluidized bed dryer. *J. Food Qual.* **2016**, *39*, 98–108. [CrossRef]
28. Bahmani, A.; Jafari, S.M.; Shahidi, S.-A.; Dehnad, D. Mass transfer kinetics of eggplant during osmotic dehydration by neural networks. *J. Food Process. Preserv.* **2016**, *40*, 815–827. [CrossRef]
29. Xie, C.; Li, X.; Shao, Y.; He, Y. Color measurement of tea leaves at different drying periods using hyperspectral imaging technique. *PLoS ONE* **2014**, *9*, 1–15. [CrossRef]
30. Ieracitano, C.; Adeel, A.; Morabito, F.C.; Hussain, A. A novel statistical analysis and autoencoder driven intelligent intrusion detection approach. *Neurocomputing* **2020**, *387*, 51–62. [CrossRef]
31. Ieracitano, C.; Mammone, N.; Hussain, A.; Morabito, F.C. A novel multi-modal machine learning based approach for automatic classification of EEG recordings in dementia. *Neural Netw.* **2020**, *123*, 176–190. [CrossRef] [PubMed]
32. Association of Official Analytical Chemists (AOAC). *Official Methods of Analysis of the Association of Official Analytical Chemists*, 15th ed.; United States Department of Agriculture: Rockville, MD, USA, 1990.
33. Aghbashlo, M.; Kianmehr, M.H.; Khani, S.; Ghasemi, M. Mathematical modelling of thin-layer drying of carrot. *Int. Agrophys.* **2009**, *23*, 313–317.
34. Crank, J. *The Mathematics of Diffusion*; Oxford University Press: Oxford, UK, 1979.
35. Chen, N.; Chen, M.; Fu, B.; Song, J. Far-infrared irradiation drying behavior of typical biomass briquettes. *Energy* **2017**, *121*, 726–738. [CrossRef]
36. Jafari, S.M.; Ganje, M.; Dehnad, D.; Ghanbari, V. Mathematical, fuzzy logic and artificial neural network modeling techniques to predict drying kinetics of onion. *J. Food Process. Preserv.* **2016**, *40*, 329–339. [CrossRef]
37. Gamea, G.R.; Essa, A.A. Solar drying characteristics of strawberry. *J. Food Drug Anal* **2007**, *78*, 456–464.
38. Antonio, V.; Uribe, E.; Lemus, R.; Miranda, M. Hot-air drying characteristics of Aloe vera (Aloe barbadensis Miller) and influence of temperature on kinetic parameters. *LWT-Food Sci. Technol.* **2007**, *40*, 1698–1707.
39. Kaur, K.; Singh, A.K. Drying kinetics and quality characteristics of beetroot slices under hot air followed by microwave finish drying. *African J. Agric. Res.* **2014**, *9*, 1036–1044.
40. Sacilik, K. Effect of drying methods on thin-layer drying characteristics of hull-less seed pumpkin (Cucurbita pepo L.). *J. Food Eng.* **2007**, *79*, 23–30. [CrossRef]
41. Dash, K.K.; Gope, S.; Sethi, A.; Doloi, M. Study on Thin Layer Drying Characteristics Star Fruit Slices. *Int. J. Agric. Food Sci. Technol.* **2013**, *4*, 679–686.
42. Hashim, N.; Daniel, O.; Rahaman, E. A Preliminary Study: Kinetic Model of Drying Process of Pumpkins (Cucurbita Moschata) in a Convective Hot Air Dryer. *Int. Conf. Agric. Food Eng. CAFEi2014* **2014**, *2*, 345–352. [CrossRef]
43. Zenoozian, M.S.; Feng, H.; Razavi, S.; Shahidi, F.; Pourreza, H.R. Image analysis and dynamic modeling of thin-layer drying of osmotically dehydrated pumpkin. *J. Food Process. Preserv.* **2008**, *32*, 88–102. [CrossRef]
44. Ayadi, M.; Ben Mabrouk, S.; Zouari, I.; Bellagi, A. Kinetic study of the convective drying of spearmint. *J. Saudi Soc. Agric. Sci.* **2014**, *13*, 1–7. [CrossRef]
45. Kose, U.; Arslan, A. Optimization of self-learning in computer engineering courses: An Intelligent software system supported by artificial neural network and vortex optimization algorithm. *Comput. Appl. Eng. Educ.* **2017**, *25*, 142–156. [CrossRef]
46. Haykin, S. *Neural Networks a Comprehensive Introduction*; Prentice Hall: Upper Saddle River, NJ, USA, 1999.
47. Fu, L.-M. *Neural Networks in Computer Intelligence*; Tata McGraw-Hill Education: New York, NY, USA, 2003.
48. Benkovi, M.; Mari, L.; Male, E.; Valinger, D.; Jurina, T.; Gajdo, J. Effects of drying on physical and chemical properties of root vegetables: Artificial neural network modelling. *Food Bioprod. Process.* **2020**, *119*, 148–160.
49. Onwude, D.I.; Hashim, N.; Janius, R.B.; Nawi, N.; Abdan, K. Modelling the convective drying process of pumpkin (Cucurbita moschata) using an artificial neural network. *Int. Food Res. J.* **2016**, *23*, S237–S243.

50. Das, M.; Akpınar, E. Investigation of pear drying performance by different methods and regression of convective heat transfer coefficient with support vector machine. *Appl. Sci.* **2018**, *8*, 2–16.
51. Vapnik, V.N. An overview of statistical learning theory. *IEEE Trans. Neural Netw.* **1999**, *10*, 988–999. [CrossRef]
52. Shevade, S.; Keerthi, S.S.; Bhattacharyya, C.; Murthy, K.R.K. Improvements to the SMO algorithm for SVM regression. *IEEE Trans. Neural Netw.* **2000**, *11*, 1188–1193. [CrossRef]
53. Xiao, H.; Pang, C.; Wang, L.; Bai, J.; Yang, W.; Gao, Z. Drying kinetics and quality of Monukka seedless grapes dried in an air-impingement jet dryer. *Biosyst. Eng.* **2010**, *105*, 233–240. [CrossRef]
54. Abbaspour-Gilandeh, Y.; Jahanbakhshi, A.; Kaveh, M. Prediction kinetic, energy and exergy of quince under hot air dryer using ANNs and ANFIS. *J. Sci. Food Agric.* **2020**, *8*, 594–611. [CrossRef]
55. Sacilik, K.; Elicin, A.K. The thin layer drying characteristics of organic apple slices. *J. Food Eng.* **2006**, *73*, 281–289. [CrossRef]
56. Telis-Romero, J.; Gabas, A.L.; Menegalli, F.C.; Telis, V.R.N. Drying of persimmon: Mathematical model for diffusivity as a simultaneous function of moisture content and shrinkage. In Proceedings of the Second Inter-American Drying Conference, Monteral, QC, Canada, 8–10 July 2001; pp. 243–251.
57. Tunde-Akintunde, T.Y.; Ogunlakin, G.O. Influence of drying conditions on the effective moisture diffusivity and energy requirements during the drying of pretreated and untreated pumpkin. *Energy Convers. Manag.* **2011**, *52*, 1107–1113. [CrossRef]
58. Falade, K.O.; Olurin, T.O.; Ike, E.A.; Aworh, O.C. Effect of pretreatment and temperature on air-drying of Dioscorea alata and Dioscorea rotundata slices. *J. Food Eng.* **2007**, *80*, 1002–1010. [CrossRef]
59. Meisami-asl, E.; Rafiee, S.; Keyhani, A. Tabatabaeefar, and others. Drying of apple slices (var. Golab) and effect on moisture diffusivity and activation energy. *Plant Omics* **2010**, *3*, 97.
60. Zogzas, N.P.; Maroulis, Z.B. Drying technology: An international journal moisture diffusivity data compilation in foodstuffs. *Dry. Technol. Int. J.* **2007**, *14*, 37–41.
61. Onwude, D.I.; Hashim, N.; Abdan, K.; Janius, R.; Chen, G. Modelling the mid-infrared drying of sweet potato: Kinetics, mass and heat transfer parameters, and energy consumption. *Heat Mass Transf.* **2018**, *54*, 2917–2933. [CrossRef]
62. Younis, M.; Abdelkarim, D.; El-abdein, A.Z. Saudi journal of biological sciences kinetics and mathematical modeling of infrared thin-layer drying of garlic slices. *Saudi J. Biol. Sci.* **2018**, *25*, 332–338. [CrossRef]
63. Zenoozian, M.S.; Devahastin, S.; Razavi, M.A.; Shahidi, F.; Poreza, H.R. Use of artificial neural network and image analysis to predict physical properties of osmotically dehydrated pumpkin. *Dry. Technol.* **2014**, *26*, 132–144. [CrossRef]
64. Ghaderi, A.; Abbasi, S.; Motevali, A.; Minaei, S. Comparison of mathematical models and artificial neural networks for prediction of drying kinetics of mushroom in microwave vacuum dryer. *Chem. Ind. Chem. Eng. Q.* **2012**, *18*, 283–293. [CrossRef]
65. Nadian, M.H.; Rafiee, S.; Aghbashlo, M.; Hosseinpour, S.; Mohtasebi, S.S. Continuous real-time monitoring and neural network modelling of apple slices color changes during hot air Drying. *Food Bioprod. Process.* **2014**, *94*, 1–40.
66. Samsudin, S.H.; Shafri, H.; Hamedianfar, A.; Mansor, S.B. Spectral feature selection and classification of roofing materials using field spectroscopy data. *J. Appl. Remote Sens.* **2015**, *9*, 95079. [CrossRef]
67. Nozad, M.; Khojastehpour, M.; Tabasizadeh, M. Characterization of hot-air drying and infrared drying of spearmint (Mentha spicata L) leaves. *J. Food Meas. Charact.* **2016**, *10*, 466–473. [CrossRef]
68. Onwude, D.; Hashim, N.; Janius, R.; Nawi, N.; Abdan, K. Color change kinetics and total carotenoid content of pumpkin. *Ital. J. Food Sci.* **2017**, *29*, 1–18.

processes

MDPI

Article

Some Physical Properties and Mass Modelling of Pepper Berries (*Piper nigrum* L.), Variety Kuching, at Different Maturity Levels

Puteri Nurain Megat Ahmad Azman [1], Rosnah Shamsudin [1,2,*], Hasfalina Che Man [3] and Mohammad Effendy Ya'acob [1]

1 Department of Process and Food Engineering, Faculty of Engineering, Universiti Putra Malaysia, Serdang 43400, Malaysia; gs53584@student.upm.edu.my (P.N.M.A.A.); m_effendy@upm.edu.my (M.E.Y.)
2 Institute of Advanced Technology, Universiti Putra Malaysia, Serdang 43400, Malaysia
3 Department of Biological and Agricultural Engineering, Faculty of Engineering, Universiti Putra Malaysia, Serdang 43400, Malaysia; hasfalina@upm.edu.my
* Correspondence: rosnahs@upm.edu.my; Tel.: +60-0397696366

Received: 9 August 2020; Accepted: 16 September 2020; Published: 19 October 2020

Abstract: Pepper berry (*Piper nigrum* L.) is known as the king of spices and has sharp, pungent flavour and aroma. In this study, the physical properties (weight, dimensions, sphericity, volume, surface area, and projected area) were measured, and the mass of pepper berries of the Kuching variety at different maturity levels (immature, mature, and ripe) was predicted using four models: linear, quadratic, s-curve, and power. When the models were based on volume and projected area, the mass could be predicted with maximum precision. The Quadratic model was best fitted for mass prediction at all mass maturity levels (immature, mature, and ripe). The results showed that mass modelling based on the actual volume of pepper berries was more applicable compared to other properties with the highest determination coefficient, 0.995, at the 1% probability level. From an economical point of view, mass prediction based on actual volume in the Quadratic form, $M = 0.828 - 0.015\ V + 7.376 \times 10^{-5} V^2$, is recommended. The findings of physical properties and mass modelling of the berries would be useful to the scientific knowledge base, which may help in developing grading, handling, and packaging systems.

Keywords: piper nigrum; dimensions; mass; maturity levels; modelling

1. Introduction

Pepper (*Piper nigrum* L.) is a perennial climber belonging to the family Piperaceae. The leafy pepper tree is comparable to an almond tree that is able to grow tall, reaching a height of 10 m or more. When the main stem matures, many side shoots start growing on it to make a dense canopy. Pepper has a sharp, pungent aroma and flavour. It is widely planted in Vietnam, Malaysia, Indonesia, Sri Lanka, and other areas. According to the Food and Agriculture Organisation of the United Nations Statistics [1], Malaysia was the world's seventh largest pepper producer in 2017. About 98% of pepper production comes from Sarawak, the largest national producer; the other 2% is produced in Johor and other states in Malaysia.

A few recommended local pepper varieties include Kuching, Semongok Emas, and Semongok Aman [2]. Kuching pepper is the most commonly grown cultivar in Sarawak and Johor, Malaysia, due to its high and stable yield. It also has a denser canopy and thinner pericarp, which is the preferred variety for white pepper, production as compared to others. Semongok Emas and Semongok Aman are often used for black pepper production due to their pungent flavour, high yield, and thicker pericarp. In the harvesting process, the selection of pepper berries depends on the maturity levels such as immature, mature, and ripe. Immature pepper berries are described as light green. Mature pepper berries are dark green and yellowish-green, and ripe pepper berries are red. Black pepper and white pepper are the most common and well-known worldwide. For producing black pepper, dark green pepper berries (mature pepper berries) with soft peppercorns and sharp, pungent aroma and flavour are used. They will undergo the drying process without removing the outer skin (pericarp). Yellowish-green and red pepper berries (mature and ripe pepper berries) are used for white pepper production due to the hardness of the peppercorn and thinner pericarp. They are produced by removing the outer skin (pericarp) after a soaking process, and the peppercorns will undergo a drying process. In general, the average diameter of mature and ripe pepper berries for all varieties is about 5 to 6 mm. Overall, the size of pepper berries is similar at all maturity levels. Furthermore, pepper, referred to as black pepper and white pepper, is used in Ayurvedic remedies for health benefits (appetite stimulant, breathing aid, cough therapy, anaemia, and others), keeping food fresh by combining with salt to cure and flavour a wide variety of meats (early food-preservation techniques), and reducing pain perception and inflammation. It is also rich in vitamins and minerals.

Although the berries are often graded by size, weight-based grading is essential in packaging and handling because it may be more economical. Therefore, the physical properties and the relationships must be determined to design and optimise a machine for handling, cleaning, conveying, and storing [3]. The most crucial properties in the design of the grading system are dimension, mass, volume, and projected area [4,5]. According to Pradhan et al. [6] and Pathak et al. [7], basic data on the physical properties of bio-materials are useful to all engineers, industrial processors, scientists, pharmacists, crop breeders, researchers, and others who may discover their advanced usage. Analysis of regression produces an equation to describe and predict the statistical relationship between one or more predictor factors and the variable of reaction. Some regression relationships, such as Linear, Quadratic, S-curve, and Power, were mostly used in previous studies. The statistical significance in the regression relationship shows that the dependent variable modifications in the dependent variables correlate with shifts. Therefore, a high coefficient of determination (R^2) indicates that the model explains a decent percentage of the variability in the dependent variable. The correlation determination technique between mass and its physical properties is more specific for automatic classification of most berries and fruits.

There is a lack of previous research on mass prediction by using model equations for berries compared to other fruits. A few differences between berries and other fruits can be observed. The berries are fleshy fruits produced from a single ovary, while other fruits can be produced from single or multiple ovaries. The berries are edible when the entire ovary wall is ripened and yet, it may not be the same for all fruits. The study of mass modelling of pepper berries has not been carried out in detail in published works. Thus, this research aims to determine some physical properties such as the dimensions (major, medium and minor axis), volume, aspect ratio, geometric mean diameter, sphericity, surface area, and projected areas of immature, mature, and ripe pepper berries. It also serves to determine the most applicable model for predicting the mass of pepper berries at different maturity levels to form an essential database for other researchers. To the authors' knowledge, two previous published works concerned the mass modelling of berries based on their physicochemical properties such as raisin berries [8] and sea buckthorn berries [9]. In other research, many valuable studies were concerned about the mass modelling of fruits based on their physicochemical properties that included oranges [10], apples [11], pomegranates [12], onions [13], fava beans [3], persimmons [5], pomelos [14], dates [15], Kinnow mandarin [16], chebula fruits [7], and banana [17]. In summary, the study findings may help in developing grading, handling, and packaging systems based on the physical properties at different maturity levels of pepper berries.

2. Materials and Methods

The Kuching variety of pepper berries was obtained from a farm located in Johor, south of Malaysia (Johor, Malaysia). The fresh pepper berries were immediately transported to the laboratory within 4 h. The pepper berries were selected and divided into three maturity levels: immature, mature, and ripe. For sample selection, colour is an essential visual parameter to differentiate pepper based on the maturity levels. Light green and red berries were labelled as immature and ripe pepper berries, respectively [2]. Dark green and yellowish-green berries were labelled as mature pepper berries [2]. One hundred samples of pepper berries from each maturity level were selected and used for physical property measurements [18,19] as shown in Table 1. All the experiments were conducted at room temperature in the laboratory.

2.1. Measurements of Physical Properties

Pepper berry mass (M) was determined with 0.01 g sensitivity of an electronic balance. Three linear dimensions, namely, the major axis (L), medium axis (T), and minor axis (W), were measured by using a digital vernier calliper with 0.01 mm sensitivity to determine the average size of the samples. The method of water displacement [1,3,12,20,21] was used to determine the actual volume (V). The uniform presumed shape of the fruit can be an aspect to determine the fruit volume accuracy [5]. The aspect ratio (AR), geometric mean diameter (D_g), and sphericity (Φ) were calculated by using the following respective formulas [5,20–22]:

$$AR = \frac{L}{W} \tag{1}$$

$$D_g = (LWT)^{\frac{1}{3}} \tag{2}$$

$$\phi = \frac{(LWT)^{\frac{1}{3}}}{L} \tag{3}$$

The surface area is defined as the total three-dimensional shape areas of all surfaces. Equation (4) was used to calculate the spheroid surface area of pepper berries.

$$SA_{sp} = 4\pi r^2 \tag{4}$$

The projected areas of pepper berries (PA_L, PA_T, and PA_W) in three perpendicular directions to the dimensions (major axis, medium axis, and minor axis) and the criteria projected area (CPA) were calculated by using Equations (5)–(8), respectively. These equations were suggested by Mohsenin [4] and Nur Salihah et al. [14] and are defined as follows:

$$PA_L = \frac{\pi LW}{4} \tag{5}$$

$$PA_T = \frac{\pi TW}{4} \tag{6}$$

$$PA_W = \frac{\pi WW}{4} \tag{7}$$

$$CPA = \frac{PA_L + PA_T + PA_W}{3} \tag{8}$$

Table 1. Some physical properties of pepper berries at different maturity levels.

	Immature			Mature			Ripe		
Parameter	Mean	Maximum Value	Minimum Value	Mean	Maximum Value	Minimum Value	Mean	Maximum Value	Minimum Value
L (mm)	4.75 ± 0.38 [c]	5.50	4.00	**5.73 ± 0.32 [a]**	6.60	5.40	5.55 ± 0.66 [b]	7.40	4.36
T (mm)	5.09 ± 0.50 [c]	5.70	4.30	**6.16 ± 0.44 [a]**	7.10	5.60	5.60 ± 0.33 [b]	7.10	4.31
W (mm)	4.74 ± 0.34 [c]	5.30	4.00	**5.87 ± 0.34 [a]**	6.70	5.40	5.67 ± 0.51 [b]	7.01	4.36
AR	**1.00 ± 0.03 [a]**	1.04	0.96	0.98 ± 0.02 [a]	1.00	0.93	0.98 ± 0.13 [a]	1.18	0.79
D_g (mm)	4.85 ± 0.38 [c]	5.50	4.10	**5.92 ± 0.35 [a]**	6.80	5.47	5.61 ± 0.46 [b]	7.17	4.67
Φ	1.02 ± 0.02 [ab]	1.05	0.97	**1.03 ± 0.02 [a]**	1.08	1.01	1.01 ± 0.08 [b]	1.43	0.72
Weight (g)	0.10 ± 0.01 [a]	0.11	0.09	0.14 ± 0.01 [a]	0.14	0.13	**0.15 ± 0.04 [a]**	0.23	0.11
V (mm^3)	96.67 ± 5.77 [b]	100.00	90.00	**120 ± 10.00 [a]**	130.00	110.00	**120 ± 21.60 [a]**	140.00	90.00
SA_{sp} (mm^2)	74.65 ± 11.43 [c]	95.03	52.81	**110.49 ± 13.66 [a]**	145.27	93.88	99.39 ± 16.66 [b]	161.51	69.20
PA_L (mm^2)	17.78 ± 2.75 [c]	22.89	12.57	**26.50 ± 3.28 [a]**	34.73	22.90	24.78 ± 4.27 [b]	40.74	17.70
PA_T (mm^2)	19.04 ± 3.08 [c]	23.73	13.51	**28.51 ± 3.93 [a]**	37.36	23.75	25.10 ± 4.51 [b]	39.09	14.96
PA_W (mm^2)	17.73 ± 2.57 [c]	22.06	12.57	**27.16 ± 3.45 [a]**	35.26	22.90	25.45 ± 4.56 [b]	38.59	14.93
CPA (mm^2)	18.19 ± 2.73 [c]	22.89	12.88	**27.39 ± 3.52 [a]**	35.78	23.18	25.11 ± 4.13 [b]	39.48	16.29

Data are expressed as the mean ± SD maximum minimum; *L*, major axis; *T*, medium axis; *W*, minor axis; *AR*, aspect ratio; D_g, geometric diameter; Φ, sphericity; weight; *V*, actual volume; SA_{sp}, spherical surface area; PA_L, projected area perpendicular to major axis; PA_T, projected area perpendicular to medium axis; PA_W, projected area perpendicular to minor axis; *CPA*, criteria projected area. Different letters in a row indicate statistically significant differences at $p < 0.001$. Means that do not share a letter are significantly different. Tukey's test was applied with 95% confidence intervals.

2.2. Regression Analysis and Mass Modelling

In order to predict the mass of pepper berries based on measured physical properties (dimensional characteristics, volume, surface area, and projected area), the considerations of model classifications are as follows:

1. Single variable regression of pepper berry mass based on dimensional characteristics of the pepper berry—major axis (L), medium axis (T), minor axis (W), and geometric mean diameter (D_g).
2. Single regression of pepper berry volume—actual volume (V).
3. Single regression of pepper berry surface area—surface area of the fruit assumed as a spheroid (SA_{sp}).
4. Single variable regression of pepper berry projected area —PA_L, PA_T, PA_W, and CPA.

Four models, namely, Linear, Quadratic, S-curve, and Power, were used and fitted with the results obtained from the experiments. These models are presented in Equations (9)–(12), respectively [3,5]:

$$M = a + bX \tag{9}$$

$$M = a + bX + cX^2 \tag{10}$$

$$M = a + \frac{b}{X} \tag{11}$$

$$M = aX^b \tag{12}$$

where M = *mass* (g); X = the value of an independent parameter (physical properties) to find the relationship of it with mass; a, b, and c = curve fitting parameters which are different in each equation.

2.3. Statistical Analysis

Data analysis and mass modelling prediction were performed by using statistical software such as SigmaPlot (Version 12.0) with significance at the 1% probability level. Coefficient of determination (R^2) and standard error of the estimate (SEE) were selected as the criteria to evaluate the applicability of the regression models. The applicable models were selected as those with higher R^2 and lower SEE values [21].

3. Results and Discussion

3.1. Physical Properties of Pepper Berries

Table 1 summarizes the results of the physical properties of pepper berries based on the maturity levels. The mean values of the major axis (L), medium axis (T), and minor axis (W) for immature pepper berries were 4.75 mm ± 0.38, 5.09 mm ± 0.50, and 4.74 mm ± 0.34, respectively. Therefore, the medium axis had the highest mean value as compared to the major and minor axes of immature pepper berries. The mature pepper berries had mean values of 5.73 mm ± 0.32, 6.16 mm ± 0.44, and 5.87 mm ± 0.34 for major axis, medium axis, and minor axis, respectively. Again, the medium axis had the highest mean value as compared to the major and minor axes of mature pepper berries. Furthermore, the mean values of the major axis, medium axis, and minor axis for ripe pepper berries were 5.55 mm ± 0.66, 5.60 mm ± 0.58, and 5.67 mm ± 0.51, respectively. For the results of dimensional characteristics, the minor axis had the highest mean value as compared to the major and medium axis of ripe pepper berries. As can be seen, mature pepper berries had the highest values of the major, medium, and minor axes among these three maturity levels.

The mean value of the aspect ratio (AR) of immature pepper berries was 1.00 with a standard deviation of 0.03. The mature pepper berries had a mean value of 0.98 with a standard deviation of 0.02 for the aspect ratio. The mean value of the aspect ratio of ripe pepper berries was 0.98 with a standard deviation of 0.13. However, the value of the aspect ratio for immature, mature, and ripe pepper berries was significantly in same range as shown in Table 1.

The mean value of the geometric mean diameter (D_g) of immature pepper berries was 4.85 mm with a standard deviation of 0.38. Mature pepper berries had a mean value of 5.92 mm with a standard deviation of 0.35 for the geometric mean diameter. Furthermore, the mean value of the geometric mean diameter of ripe pepper berries was 5.61 mm with a standard deviation of 0.46. Thus, the mean value of the geometric mean diameter of mature pepper berries was the highest among all three maturity levels.

Sphericity can be described a solid shape formed relative to that of the same volume of a sphere. The ideal shape of the sphere is a solid with a high sphericity value [4,18]. A sphericity value of 1 is an ideal sphere. The mean sphericity value for immature pepper berries was 1.02 with a standard deviation of 0.02. The mature pepper berries had a mean value of 1.03 with a standard deviation of 0.02 for sphericity. As for ripe pepper berries, the mean value of sphericity was 1.01 with a standard deviation of 0.08. Thus, from the mean values of sphericity for all three maturity levels, it can be considered that the shapes of immature, mature, and ripe pepper berries were an ideal sphere. The mean weight of immature pepper berries was 0.10 ± 0.01 g. The mature pepper berries had a mean weight of 0.14 ± 0.01 g. The mean weight of ripe pepper berries was 0.15 g ± 0.04 g. However, the weight of immature, mature, and ripe pepper berries was significantly in the same range as shown in Table 1.

Based on Table 1, other physical properties of pepper berries included actual volume. The immature pepper berries had mean actual volumes of 96.67 mm^3 ± 5.77. The mean actual volume of mature pepper berries was 120 mm^3 ± 10.00. Furthermore, the ripe mature pepper berries had mean values of 120 mm^3 ± 21.60 for the actual volume. Overall, the mature and ripe pepper berries had the highest mean actual volumes when compared to the immature pepper berries.

The mean value of the spheroid surface area of immature pepper berries was 74.65 mm^2 ± 11.43. The mature pepper berries had a mean value of 110.49 mm^2 ± 13.66 for the spheroid surface area. The mean value of the spheroid surface area of ripe pepper berries was 99.39 mm^2 ± 16.66. Thus, the highest mean value of the surface area for the spheroid was obtained for mature pepper berries as shown in Table 1.

Table 1 shows the results of mean projected areas, perpendicular to the major axis (PA_L), medium axis (PA_T), and minor axis (PA_W). The results obtained for immature pepper berries were 17.78 mm^2 ± 2.75 (PA_L), 19.04 mm^2 ± 3.08 (PA_T), and 17.73 mm^2 ± 2.57 (PA_W). The mean values of PA_L, PA_T, and PA_W for mature pepper berries were 26.50 mm^2 ± 3.28, 28.51 mm^2 ± 3.93, and 27.16 mm^2 ± 3.45, respectively. As for the ripe pepper berries, the mean values were 24.78 mm^2 ± 4.27, 25.10 mm^2 ± 4.51, and 25.45 mm^2 ± 4.56 for PA_L, PA_T, and PA_W, respectively. Therefore, the mature pepper berries had the highest mean values of PA_L, PA_T, and PA_W when compared to the other two maturity levels of pepper berries. The criteria of projected area were determined by using the results of PA_L, PA_T, and PA_W as indicated in Equation (11). The mean values of projected area criteria for immature, mature, and ripe pepper berries were 18.19 mm^2 ± 2.73, 27.39 mm^2 ± 3.52, and 25.11 mm^2 ± 4.13, respectively. Thus, the projected area criteria for mature pepper berries had the highest mean value in Table 1.

Projected area values are fundamental information in the design and development of machine vision-based grading systems [16]. Projected area is also useful for estimating the respiration rate, maturity index, and gas permeability to predict optimum harvest time, water loss, and heat and mass transfer during drying and cooling [7,16]. However, some ellipsoidal shapes of ripe pepper berries were observed among the samples. Also, the measured distinction in the physical properties may be due to the inherent variation in fruit dimensional characteristics.

3.2. Mass Modelling

The average dimensions, volumes, weight, surface areas, and projected areas of pepper berries obtained were used in mass modelling. Tables 2–4 show the best models obtained and their coefficients of determination, R^2, and SEE to predict mass by using the measured average dimensions, volumes, weight, surface areas, and projected areas of pepper berries. The correlations of physical properties with pepper berry mass as shown from the results obtained were significant at the 0.01 probability level. The regression mass was evaluated by using the coefficient of determination (R^2), where the best fit model was shown with a higher R^2 value (near 1.00).

3.3. Models Based on Dimensions

According to Table 2, the major axis (L), medium axis (T), minor axis (W), and geometric mean diameter (D_g) showed that the Quadratic model was the best-fit model to calculate and evaluate the mass of immature, mature, and ripe pepper berries. Table 2 shows the fitted models based on dimensions such as L, T, W, and D_g with the values of R^2 and SEE.

For immature pepper berries, D_g had a highest value of R^2 and the lowest value of SEE, which were 0.938 and 0.002, respectively, as indicated in Table 2. Equation (13) shows the equation of the Quadratic model obtained.

$$M = 0.225 - 0.071\, D_g + 0.009\, {D_g}^2 \tag{13}$$

According to Table 2, the mature pepper berries had W and D_g values with the highest R^2 (0.960) and the lowest SEE (0.001). The model equation obtained for these parameters was Quadratic. Equations (14) and (15) were determined for the parameters W and D_g.

$$M = 0.260 - 0.053\, W + 0.005\, W^2 \tag{14}$$

$$M = 0.196 - 0.032\, D_g + 0.004\, {D_g}^2 \tag{15}$$

L had the highest R^2 (0.980) and the lowest SEE (0.007) values for ripe pepper berries when compared to the others as given in the Quadratic model equation (Equation (16)).

$$M = 0.351 - 0.107\, L + 0.012\, L^2 \tag{16}$$

A similar model for onion and a Malaysian variety of pomelo fruit in another study was suggested and reported by Ghabel et al. [13] and Nur Salihah et al. [14], where the best model for mass determination based on L was a Quadratic model following Equations (17) and (18):

$$M = 36.137 - 1.64\, L + 0.35\, L^2,\ R^2 = 0.96 \tag{17}$$

$$M = 28500.33 - 417.191\, L + 1.587\, L^2,\ R^2 = 0.993 \tag{18}$$

For the entire dimensions, the S-curve model was reported to have lower R^2 values as compared to other fitted models. The lower R^2 values could be due to the non-uniform mass of pepper berries corresponding to their size. Thus, the sizing of pepper berries based on length is recommended.

Table 2. Mass models of pepper berries based on dimensions.

Dependent Parameter	Independent Parameter	Model Equation	Maturity Levels	Regression Constant			Statistical Parameters		The Best Fitted Model
				a	b	c	R^2	*SEE*	
M (g)	*L* (mm)	Linear	Immature	0.021	0.016	-	0.828	0.003	Quadratic
			Mature	0.058	0.013	-	**0.949**	**0.001**	
			Ripe	−0.056	0.036	-	0.894	0.014	
M (g)	*L* (mm)	Quadratic	Immature	0.130	−0.031	0.005	0.852	0.003	
			Mature	−0.013	0.037	−0.002	0.952	0.001	
			Ripe	0.351	−0.107	0.012	**0.980**	**0.007**	
M (g)	*L* (mm)	S-curve	Immature	0.168	−0.338	-	0.781	0.004	
			Mature	0.218	−0.478	-	**0.951**	**0.001**	
			Ripe	0.345	−1.085	-	0.796	0.020	
M (g)	*L* (mm)	Power	Immature	0.028	0.786	-	0.824	0.003	
			Mature	0.049	0.580	-	**0.950**	**0.001**	
			Ripe	0.012	1.432	-	0.911	0.013	
M (g)	*T* (mm)	Linear	Immature	0.033	0.012	-	0.862	0.003	Quadratic
			Mature	0.076	0.009	-	**0.903**	**0.002**	
			Ripe	−0.052	0.035	-	0.795	0.020	
M (g)	*T* (mm)	Quadratic	Immature	0.239	−0.071	0.008	0.925	0.002	
			Mature	0.286	−0.057	0.005	**0.952**	**0.001**	
			Ripe	0.419	−0.133	0.015	0.883	0.017	
M (g)	*T* (mm)	S-curve	Immature	0.155	−0.296	-	0.821	0.003	
			Mature	0.194	−0.370	-	**0.871**	**0.002**	
			Ripe	0.334	−1.032	-	0.711	0.024	
M (g)	*T* (mm)	Power	Immature	0.033	0.653	-	0.856	0.003	
			Mature	0.060	0.439	-	**0.895**	**0.002**	
			Ripe	0.012	1.419	-	0.809	0.019	

Table 2. *Cont.*

Dependent Parameter	Independent Parameter	Model Equation	Maturity Levels	Regression Constant			Statistical Parameters		The Best Fitted Model
				a	b	c	R^2	*SEE*	
M (g)	*W* (mm)	Linear	Immature	0.012	0.018	-	0.794	0.003	Quadratic
			Mature	0.062	0.012	-	**0.927**	**0.001**	
			Ripe	−0.065	0.038	-	0.807	0.020	
M (g)	*W* (mm)	Quadratic	Immature	0.306	−0.110	0.014	0.895	0.003	
			Mature	0.260	−0.053	0.005	**0.960**	**0.001**	
			Ripe	0.557	−0.186	0.020	0.926	0.014	
M (g)	*W* (mm)	S-curve	Immature	0.172	−0.357	-	0.730	0.004	
			Mature	0.208	−0.436	-	**0.895**	**0.002**	
			Ripe	0.346	−1.094	-	0.722	0.023	
M (g)	*W* (mm)	Power	Immature	0.024	0.882	-	0.790	0.003	
			Mature	0.051	0.543	-	**0.920**	**0.001**	
			Ripe	0.010	1.523	-	0.826	0.019	
M (g)	D_g (mm)	Linear	Immature	0.016	0.017	-	0.878	0.003	Quadratic
			Mature	0.063	0.012	-	**0.945**	**0.001**	
			Ripe	−0.065	0.037	-	0.821	0.019	
M (g)	D_g (mm)	Quadratic	Immature	0.225	−0.071	0.009	0.938	0.002	
			Mature	0.196	−0.032	0.004	**0.960**	**0.001**	
			Ripe	0.479	−0.152	0.016	0.886	0.017	
M (g)	D_g (mm)	S-curve	Immature	0.171	−0.360	-	0.822	0.003	
			Mature	0.209	−0.445	-	**0.922**	**0.001**	
			Ripe	0.361	−1.195	-	0.767	0.021	
M (g)	D_g (mm)	Power	Immature	0.026	0.840	-	0.874	0.003	
			Mature	0.051	0.541	-	**0.940**	**0.001**	
			Ripe	0.011	1.474	-	0.832	0.018	

Data are expressed as *L*, major axis; *T*, medium axis; *W*, minor axis; D_g, geometric diameter; R^2, coefficient of determination; *SEE*, standard error of estimate; a, b, c, constants of curve fittings.

Table 3. Mass models of pepper berries based on volume and surface area.

Dependent Parameter	Independent Parameter	Model Equation	Maturity Levels	Regression Constant			Statistical Parameters		The Best Fitted Model
				a	b	c	R^2	*SEE*	
M (g)	V (mm^3)	Linear	Immature	0.108	−0.015	-	0.636	0.005	
			Mature	0.049	0.001	-	0.806	0.002	
			Ripe	−0.095	0.002	-	**0.816**	**0.027**	
M (g)	V (mm^3)	Quadratic	Immature	0.125	−0.190	0.163	0.925	0.002	
			Mature	0.795	−0.012	5.164×10^{-5}	0.988	0.001	
			Ripe	0.828	−0.015	7.376×10^{-5}	**0.995**	**0.006**	Quadratic
M (g)	V (mm^3)	S-curve	Immature	0.092	0.002	-	0.692	0.004	
			Mature	0.217	−9.947	-	**0.762**	**0.002**	
			Ripe	0.375	−24.784	-	0.727	0.033	
M (g)	V (mm^3)	Power	Immature	0.096	6.619×10^{-18}	-	0.000	0.008	
			ture	0.006	0.639	-	0.798	0.002	
			Ripe	2.626×10^{-5}	1.819	-	**0.849**	**0.024**	
M (g)	SA_{sp} (mm^2)	Linear	Immature	0.055	6×10^{-4}	-	0.899	0.002	
			Mature	0.100	3×10^{-4}	-	0.952	0.001	
			Ripe	0.020	0.001	-	**0.955**	**0.009**	
M (g)	SA_{sp} (mm^2)	Quadratic	Immature	0.097	-6×10^{-4}	7.825×10^{-6}	0.936	0.002	
			Mature	0.124	-1×10^{-4}	1.715×10^{-6}	0.960	0.001	
			Ripe	0.117	−0.001	7.80×10^{-6}	**0.983**	**0.006**	Quadratic
M (g)	SA_{sp} (mm^2)	S-curve	Immature	0.132	−2.596	-	0.789	0.003	
			Mature	0.172	−4.186	-	**0.905**	**0.002**	
			Ripe	0.280	−12.808	-	0.827	0.018	
M (g)	SA_{sp} (mm^2)	Power	Immature	0.016	0.419	-	0.874	0.003	
			Mature	0.038	0.270	-	0.940	0.001	
			Ripe	0.002	0.885	-	**0.950**	**0.010**	

Data are expressed as V, actual volume; SA_{sp}, spherical surface area; R^2, coefficient of determination; *SEE*, standard error of estimate; a, b, c, constants of curve fittings.

Table 4. Mass models of pepper berries based on projected area.

Dependent Parameter	Independent Parameter	Model Equation	Maturity Levels	Regression Constant			Statistical Parameters		The Best Fitted Model
				a	b	c	R^2	SEE	
M (g)	PA_L (mm^2)	Linear	Immature	0.055	0.002	-	0.831	0.003	Quadratic
			Mature	0.098	0.001	-	0.912	0.001	
			Ripe	0.042	0.004	-	0.792	0.020	
M (g)	PA_L (mm^2)	Quadratic	Immature	0.086	−0.001	1×10^{-4}	**0.856**	**0.003**	
			Mature	0.111	0.001	1.466×10^{-5}	**0.913**	**0.002**	
			Ripe	0.158	−0.005	1×10^{-4}	**0.835**	**0.020**	
M (g)	PA_L (mm^2)	S-curve	Immature	0.132	−0.614	-	0.729	0.004	
			Mature	0.175	−1.063	-	0.883	0.002	
			Ripe	0.252	−2.626	-	0.700	0.024	
M (g)	PA_L (mm^2)	Power	Immature	0.029	0.422	-	0.807	0.003	
			Mature	0.054	0.280	-	0.906	0.002	
			Ripe	0.013	0.723	-	0.782	0.021	
M (g)	PA_T (mm^2)	Linear	Immature	0.055	0.002	-	0.887	0.003	Quadratic
			Mature	0.102	0.001	-	0.941	0.001	
			Ripe	0.040	0.004	-	0.810	0.019	
M (g)	PA_T (mm^2)	Quadratic	Immature	0.114	−0.004	2×10^{-4}	**0.946**	**0.002**	
			Mature	0.146	−0.002	4.808×10^{-5}	**0.971**	**0.001**	
			Ripe	0.199	−0.009	2×10^{-4}	**0.930**	**0.013**	
M (g)	PA_T (mm^2)	S-curve	Immature	0.131	−0.647	-	0.783	0.004	
			Mature	0.169	−0.980	-	0.874	0.001	
			Ripe	0.239	−2.167	-	0.642	0.027	
M (g)	PA_T (mm^2)	Power	Immature	0.029	0.412	-	0.860	0.003	
			Mature	0.058	0.250	-	0.921	0.001	
			Ripe	0.013	0.741	-	0.790	0.020	

Table 4. *Cont.*

Dependent Parameter	Independent Parameter	Model Equation	Maturity Levels	Regression Constant			Statistical Parameters		The Best Fitted Model
				a	b	c	R^2	SEE	
M (g)	PA_W (mm^2)	Linear	Immature	0.052	0.003	-	0.820	0.003	
			Mature	0.099	0.001	-	0.934	0.001	
			Ripe	0.039	0.004	-	0.815	0.019	
M (g)	PA_W (mm^2)	Quadratic	Immature	0.112	−0.005	2×10^{-4}	**0.888**	**0.003**	
			Mature	0.139	−0.002	4.768×10^{-5}	**0.955**	**0.001**	
			Ripe	0.207	−0.010	3×10^{-4}	**0.942**	**0.012**	Quadratic
M (g)	PA_W (mm^2)	S-curve	Immature	0.132	−0.619	-	0.695	0.004	
			Mature	0.172	−1.011	-	0.872	0.002	
			Ripe	0.240	−2.168	-	0.652	0.026	
M (g)	PA_W (mm^2)	Power	Immature	0.027	0.441	-	0.790	0.003	
			Mature	0.055	0.271	-	0.916	0.001	
			Ripe	0.013	0.744	-	0.795	0.020	
M (g)	*CPA*	Linear	Immature	0.054	0.002	-	0.832	0.003	
			Mature	0.099	0.001	-	0.950	0.001	
			Ripe	0.034	0.004	-	0.833	0.018	
M (g)	*CPA*	Quadratic	Immature	0.102	−0.003	2×10^{-4}	**0.880**	**0.003**	
			Mature	0.130	−0.001	3.55×10^{-5}	**0.962**	**0.001**	
			Ripe	0.221	−0.011	3×10^{-4}	**0.966**	**0.009**	Quadratic
M (g)	*CPA*	S-curve	Immature	0.132	−0.632	-	0.716	0.004	
			Mature	0.172	−1.037	-	0.896	0.002	
			Ripe	0.250	−2.445	-	0.683	0.025	
M (g)	*CPA*	Power	Immature	0.028	0.431	-	0.805	0.003	
			Mature	0.055	0.272	-	0.935	0.001	
			Ripe	0.011	0.784	-	0.816	0.019	

Data are expressed as PA_L, projected area perpendicular to major axis; PA_T, projected area perpendicular to medium axis; PA_W, projected area perpendicular to minor axis; *CPA*, criteria projected area; R^2, coefficient of determination; *SEE*, standard error of estimate; a, b, c, constants of curve fittings.

3.4. Models Based on Volume

In Table 3 which shows the mass prediction results of the pepper berries based on actual volume (V), the Quadratic model based on V (Equation (19)) was found to be the best fit when compared to the other models. It had the highest R^2 of 0.995 and the lowest SEE of 0.006 for ripe pepper berries.

$$M = 0.828 - 0.015\ V + 7.376 \times 10^{-5} V^2 \tag{19}$$

For immature pepper berries, the Quadratic model based on V was suitable with the highest values of R^2 and SEE, which were 0.925 and 0.002, respectively, as shown in Table 3. Equation (20) was obtained for the pepper berries at the immature maturity level.

$$M = 0.081 - 5.726 \times 10^{-6} V + 3.703 \times 10^{-6} V^2 \tag{20}$$

The mature pepper berries had a Quadratic model based on V as the best model with the highest R^2 of 0.988 and the lowest SEE of 0.001, shown in Table 3. The Quadratic model equation (Equation (21)) is as follows:

$$M = 0.795 - 0.012\ V + 5.164 \times 10^{-5} V^2 \tag{21}$$

Thus, the ripe pepper berries had the highest value of R^2 and lowest SEE for actual volume among the maturity levels. Therefore, a quadratic form was shown as the suggested mass model-based volume, similar to the prediction of the Fava bean mass with R^2 = 0657 [3].

3.5. Models Based on Surface Area

As shown in Table 3 for the results of mass prediction of pepper berries based on surface area (SA_{sp}), the Quadratic model was the best based on the highest value of R^2 compared to the other models. For the best fit model, the Quadratic model based on SA_{sp} of ripe pepper berries had the highest value of R^2 and lowest SEE of the surface area-assumed shape; the respective values were 0.984 and 0.006 (as shown in Equation (22)). It was also the best fit among the models of other maturity levels of pepper berries.

$$M = 0.118 - 0.001\ SA_{sp} + 7.900 \times 10^{-6}\ {SA_{sp}}^2 \tag{22}$$

The Quadratic model was suitable for SA_{sp} which had the highest R^2 of 0.936 and lowest SEE of 0.002 for immature pepper berries (as shown in Equations (23)). The suitable model of immature pepper berries is shown in the following equation:

$$M = 0.097 - 6 \times 10^{-4}\ SA_{sp} + 7.825 \times 10^{-6}\ {SA_{sp}}^2 \tag{23}$$

For mature pepper berries, the suitable model equation was Quadratic. Based on Table 3, the parameter SA_{sp} had the highest value of R^2 and lowest SEE, which were 0.960 and 0.001, respectively. The model equation was formed as indicated in Equation (24).

$$M = 0.124 - 1 \times 10^{-4}\ SA_{sp} + 1.715 \times 10^{-6}\ {SA_{sp}}^2 \tag{24}$$

Among these maturity levels, the ripe pepper berries had the highest value of R^2 and lowest SEE for the spheroid surface area in Quadratic form. Therefore, the Quadratic form was shown as the suggested mass model.

3.6. Models Based on Projected Area

Among the models based on the projected area (PA_L, PA_T, PA_W, and CPA), the Quadratic model comprising PA_T was the best fit with the highest R^2 of 0.71 and lowest SEE of 0.001 for mature pepper berries as shown in Table 4. Equation (25) shows the model equation obtained. The Quadratic model based on PA_T (Equation (26)) was suitable, with R^2 of 0.946 and SEE of 0.002, with respect to immature pepper berries. The Quadratic model based on PA_W (Equation (27)) also achieved a suitable R^2 of 0.942 with SEE of 0.012 for ripe pepper berries.

$$M = 0.123 - 0.002PA_T + 4.808 \times 10^{-5} {PA_T}^2 \tag{25}$$

$$M = 0.114 - 0.004PA_T + 2 \times 10^{-4} {PA_T}^2 \tag{26}$$

$$M = 0.207 - 0.010PA_W + 3 \times 10^{-4} {PA_W}^2 \tag{27}$$

Furthermore, the Quadratic model of the projected area based on the criteria projected area (CPA) shown in Table 4 was preferred as the best model to calculate the mass of pepper berries. Due to the high R^2 value of 0.966, a model based on CPA (Equation (28)) for ripe pepper berries was found as the best fit with SEE of 0.009. The mature pepper berries had values of R^2 and SEE (Equation (29)) which were 0.962 and 0.001, respectively. The immature pepper berries had the lowest R^2 and SEE values for the Quadratic model based on CPA (Equation (30)), which were 0.880 and 0.003, respectively, compared to other maturity levels of pepper berries.

$$M = 0.221 - 0.011CPA + 3 \times 10^{-4} CPA^2 \tag{28}$$

$$M = 0.130 - 0.001CPA + 3.55 \times 10^{-5} CPA^2 \tag{29}$$

$$M = 0.102 - 0.003CPA + 2 \times 10^{-4} CPA^2 \tag{30}$$

Thus, all three projected areas are necessary to be specified and applied in grading the pepper berries by using this model.

4. Conclusions

In the current study, the mass of ripe pepper berries based on volume (Quadratic model) is the recommended equation, as the nonlinear form $M = 0.828 - 0.015V + 7.376 \times 10^{-5} V^2$ had the highest R^2, 0.995, and lowest SEE, 0.006, when compared to other maturity levels of pepper berries. This shows a very good relationship between the mass and actual volume of pepper berries. The model predicting the mass of pepper berries considered as spheroid was found to be the most applicable (Quadratic model is recommended). Finally, the Quadratic model is applicable to all properties due to its economical viewpoint. The mass model of pepper berries based on actual volume in the obtained results is recommended for designing and optimizing machines for handling, cleaning, conveying, and storing.

Author Contributions: P.N.M.A.A. conducted the experiments, collected and analyzed the data of results, and wrote the manuscript. R.S. supervised the research and revised the manuscript. H.C.M. and M.E.Y. supervised the experiments. All authors have read and agreed to the published version of the manuscript.

Funding: This research received no external funding.

Acknowledgments: The authors would like to express their appreciation to Universiti Putra Malaysia for the financial and technical support given during this research work.

Conflicts of Interest: The authors declare no conflict of interest.

References

1. FAOSTAT. Available online: http://www.fao.org/faostat/en/#rankings/countries_by_commodity (accessed on 18 June 2020).
2. Malaysian Pepper Board (MPB). Available online: https://www.mpb.gov.my/mpb2/index.php/my/ (accessed on 18 June 2020).
3. Lorestani, A.N.; Ghari, M. Mass modeling of Fava bean (*Vicia faba* L.) with some physical characteristics. *Sci. Hortic.* **2012**, *133*, 6–9. [CrossRef]
4. Mohsenin, N.N. Physical properties of plant and animal materials. Structure, physical characteristics and mechanical properties. *N. Y. Gordon Breach Sci. Publ.* **1986**, *31*, 987.
5. Feizollah, S.; Satar, R. Mass modeling of persimmon fruit with some physical characteristics. *Agric. Eng. Int. CIGR J.* **2014**, *16*, 289–293.
6. Pradhan, R.C.; Meda, V.; Naik, S.N.; Tabil, L. Physical properties of canadian grown flaxseed in relation to its processing. *Int. J. Food Prop.* **2010**, *13*, 732–743. [CrossRef]
7. Pathak, S.S.; Pradhan, R.C.; Mishra, S. Physical characterization and mass modeling of dried Terminalia chebula fruit. *J. Food Process Eng.* **2019**, *42*, e12992. [CrossRef]
8. Karimi, N.; Arabhosseini, A.; Kianmehr, M.H.; Khazaei, J. Modelling of raisin berries by some physical and statistical characteristics. *Int. Agrophys* **2011**, *25*, 141–147.
9. Jaiswal, S.G.; Dole, B.R.; Satpathy, S.K.; Naik, S.N. Physical attributes and modelling of trans-himalayan seabuckthorn berries. *Curr. Res. Nutr. Food Sci.* **2017**, *5*, 391–397. [CrossRef]
10. Tabatabaeefar, A.; Vefagh-Nematolahee, A.; Rajabipour, A. Modeling of Orange Mass Based on Dimensions. *J. Agric. Sci. Technol.* **2000**, *2*, 299–305.
11. Tabatabaeefar, A.; Rajabipour, A. Modeling the mass of apples by geometrical attributes. *Sci. Hortic.* **2005**, *105*, 373–382. [CrossRef]
12. Khoshnam, F.; Tabatabaeefar, A.; Varnamkhasti, M.G.; Borghei, A. Mass modeling of pomegranate (*Punica granatum* L.) fruit with some physical characteristics. *Sci. Hortic.* **2007**, *114*, 21–26. [CrossRef]
13. Ghabel, R.; Rajabipour, A.; Ghasemi-Varnamkhasti, M.M.O. Modeling the mass of Iranian export onion (*Allium cepa* L.) varieties using some physical characteristics. *Res. Agric. Eng.* **2010**, *56*, 33–40. [CrossRef]
14. Nur Salihah, B.; Rosnah, S.; Norashikin, A.A. Mass modeling of Malaysian varieties pomelo fruit (*Citrus grandis* L. Osbeck) with some physical characteristics. *Int. Food Res. J.* **2015**, *22*, 488–493.
15. Khodabakhshian, R.; Emadi, B. Mass model of date fruit (cv. Mazafati) based on its physiological properties. *Int. Food Res. J.* **2016**, *23*, 2057–2062.
16. Mahawar, M.K.; Bibwe, B.; Jalgaonkar, K.; Ghodki, B.M. Mass modeling of kinnow mandarin based on some physical attributes. *J. Food Process Eng.* **2019**, *42*, e13079. [CrossRef]
17. Zainal A'Bidin, F.N.; Shamsudin, R.; Mohd Basri, M.S.; Mohd Dom, Z. Mass Modelling and Effects of Fruit Position on Firmness and Adhesiveness of Banana Variety Nipah. *Int. J. Food Eng.* **2020**, *1*. (ahead of print).
18. Rosnah, S.; Chan, S. Enzymatic rettings of green pepper berries for white pepper production. *Int. Food Res. J.* **2014**, *21*, 237–245.
19. Murthy, C.T.; Bhattacharya, S. Moisture dependant physical and uniaxial compression properties of black pepper. *J. Food Eng.* **1998**, *37*, 193–205. [CrossRef]
20. Akar, R.; Aydin, C. Some physical properties of gumbo fruit varieties. *J. Food Eng.* **2005**, *66*, 387–393. [CrossRef]
21. Aydin, C.; Özcan, M.M. Determination of nutritional and physical properties of myrtle (*Myrtus communis* L.) fruits growing wild in Turkey. *J. Food Eng.* **2007**, *79*, 453–458. [CrossRef]
22. Karababa, E. Physical properties of popcorn kernels. *J. Food Eng.* **2006**, *72*, 100–107. [CrossRef]

processes

MDPI

Article

Susceptibility of *Tribolium castaneum* (Coleoptera: Tenebrionidae) to the Fumigation of Two Essential *Satureja* Oils: Optimization and Modeling

Asgar Ebadollahi [1,*], Ebrahim Taghinezhad [1], William N. Setzer [2,3] and Guangnan Chen [4,*]

1 Moghan College of Agriculture and Natural Resources, University of Mohaghegh Ardabili, Ardabil 56199-36514, Iran; taghinezhad@uma.ac.ir
2 Department of Chemistry, University of Alabama in Huntsville, Huntsville, AL 35899, USA; setzerw@uah.edu
3 Aromatic Plant Research Center, 230 N 1200 E, Suite 100, Lehi, UT 84043, USA
4 Faculty of Health, Engineering and Sciences, University of Southern Queensland, Toowoomba, QLD 4350, Australia
* Correspondence: ebadollahi@uma.ac.ir (A.E.); Guangnan.Chen@usq.edu.au (G.C.)

Abstract: Due to the numerous side effects of synthetic pesticides, including environmental pollution, threats to human health, harmful effects on non-target organisms and pest resistance, the use of alternative healthy, available and efficient agents in pest management strategies is necessary. In this paper, the susceptibility of the cosmopolitan, polyphagous, stored-product pest *Tribolium castaneum* (red flour beetle) to the fumigation of the essential oils of two important medicinal and food additive plants, *Satureja hortensis* and *S. intermedia*, was investigated. The insecticidal properties of the essential oils were modeled and optimized using response surface methodology. It was found that a maximum significant mortality of 94.72% and 92.97% could be achieved within 72 h with the applications of 55.15 μL/L of *S. hortensis* (with the linear model) and 58.82 μL/L of *S. intermedia* (with the quadratic model), respectively. There were insecticidal terpenes and phenylpropanoids in both essential oils, including thymol (50.8%), carvacrol (11.2%) and *p*-cymene (13.4%), in the *S. intermedia* and estragole (68.0%) and methyl eugenol (5.6%) in the *S. hortensis*. It was suggested that the essential oils of *S. hortensis* and *S. intermedia* could be offered as promising pesticidal agents against *T. castaneum* for further studies in the management of such pests instead of detrimental synthetic pesticides.

Keywords: biorational pesticides; chemical profile; fumigant toxicity; modeling; optimization; *S. hortensis*; *S. intermedia*

Citation: Ebadollahi, A.; Taghinezhad, E.; Setzer, W.N.; Chen, G. Susceptibility of *Tribolium castaneum* (Coleoptera: Tenebrionidae) to the Fumigation of Two Essential *Satureja* Oils: Optimization and Modeling. *Processes* **2021**, *9*, 1243. https://doi.org/10.3390/pr9071243

Academic Editor: David Fernández-Calviño

Received: 18 June 2021
Accepted: 13 July 2021
Published: 19 July 2021

1. Introduction

The globally distributed insect pest, the red flour beetle *Tribolium castaneum* Herbst (Coleoptera: Tenebrionidae), attacks several stored grain products such as beans, cereals, chocolate, flour, meal, nuts, seeds and spices [1]. The adults of *T. castaneum* are long-lived; they can live for up to three years [2]. In addition to qualitative nutritional damage, contaminated products also have reduced quality due to the creation of an unpleasant odor by the secretion of benzoquinone compounds from abdominal glands and remains of various molting stages and cadavers [3]. Furthermore, *T. castaneum* can transmit phytopathogenic microbial agents such as *Aspergillus*, *Pseudomonas* and *Staphylococcus* to infested storage products [4,5]. Bosly and Kawanna [6] demonstrated that *T. castaneum* can actually carry and distribute some *Aspergillus* species, particularly *A. flavus*, the main producer of carcinogenic aflatoxins in agricultural products [7], in stored flour.

Although the use of synthetic pesticides is the principal approach in pest management, the widespread use of these compounds has resulted in numerous side effects, such as environmental contaminations, the accumulation of hazardous residues in food and high toxicity to non-target organisms [8–10]. In addition, the resistance of pests to frequently used chemical pesticides is increasing. For example, the resistance of *T. castaneum* to phosphine, as one of the commonly used fumigants against stored grain insect pests, was reported [11–13]. Therefore, the introduction of low-risk, available and effective pesticides is essential to alternate with synthetic chemicals in the management of such pests.

The effectiveness of plant-derived essential oils in insect pest control has been demonstrated in numerous studies in recent years [14,15]. Along with the prospective insecticidal effects of essential oils on insect pests from various species, genera, families and orders, their biodegradable nature, safety compared to synthetic chemicals and multiple modes of action in targeted pests indicate that they can be suitable alternatives to detrimental chemical pesticides [16,17]. In fact, the volatile oils obtained from different species of the *Satureja* genus were enriched with terpenes such as 1,8-cineole, γ-terpinene, thymol and β-caryophyllene, which were grouped into four primary classifications: monoterpene hydrocarbons, oxygenated monoterpenes, sesquiterpene hydrocarbons and oxygenated sesquiterpenes [18–20]. Promising insecticidal effects of *Satureja* essential oils against insect pests of storage products were also reported. These included, for example, insecticidal activities of *S. bachtiarica* Bung and *S. khuzestanica* Jamzad essential oils against *T. castaneum* [21], *S. hortensis* L. essential oil against the bean weevil (*Bruchus dentipes* (Baudi)) [19] and the cowpea weevil (*Callosobruchus maculatus* (Fabricius)) [22] and *S. spicigera* Boiss essential oil against the granary weevil (*Sitophilus granarius* (L.)) [23] and the maize weevil (*Sitophilus zeamais* Motschulsky) [24].

Response surface methodology (RSM) is a set of mathematical and statistical tools for the optimization of independent factors and modeling of dependent factors [25]. The optimization of agricultural and chemical processes requires the simultaneous optimization of several objective functions [26]. Therefore, RSM has been used in numerous agricultural and pharmacological fields [27–30]. For example, RSM was used to find mathematical models and optimized conditions for the toxicity of *Eucalyptus globulus* Labill and *Teucrium polium* L. essential oils against *T. castaneum* [31,32].

Given the importance of using natural and low-risk compounds in pest management, the main purpose of the present study is to investigate the possibility of using *S. intermedia* C. A. Mey and *S. hortensis* L. essential oils in the control of *T. castaneum*. The insecticidal properties of essential oils are modeled and optimized using response surface methodology. Due to changes in essential oil composition under different environmental and geographical conditions [18], the chemical profiles of the *S. intermedia* and *S. hortensis* essential oils were also assessed.

2. Materials and Methods

2.1. Plant Materials and Extraction of the Essential Oils

S. intermedia was gathered from its wild populations in the Heiran regions in Ardebil Province, Iran (38°230′ N, 48°350′ E and an elevation of 910 m). *S. hortensis* was collected from Parsabad, Ardebil Province, Iran (39°38′ N, 47°52′ E and an elevation of 52 m). The *Satureja* species were identified based on the work of Jamzad [33]. The fresh leaves and flowers of *S. intermedia* and *S. hortensis* were used for essential oil extraction. All specimens were allowed to dry in the shade over a 10-day period. For each plant sample, 100 g of plant material was added to a 2-L flask of the Clevenger apparatus and subjected to 3 h hydrodistillation. The extracted essential oils were then individually poured into glass vials, covered with aluminum foil, dried over anhydrous Na_2SO_4 and kept in the refrigerator at 4 °C until use.

2.2. Chemical Profiles of the Essential Oils

The *Satureja* essential oils were analyzed by gas chromatography–mass spectrometry using an Agilent 7890B GC with an Agilent 5977A mass selective detector (MSD), operated in the EI mode (electron energy = 70 eV, scan range = 50–550 amu and scan rate = 3.99 scans/s). The GC column was an HP-5ms fused silica capillary column (30 m in length, 0.25 mm internal diameter, (5% phenyl)-polymethylsiloxane stationary phase and 0.25 μm film thickness). Helium was the carrier gas, with a 52.8 kPa column head pressure and 1.0 mL/min flow rate. The inlet temperature was 200 °C, and the interface temperature was 280 °C. The GC oven temperature was programed to hold for 1 min at 50 °C and then increase at a rate of 6 °C/min to a final temperature of 290 °C, with a total run time of 50 min. A 10% *w/v* solution of the sample in methanol was prepared, and 1 µL was injected using a 100:1 split ratio. Identification of the oil components was based on their retention indices, determined by reference to a homologous series of *n*-alkanes and by comparison of their mass spectral fragmentation patterns with those reported in the databases [34–36].

2.3. Insect Rearing

Adult *T. castaneum* specimens were collected from contaminated wheat grains in warehouses in Parsabad, Ardabil Province, Iran (39°38′ N, 47°52′ E and an elevation of 52 m). The adults were reared on wheat grains in cylindrical glass containers (720 mL) covered by a fine mesh cloth for ventilation. Contaminated grains were kept in the incubator at 27 ± 2 °C and 60 ± 5% RH in the dark. Newly emerged adult insects 1–10 days old were selected for the bioassays.

2.4. Fumigant Toxicity

A series of concentrations from 20.59 to 58.82 µL/L and from 21.00 to 55.15 µL/L was selected to assess the fumigant toxicity of *S. intermedia* and *S. hortensis* essential oils, respectively. Filter papers (Whatman No. 1; 2 × 2 cm) separately treated with the concentrations of the essential oils were glued to the inner surface of the screw caps of fumigant glass containers (340 mL). Twenty unsexed adults (1–10 days old) were transferred to containers before screwing their caps. Then, the caps were sealed to be impermeable to air with parafilm. The bioassay was repeated four times, and in the control groups, all steps were performed except for adding essential oils. The fumigant containers were kept in an incubator with 27 ± 2 °C and 60 ± 5% RH, and insect mortality was recorded after 24, 48 and 72 h of exposure. The mortality in the control groups was corrected by the following formula: *Pt* = [(*Po* − *Pc*)/(100 − *Pc*)] × 100, in which *Pt* is the corrected mortality percentage, *Po* is the mortality of the insects treated by essential oil concentrations and *Pc* is the mortality of the insects in the control groups [37].

2.5. Modeling and Optimization by RSM

Using RSM under the historical data design by the statistical software Design Expert 8.0.6 (Stat-Ease, Inc., Minneapolis, MN, USA), various concentrations of essential oils were analyzed in five levels, with the exposure times in three levels as the independent variables and with the mortality of *T. castaneum* as the dependent variable.

Multiple linear regression analysis for the interactions of the independent and dependent variables was used to achieve the statistically appropriate mathematical model in the following second-order polynomial equation [38]:

$$y = \beta o + \sum_{i=1}^{k} \beta_i X_i + \sum_{i=1}^{k} \beta_j X_j + \sum_{i=1}^{k} \sum_{j=1}^{k} \beta_{ij} X_i X_j + \sum_{i=1}^{k} \beta_{jj} X_j^2 \quad (1)$$

where y is the mortality of *T. castaneum* (dependent variable), X_i and X_j are the exposure times and essential oil concentrations, respectively (independent variables), k is a number of independent variables (2), β_o is the intercept of the model, β_i and β_j are the coefficients of the linear parameters and β_{ij} specifies the quadratic parameter coefficient. The statistical relationship between the independent and dependent variables was evaluated by the correlation coefficient of determination (R^2), adjusted R^2 and predicted R^2. Optimization of the insect pest mortality caused by the fumigation of *S. intermedia* and *S. hortensis* essential oils was performed to the maximum desirability using Design Expert software with RSM. The statistical significance of the independent variables on the response variables was examined at a 95% confidence level ($p < 0.05$), and only the variables with a significant effect on the response variable were used in the proposed regression equation.

3. Results

The results of the GC-MS analyses of the essential oils of *S. intermedia* and *S. hortensis* are presented in Table 1. It can be seen that the essential oil of *S. intermedia* was rich in the phenolic monoterpenes thymol (50.8%) and carvacrol (11.2%) along with monoterpenes *p*-cymene (13.4%) and 1,8-cineole (4.3%). The essential oil of *S. hortensis*, on the other hand, was dominated by phenylpropanoids, especially estragole (68.0%) as well as methyl eugenol (5.6%) and (*E*)-*p*-methoxycinnamaldehyde (4.4%) (Table 1 and Figure 1).

Table 1. Chemical profiles of the essential oils isolated from the aerial parts of *S. intermedia* and *S. hortensis*.

RIcalc	RIdb	Compound	Percent Composition	
			S. intermedia	*S. hortensis*
933	933	α-Pinene	1.3	1.3
978	982	1-Octen-3-ol	0.9	—
1005	999	3-Octanol	0.4	—
1025	1024	Limonene	—	0.9
1027	1025	p-Cymene	13.4	—
1031	1032	1,8-Cineole	4.3	—
1031	1032	(Z)-β-Ocimene	—	1.2
1048	1044	(E)-β-Ocimene	—	0.9
1054	1057	γ-Terpinene	1.0	—
1097	1091	Rosefuran	—	0.2
1101	1101	Linalool	0.2	0.2
1105	1102	6-Methylhepta-3,5-dien-2-one	—	0.2
1129	1127	allo-Ocimene	—	1.3
1137	1138	cis-p-Mentha-2,8-dien-1-ol	—	0.2
1143	1142	(E)-Myroxide	—	0.2
1149	1145	trans-Verbenol	—	0.1
1174	1171	p-Mentha-1,5-dien-8-ol	—	0.2
1174	1173	Borneol	0.2	0.1
1181	1180	Terpinen-4-ol	0.7	0.1
1203	1201	Estragole (=Methyl chavicol)	—	68.0
1211	1211	β-Cyclocitral	—	0.4
1223	1223	trans-Carveol	—	0.1
1237	1239	Thymyl methyl ether	1.8	—
1248	1244	Carvacryl methy ether	3.1	—
1256	1252	Chavicol	—	0.2
1259	1257	p-Anisaldehyde	—	0.5
1288	1289	Thymol	50.8	0.3
1296	1298	Carvacrol	11.2	0.8
1361	1356	Eugenol	—	0.1
1362	1365	Carvacryl acetate	0.1	—
1407	1405	Methyl eugenol	—	5.6
1424	1424	(E)-β-Caryophyllene	1.1	—
1478	1478	γ-Muurolene	0.2	—
1490	1490	(E)-β-Ionone	—	0.3
1497	1491	Viridiflorene	0.4	—
1500	1500	α-Muurolene	0.1	—
1508	1508	β-Bisabolene	0.9	—
1516	1514	γ-Cadinene	0.2	—
1524	1518	δ-Cadinene	0.4	—
1540	1541	(E)-α-Bisabolene	0.1	—
1544	1544	α-Calacorene	0.1	—
1552	1554	Thymohydroquinone	0.4	—
1575	1567	(E)-p-Methoxycinnamaldehyde	—	4.4
1576	1576	Spathulenol	1.0	0.5
1587	1583	allo-Spathulenol	—	3.5
1589	1587	Caryophyllene oxide	0.8	0.8
1611	1614	1,10-di-epi-Cubenol	—	0.4
1613	1613	Humulene epoxide II	0.1	—

Table 1. *Cont.*

RIcalc	RIdb	Compound	Percent Composition	
			S. intermedia	*S. hortensis*
1629	1631	1-epi-Cubenol	tr	—
1635	1638	cis-Cadin-4-en-7-ol	0.1	—
1636	1629	iso-Spathulenol	—	0.2
1639	1636	Caryophylla-4(12),8(13)-dien-5β-ol	0.1	—
1641	1640	τ-Cadinol	0.1	—
1646	1645	α-Muurolol (=δ-Cadinol)	tr	—
1655	1655	α-Cadinol	0.1	—
1658	1656	14-Hydroxy-9-epi-(Z)-caryophyllene	0.1	—
1667	1668	ar-Turmerone	0.5	0.2
1667	1668	β-Turmerone	tr	—
1672	1666	14-Hydroxy-9-epi-(E)-caryophyllene	0.2	—
1675	1677	Cadalene	tr	—
1681	1679	epi-α-Bisabolol	tr	—
1687	1683	Germacra-4(15),5,10(14)-trien-1α-ol	tr	—
1699	1701	Curlone	0.1	—
1841	1841	Phytone	0.1	0.4
1926	1925	Methyl palmitate	—	0.2
2109	2109	Phytol	—	1.2
		Monoterpene hydrocarbons	15.7	5.7
		Oxygenated monoterpenoids	72.8	2.8
		Sesquiterpene hydrocarbons	3.5	0.0
		Oxygenated sesquiterpenoids	3.1	5.6
		Benzenoid aromatics	0.0	78.7
		Others	1.4	2.3
		Total identified	96.4	95.0

RIcalc = Retention index, determined with respect to a homologous series of n-alkanes on an HP-5 ms column; RIdb = Retention index from the databases.

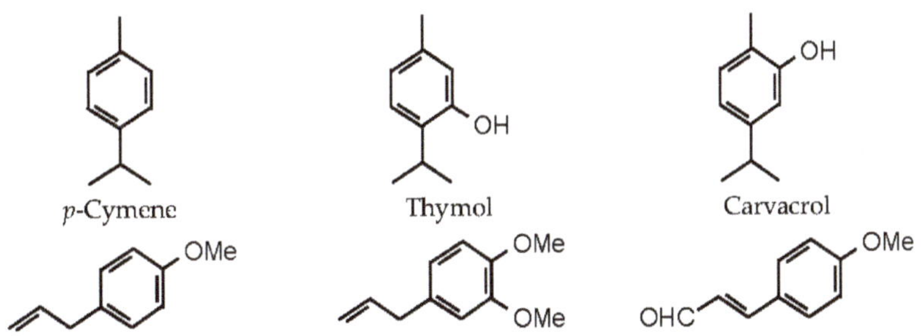

Figure 1. The structures of the major components in the essential oils of *Satureja intermedia* and *S. hortensis*.

Table 2 shows the results of the analysis of variance (ANOVA), which indicated that the mortality of *T. castaneum* adults was significantly affected by A (exposure time) and B (essential oil concentration) in the *S. hortensis* treatment and by A, B, and B^2 in the *S. intermedia* treatment ($p < 0.01$). The lack-of-fit test for both essential oils was non-significant, demonstrating the validation of treatments (Table 2).

In Table 3, the fitting effect of different levels of the essential oil concentration and exposure time on the mortality amount is presented. The model's fitting was evaluated on the basis of the coefficient of determination (R^2), adjusted R^2 and prediction R^2 as well as the coefficient of variation (CV). It can be seen that all the R^2 values were high (>0.94), meaning that the response surface methodology models were suitable. Furthermore, the coefficient of variation for almost all parameters was below 8.4%. This means that the results demonstrated good accuracy and precision with the reliability of experiments. As is shown in Table 3, it was found that the linear effects of the dependent variables on the mortality amount were significant ($p < 0.05$) under the essential oil of *S. hortensis*. The effect on the mortality value under the essential oil of *S. intermedia* was significant according to the quadratic equation ($p < 0.05$). The mortality was strongly influenced by the essential oil concentrations, based on the higher coefficient for B in the equation of Table 3.

Table 2. Results of analysis of variance for prediction of the fumigant toxicity of *S. intermedia* and *S. hortensis* essential oils against *T. castaneum*.

Essential Oil	Source	Sum of Squares	df	Mean Square	F Value	*p*-Value
S. intermedia	Model	21,460.07	3	7153.36	287.26	<0.0001
	A	4305.62	1	4305.62	172.90	<0.0001
	B	16,798.40	1	16,798.40	674.58	<0.0001
	B^2	518.89	1	518.89	20.84	<0.0001
	Residual	1394.51	56	24.90		
	Lack of Fit	250.76	11	22.80	0.90	0.5504 NS
	Pure Error	1143.75	45	25.42		
	Model	25,853.86	2	12,926.93	760.04	<0.0001
S. hortensis	A	4622.50	1	4622.50	271.78	<0.0001
	B	21,231.36	1	21,231.36	1248.30	<0.0001
	Residual	969.47	57	17.01		
	Lack of Fit	231.97	12	19.33	1.18	0.3262 NS
	Pure Error	737.50	45	16.39		
	Cor Total	26,823.33	59			

A: exposure time (h); B: essential oil concentrations (μL/L); NS: non-significant.

Table 3. RSM modeling results for predicting the mortality percentage under different concentrations of essential oils and exposure times.

Essential Oil	Equation	R^2 Value	Adj R^2	Pred R^2	C.V. (%)
S. intermedia	$-34.33772 + 0.43229A + 2.77692B - 0.019410B^2$	0.9390	0.9357	0.9297	8.40
S. hortensis	$-23.11885 + 0.44792A + 1.55199B$	0.9639	0.9626	0.9600	7.59

A: exposure time (h); B: essential oil concentrations (μL/L).

Figure 2 presents the effect of different concentrations of essential oils (in five levels) and exposure times (in three intervals) on the mortality of *T. castaneum* adults. It can be seen that the susceptibility or the mortality of the insect pest was increased by increasing the concentrations of both the essential oils and the exposure times. According to the color points in Figure 2, the mortality of *T. castaneum* was increased from 15% to 95% and from 20% to 100% by increasing the concentrations of the *S. hortensis* and *S. intermedia* essential oils from 21.00 to 55.15 μL/L and from 20.59 to 58.82 μL/L, respectively, and the exposure times from 24 to 72 h.

Plots of the residuals versus the predicted mortality of *T. castaneum* adults caused by the fumigation of *S. intermedia* and *S. hortensis* essential oils are shown in Figure 3. It can be seen that the introduced models for predicting *T. castaneum* mortality were accurate and consistent with the variation hypothesis.

The optimized conditions for the significant maximum mortality of *T. castaneum* adults treated by the essential oils of *S. intermedia* and *S. hortensis* after different exposure times are shown in Table 4. After 72 h of exposure time, the maximum mortality of the pest (92.973%) may be achieved with 58.820 μL/L of *S. intermedia* essential oil with a desirability of 0.9121, while a concentration of 55.150 μL/L would be sufficient to attain the maximum mortality (94.72%) with *S. hortensis* essential oil with desirability of 0.997%. This indicated that the *T. castaneum* adults were more susceptible to the essential oil of *S. hortensis* than the *S. intermedia* (Table 4).

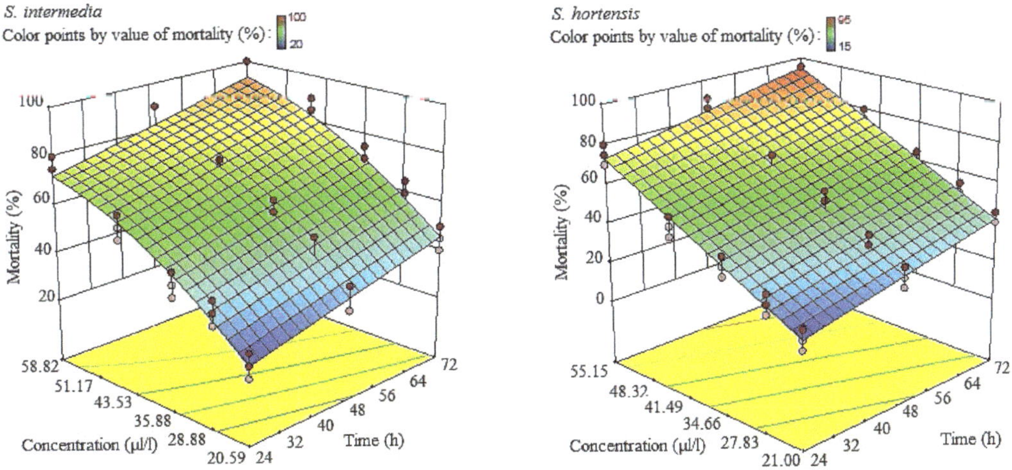

Figure 2. Three-dimensional diagrams of the mortality of *T. castaneum* caused by the fumigation of *S. intermedia* and *S. hortensis* essential oils.

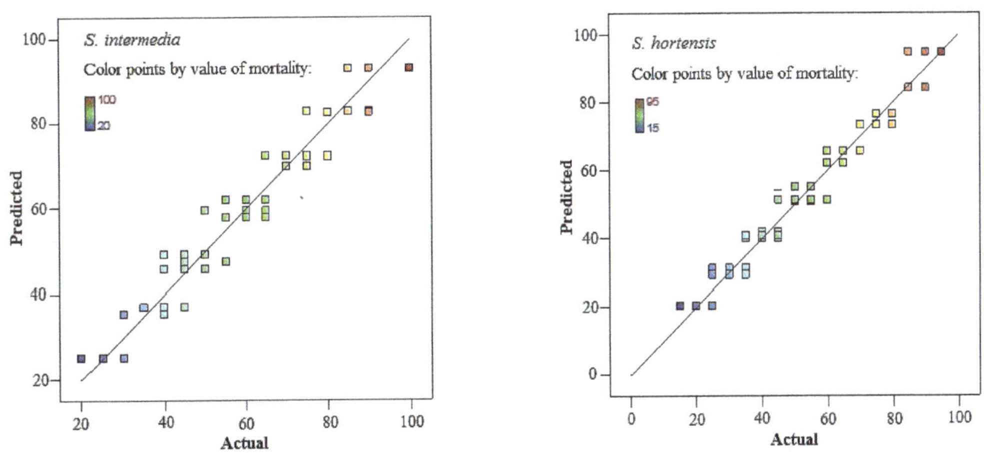

Figure 3. Plots of residual vs. predicted mortality of *T. castaneum* caused by the fumigation of *S. intermedia* and *S. hortensis* essential oils.

Table 4. Optimization of the mortality of *T. castaneum* caused by the fumigation of *S. intermedia* and *S. hortensis* essential oils.

Essential Oil	Mortality (%) *	Time (h)	Concentration (µL/L)	Desirability
S. intermedia	50.000	31.569	33.127	1.000
	92.973 *	72.000	58.820	0.912
S. hortensis	50.000	61.952	29.233	1.000
	94.723 *	72.000	55.150	0.997

* The maximum significant mortality percentage based on high desirability, calculated by Design Expert software.

4. Discussion

In the *S. intermedia* essential oil, 95.1% of the identified compounds were terpenes, in which oxygenated monoterpenoids such as thymol (50.8%) and carvacrol (11.2%) had high concentrations. In contrast, the terpenes had only 14.1% of the total *S. hortensis* essential oil, and this essential oil was rich in benzenoid aromatics such as estragole (68.0%), methyl eugenol (5.6%) and (*E*)-*p*-methoxycinnamaldehyde (4.4%). Therefore, the essential oils studied in the present study were completely different in terms of chemical components. It was reported that γ-terpinene (42.3%), carvacrol (32.8%), *p*-cymene (8.1%), β-pinene (2.3%), β-caryophyllene (2.2%), α-thujene (1.9%) and α-pinene (1.5%) were the main components of the essential oil isolated from the aerial parts of *S. intermedia* at the flowering stage from Romania [39]. In the present study, only carvacrol (0.8%) and α-pinene (1.3%) in low concentrations were found in *S. hortensis* at the pre-flowering stage. In contrast, estragole and methyl eugenol, identified in high amounts in the present study, were not detected in the essential oil investigated by Chambre et al. [39]. The chemical composition of *S. intermedia*, as one of the *Satureja* species endemic to Iran, was assessed [40], and γ-terpinene (37.1%), thymol (30.2%), *p*-cymene (16.2%), limonene (3.9%), α-terpinene (3.3%) and β-myrcene (2.5%) were reported as the main components. In the present study, γ-terpinene, thymol and *p*-cymene were also identified, but there were no traces of limonene, α-terpinene or β-myrcene. Accordingly, differences in the chemical compositions of essential oils may be related to the different species and stages of plants as well as the temperature, humidity, height and other geographical conditions of the cultivation locales [18,41].

The insecticidal effects of the essential oils extracted from *S. hortensis* and *S. intermedia* against stored-product insect pests were also reported in some recent studies. The fumigant toxicity of *S. hortensis* essential oil against *C. maculatus*, *S. zeamais* and the Mediterranean flour moth (*Ephestia kuehniella* Zeller) was studied [22,24,42]. Additionally, the fumigant toxicity of *S. intermedia* essential oils against the lesser grain borer (*Rhyzopertha dominica* (Fabricius)) and the khapra beetle (*Trogoderma granarium* Everts) was observed in one of our previous works [43]. These results are consistent with the current findings about the insecticidal potential of essential oils of the *S. hortensis* and *S. intermedia* species. Overall, the insecticidal effects of plant essential oils are related to their active compounds such as terpenes and phenylpropanoids [44,45]. As is shown in Table 5, the strong insecticidal effects of the components were identified in high percentages in the *S. intermedia* and *S. hortensis* essential oils, including 1,8-cineole, carvacrol, estragole, methyl eugenol, *p*-cymene, thymol and α-pinene (in both essential oils). The insecticidal activity of *S. intermedia* and *S. hortensis* essential oils may be highly associated with these active compounds.

Table 5. Review of insecticidal effects for the main compounds identified in *S. intermedia* and *S. hortensis* essential oils.

Compound	Reported Insecticidal Effects
p-Cymene	Significant toxicity against the third and fourth instar larvae and pupae of mosquito *Culex quinquefasciatus* Say [46].
1,8-Cineole	High contact toxicity to the third instar larvae of the diamondback moth (*Plutella xylostella* (L.)), which was synergistically increased by terpene pulegone [47]. Significant fumigant toxicity against the adults of rice weevil (*Sitophilus oryzae* (L.)) and *T. castaneum* [48].
Thymol	High contact toxicity to the third instar larvae of the diamondback moth, which was synergistically increased by the terpene pulegone [47]. The 100% mortality of adults of *S. granarius* by the fumigation of 163.3 μL/L after 96 h [49].
Carvacrol	The 100% mortality of adults of *S. granarius* by the fumigation of 166.7 μL/L after 96 h [49]. Significant toxicity against the third and fourth instar larvae and pupae of *C. quinquefasciatus* [46].
Estragole	Significant fumigant toxicity against the adults of *T. castaneum* after 24 h [44]. Significant fumigant toxicity and repellent action against the adults of *S. zeamais* after 168 and 1 h, respectively [45].
Methyl eugenol	Strong fumigant toxicity against *S. zeamais* [50]. Toxic to the third instar larvae of cigarette beetle (*Lasioderma serricorne* (F.)) in the feeding test after 168 h [51].
α-Pinene (presence in both essential oils)	The 100% mortality of adults of *S. granarius* by the fumigation of 72.5 μL/L after 96 h [49]. Significant toxicity against the third and fourth instar larvae and pupae of *C. quinquefasciatus* [46].

Furthermore, according to the study of Kim et al. [44], estragole had more fumigant toxicity against *S. zeamais* (LC_{50} = 0.004 mg/cm^3) and *T. castaneum* (LC_{50} = 0.013 mg/cm^3) adults than the terpenes limonene, linalool, β-myrcene, α-pinene and α-humulene. Therefore, having a high percentage of estragole in *S. hortensis* essential oil can be the main reason for its higher toxicity than *S. intermedia*. However, the synergistic and antagonistic effects between all essential oil components should be considered. For example, the toxicity of 1,8-cineole and thymol against the third instar larvae of diamondback moths (*Plutella xylostella* (L.)) were synergistically enhanced in combination with another terpene, pulegone, while the pulegone alone had no significant toxicity to the pest [47].

According to our recent studies, the *T. castaneum* adults were susceptible to the fumigation of *S. hortensis* and *S. intermedia* essential oils [43,52]. However, the optimization and modeling of the fumigant toxicity of these essential oils, through RSM, are reported for the first time in the current study. The use of RSM in the optimization and modelling of the insecticidal efficiency of essential oils can be found in a few other investigations. For example, optimization of the fumigant toxicity of *Thymus vulgaris* L. essential oil indicated that a concentration of 25.86 μL/L and a 59.00-h exposure time were sufficient to achieve 50% mortality of *R. dominica* adults [53]. In another work on the essential oil of *Thymus kotschyanus* Boiss. & Hohen, the optimized condition for 50% mortality against *R. dominica* adults was 24.62 μL/L and a 57.98-h exposure time [54]. Based on the above-mentioned results, the essential oil of *T. kotschyanus* was more toxic than the *T. vulgaris* oil. Regarding the toxicity of *S. hortensis* and *S. intermedia* essential oils against *T. castaneum*, 29.23 and 33.13 μL/L of essential oils after 61.95 and 31.57 h, respectively, can result in 50% mortality of the pest. In one of our other studies, modeling and optimization of the fumigant toxicity of *Teucrium polium* L. essential oil against *T. castaneum* adults was evaluated through RSM, and it was found that 97.97% mortality of the insect pest could be attained by a 20 μL/L essential oil concentration after 72 h with the model +0.71 − 0.047A − 8.84E − 3B + 3.89E − 4AB + 3.27E − 3A^2 + 8.38E − 5B^2 (A and B are the time and essential oil concentration, respectively) [31]. In comparison with these results, and after 72 h of exposure time, 55.15 and 58.82 μL/L of *S. hortensis* and *S. intermedia* essential oils could result in 94.72% and 92.97% mortality of *T. castaneum*, respectively. The differences may be associated with different tested essential oils and subjected insect pests. In addition, the above-mentioned and present findings indicated that RSM is a useful method for the optimization and modeling of the insecticidal effects of essential oils.

The disadvantage of such essential oils (i.e., low persistence in environmental conditions) can be improved by micro-and nano-encapsulation formulations based on controlled release techniques [55,56].

5. Conclusions

The present study has shown that the essential oils isolated from the aerial parts of *S. intermedia* and *S. hortensis* had strong fumigant toxicity against the adults of the red flour beetle *T. castaneum*. GC-MS analyses indicated that there was a high percentage of insecticidal components such as estragole and methyl eugenol in the *S. hortensis* essential oil and thymol, carvacrol, *p*-cymene and 1,8-cineole in the *S. intermedia* essential oil. However, additional studies are required to determine which components have efficient insecticidal activity. Based on RSM, it was found that the best model for the prediction of mortality was linear and the quadratic equation for the essential oil of *S. hortensis and S. intermedia*, respectively. Up to 94.72% and 92.97% mortality of the insect pest, as the maximum significant mortality, can be attained with 55.15 and 58.82 μL/L of *S. hortensis* and *S. intermedia* essential oils, respectively, after 72 h. The essential oils of *S. hortensis* and *S. intermedia* can be proposed for further research in the management of *T. castaneum* and probably other stored-product insect pests.

Author Contributions: Conceptualization, A.E. and E.T.; methodology, A.E. and E.T.; software, A.E. and E.T.; formal analysis, E.T. and W.N.S.; investigation, A.E.; writing—original draft preparation, A.E. and E.T.; writing—review and editing, A.E., E.T., W.N.S. and G.C.; funding acquisition, G.C. All authors have read and agreed to the published version of the manuscript.

Funding: This research was funded by the University of Mohaghegh Ardabili (grant number 99/d/9/28635).

Institutional Review Board Statement: Not applicable.

Informed Consent Statement: Not applicable.

Data Availability Statement: All data can be made available upon request to the authors.

Acknowledgments: This research project was supported by the University of Mohaghegh Ardabili in Ardabil, Iran, which is greatly appreciated. W.N.S. participated in this work as part of the activities of the Aromatic Plant Research Center (APRC, https://aromaticplant.org/, accessed on 16 July 2021).

Conflicts of Interest: The authors declare no conflict of interest.

References

1. Rees, D. *Insects of Stored Grain: A Pocket Reference*, 2nd ed.; CSIRO Publishing: Collingwood, Australia, 2007; pp. 53–56.
2. Granousky, T.A. Stored product pests. In *Hand Book of Pest Control, the Behavior, Life History and Control of Household Pests*, 1st ed.; Moreland, D., Ed.; Mallis Handbook and Technical Training Co.: Cleveland, OH, USA, 1997; pp. 635–728.
3. Li, J.; Lehmann, S.; Weißbecker, B.; Naharros, I.O.; Schütz, S.; Joop, G.; Wimmer, E.A. Odoriferous defensive stink gland transcriptome to identify novel genes necessary for quinone synthesis in the red flour beetle, *Tribolium castaneum*. *PLoS Genet.* **2013**, *9*, e1003596. [CrossRef]
4. Agarwal, A.; Agashe, D. The red flour beetle *Tribolium castaneum*: A model for host-microbiome interactions. *PLoS ONE* **2020**, *15*, e0239051. [CrossRef] [PubMed]
5. Tuama, S.J.; Ali, S.T.; Hthdy, Z. Effect of *Tribolium castaneum* in qualitative and quantitative contamination with fungi. *Syst. Rev. Pharm.* **2020**, *11*, 1636–1640. [CrossRef]
6. Bosly, H.A.; Kawanna, M.A. Fungi species and red flour beetle in stored wheat flour under Jazan region conditions. *Toxicol. Ind. Health* **2014**, *30*, 304–310. [CrossRef] [PubMed]
7. Okoth, S.; De Boevre, M.; Vidal, A.; Di Mavungu, D.J.; Landschoot, S.; Kyallo, M.; Njuguna, J.; Harvey, J.; De Saeger, S. Genetic and toxigenic variability within *Aspergillus flavus* population isolated from maize in two diverse environments in Kenya. *Front. Microbiol.* **2018**, *9*, 57. [CrossRef] [PubMed]
8. Goulson, D. Ecology: Pesticides linked to bird declines. *Nature* **2014**, *511*, 295–296. [CrossRef]
9. Reiler, E.; Jørs, E.; Bælum, J.; Huici, O.; Alvarez Caero, M.M.; Cedergreen, N. The influence of tomato processing on residues of organochlorine and organophosphate insecticides and their associated dietary risk. *Sci. Total Environ.* **2015**, *528*, 262–269. [CrossRef]
10. Nicolopoulou-Stamati, P.; Maipas, S.; Kotampasi, C.; Stamatis, P.; Hens, L. Chemical pesticides and human health: The urgent need for a new concept in agriculture. *Front. Public Health* **2016**, *4*, 148. [CrossRef]
11. Opit, G.P.; Phillips, T.W.; Aikins, M.J.; Hasan, M.M. Phosphine resistance in *Tribolium castaneum* and *Rhyzopertha dominica* from stored wheat in Oklahoma. *J. Econ. Entomol.* **2012**, *105*, 1107–1114. [CrossRef]
12. Hubhachen, Z.; Jiang, H.; Schlipalius, D.; Park, Y.; Guedes, R.N.C.; Oppert, B.; Opit, G.; Phillips, T.W. A CAPS marker for determination of strong phosphine resistance in *Tribolium castaneum* from Brazil. *J. Pest. Sci.* **2020**, *93*, 127–134. [CrossRef]
13. Wakil, W.; Kavallieratos, N.G.; Usman, M.; Gulzar, S.; El-Shafie, H.A.F. Detection of phosphine resistance in field populations of four key stored-grain insect pests in Pakistan. *Insects* **2021**, *12*, 288. [CrossRef]
14. Isman, M.B.; Grieneisen, M.L. Botanical insecticide research: Many publications, limited useful data. *Trends Plant Sci.* **2014**, *19*, 140–145. [CrossRef]
15. Basaid, K.; Chebli, B.; Mayad, E.H.; Furze, J.N.; Bouharroud, R.; Krier, F.; Barakate, M.; Paulitz, T. Biological activities of essential oils and lipopeptides applied to control plant pests and diseases: A review. *Int. J. Pest. Manag.* **2020**, *67*, 155–177. [CrossRef]
16. Regnault-Roger, C.; Vincent, C.; Arnasson, J.T. Essential oils in insect control: Low-risk products in a high-stakes world. *Ann. Rev. Entomol.* **2012**, *57*, 405–425. [CrossRef]
17. Campos, E.V.; Proença, P.L.; Oliveira, J.L.; Bakshi, M.; Abhilash, P.; Fraceto, L.F. Use of botanical insecticides for sustainable agriculture: Future perspectives. *Ecol. Indic.* **2019**, *105*, 483–495. [CrossRef]
18. Sefidkon, F.; Jamzad, Z. Chemical composition of the essential oil of three Iranian *Satureja* species (*S. mutica, S. macrantha* and *S. intermedia*). *Food Chem.* **2005**, *91*, 1–4. [CrossRef]
19. Tozlu, E.; Cakir, A.; Kordali, S.; Tozlu, G.; Ozer, H.; Akcin, T.A. Chemical compositions and insecticidal effects of essential oils isolated from *Achillea gypsicola, Satureja hortensis, Origanum acutidens* and *Hypericum scabrum* against broadbean weevil (*Bruchus dentipes*). *Sci. Hortic.* **2011**, *130*, 9–17. [CrossRef]

20. Ghorbanpour, M.; Hadian, J.; Hatami, M.; Salehi-Arjomand, H.; Aliahmadi, A. Comparison of chemical compounds and antioxidant and antibacterial properties of various *Satureja* species growing wild in Iran. *J. Med. Plants* **2016**, *3*, 58–72.
21. Taban, A.; Saharkhiz, M.J.; Hooshmandi, M. Insecticidal and repellent activity of three Satureja species against adult red flour beetles, *Tribolium castaneum* (Coleoptera: Tenebrionidae). *Acta Ecol. Sin.* **2017**, *37*, 201–206. [CrossRef]
22. Heydarzade, A.; Moravvej, G.H. Contact toxicity and persistence of essential oils from *Foeniculum vulgare*, *Teucrium polium* and *Satureja hortensis* against *Callosobruchus maculatus* (Fabricius) (Coleoptera: Bruchidae) adults. *Turkish J. Entomol.* **2012**, *36*, 507–518.
23. Yildirim, E.; Krdali, S.; Yazici, G. Insecticidal effects of essential oils of eleven plant species from Lamiaceae on *Sitophilus granarius* (L.) (Coleoptera: Curculionidae). *Rom. Biotechnol. Lett.* **2011**, *16*, 6702–6709.
24. Kordali, S.; Emsen, B.; Yildirim, E. Fumigant toxicity of essential oils from fifteen plant species against *Sitophilus zeamais* Motschulsky (Coleoptera: Curculionidae). *Egypt. J. Biol. Pest Control* **2013**, *23*, 241–246.
25. Riswanto, F.D.O.; Rohman, A.; Pramono, S.; Martono, S. Application of response surface methodology as mathematical and statistical tools in natural product research. *J. Appl. Pharm. Sci.* **2016**, *9*, 125–133. [CrossRef]
26. Darvishi, H.; Farhudi, Z.; Behroozi-Khazaei, N. Multi-objective optimization of savory leaves drying in continuous infrared-hot air dryer by response surface methodology and desirability function. *Comput. Electron. Agric.* **2020**, *168*, 105112. [CrossRef]
27. Kaur, G.; Kumar, V.; Goyal, A.; Tanwar, B.; Kaur, J. Optimization of nutritional beverage developed from radish, sugarcane and herbal extract using response surface methodology. *Nutr. Food Sci.* **2018**, *48*, 733–743. [CrossRef]
28. Kumari, M.; Gupta, S.K. Response surface methodological (RSM) approach for optimizing the removal of trihalomethanes (THMs) and its precursor's by surfactant modified magnetic nanoadsorbents (sMNP)-An endeavor to diminish probable cancer risk. *Sci. Rep.* **2019**, *9*, 18339. [CrossRef] [PubMed]
29. Ning, J.; Yue, S. Optimization of preparation conditions of eucalyptus essential oil microcapsules by response surface methodology. *J. Food Process. Preserv.* **2019**, *43*, 14188. [CrossRef]
30. Pongsumpun, P.; Iwamoto, S.; Siripatrawan, U. Response surface methodology for optimization of cinnamon essential oil nanoemulsion with improved stability and antifungal activity. *Ultras Sonochem.* **2020**, *60*, 104604. [CrossRef] [PubMed]
31. Ebadollahi, A.; Taghinezhad, E. Modeling and optimization of the insecticidal effects of *Teucrium polium* L. essential oil against red flour beetle (*Tribolium castaneum* Herbst) using response surface methodology. *Inf. Process. Agric.* **2020**, *7*, 286–293. [CrossRef]
32. Ebadollahi, A.; Taghinezhad, E. Modeling of the toxicity of *Eucalyptus globulus* Labill essential oil against red flour beetle, *Tribolium castaneum* Herbst. *J. Appl. Sci. Environ. Manag.* **2020**, *24*, 2043–2047. [CrossRef]
33. Jamzad, Z. *Thymus and Satureja Species of Iran*, 1st ed.; Research Institute of Forests and Rangelands: Tehran, Iran, 2009; pp. 22–31.
34. Adams, R.P. *Identification of Essential Oil Components by Gas. Chromatography/Mass Spectrometry*, 4th ed.; Allured Publishing Corporation: Carol Stream, IL, USA, 2007.
35. Mondello, L. *FFNSC 3*; Shimadzu Scientific Instruments: Columbia, MD, USA, 2016.
36. NIST. *NIST17*; National Institute of Standards and Technology: Gaithersburg, MD, USA, 2017.
37. Islam, R.; Khan, R.I.; Al-Reza, S.M.; Jeong, Y.T.; Song, C.H.; Khalequzzaman, M. Chemical composition and insecticidal properties of *Cinnamomum aromaticum* (Nees) essential oil against the stored product beetle *Callosobruchus maculatus* (F.). *J. Sci. Food Agric.* **2009**, *89*, 1241–1246. [CrossRef]
38. Kaveh, M.; Abbaspour-Gilandeh, Y.; Taghinezhad, E.; Witrowa-Rajchert, D.; Nowacka, M. The quality of infrared rotary dried terebinth (*Pistacia atlantica* L.)-optimization and prediction approach using response surface methodology. *Molecules* **2021**, *26*, 1999. [CrossRef] [PubMed]
39. Chambre, D.R.; Moisa, C.; Lupitu, A.; Copolovici, L.; Pop, G.; Copolovici, D.M. Chemical composition, antioxidant capacity, and thermal behavior of *Satureja hortensis* essential oil. *Sci. Rep.* **2020**, *10*, 21322. [CrossRef] [PubMed]
40. Sharifi-Rad, J.; Sharifi-Rad, M.; Hoseini-Alfatemi, S.M.; Iriti, M.; Sharifi-Rad, M.; Sharifi-Rad, M. Composition, cytotoxic and antimicrobial activities of Satureja intermedia C. A. Mey essential oil. *Int. J. Mol. Sci.* **2015**, *16*, 17812–17825. [CrossRef]
41. Khalil, N.; El-Jalel, L.; Yousif, M.; Gonaid, M. Altitude impact on the chemical profile and biological activities of *Satureja thymbra* L. essential oil. *BMC Complement. Med. Ther.* **2020**, *20*, 186. [CrossRef]
42. Najafzadeh, R.; Ghasemzadeh, S.; Mirfakhraie, S. Effect of essential oils from *Nepeta crispa*, *Anethum graveolens* and *Satureja hortensis* against the stored-product insect "*Ephestia kuehniella* (Zeller)". *J. Med. Plants By-Prod.* **2019**, *2*, 163–169. [CrossRef]
43. Ebadollahi, A.; Setzer, W.N. Evaluation of the toxicity of *Satureja intermedia* C. A. Mey essential oil to storage and greenhouse insect pests and a predator ladybird. *Foods* **2020**, *9*, 712. [CrossRef]
44. Kim, S.-L.; Lee, D.-W. Toxicity of basil and orange essential oils and their components against two coleopteran stored products insect pests. *J. Asia Pac. Entomol.* **2014**, *17*, 13–17. [CrossRef]
45. Rosa, J.S.; Oliveira, L.; Sousa, R.M.O.F.; Escobar, C.B.; Fernandes-Ferreira, M. Bioactivity of some Apiaceae essential oils and their constituents against *Sitophilus zeamais* (Coleoptera: Curculionidae). *Bull. Entomol. Res.* **2020**, *110*, 406–416. [CrossRef]
46. Andrade-Ochoa, S.; Correa-Basurto, J.; Rodríguez-Valdez, L.M.; Sanchez-Torres, L.E.; Nogueda-Torres, B.; Nevarez-Moorillon, G.V. In Vitro and in silico studies of terpenes, terpenoids and related compounds with larvicidal and pupicidal activity against *Culex quinquefasciatus* Say (Diptera: Culicidae). *Chem. Cent. J.* **2018**, *12*, 53. [CrossRef]
47. Kumrungsee, N.; Pluempanupat, W.; Koul, O.; Bullangpoti, V. Toxicity of essential oil compounds against diamondback moth, *Plutella xylostella*, and their impact on detoxification enzyme activities. *J. Pest Sci.* **2014**, *87*, 721–729. [CrossRef]

48. Abdelgaleil, S.A.M.; Mohamed, M.I.E.; Badawy, M.E.I.; El-aramis, S.A.A. Fumigant and contact toxicities of monoterpenes to Sitophilus oryzae (L.) and Tribolium castaneum (Herbst) and their inhibitory effects on acetylcholinesterase activity. *J. Chem. Ecol.* **2009**, *35*, 518–525. [CrossRef]
49. Kordali, S.; Usanmaz, A.; Bayrak, N.; Çakır, A. Fumigation of volatile monoterpenes and aromatic compounds against adults of *Sitophilus granarius* (L.) (Coleoptera: Curculionidae). *Rec. Nat. Prod.* **2017**, *11*, 362–373.
50. Wang, C.F.; Yang, K.; Zhang, H.M.; Cao, J.; Fang, R.; Liu, Z.L.; Du, S.S.; Wang, Y.Y.; Deng, Z.W.; Zhou, L. Components and insecticidal activity against the maize weevils of *Zanthoxylum schinifolium* fruits and leaves. *Molecules* **2011**, *16*, 3077–3088. [CrossRef]
51. Imai, T.; Masuda, R. Insecticidal activities of methyleugenol and β-asarone, from the herbal medicines Saishin and Sekishōkon, and other alkoxy-propenyl-benzene derivatives against the cigarette beetle *Lasioderma serricorne* (Coleoptera: Anobiidae). *Appl. Entomol. Zool.* **2017**, *52*, 183–188. [CrossRef]
52. Ebadollahi, A. Estragole-rich essential oil of summer savory (*Satureja hortensis* L.) as an eco-friendly alternative to the synthetic insecticides in management of two stored-products insect pests. *Acta Agric. Slov.* **2020**, *115*, 307–314. [CrossRef]
53. Ebadollahi, A.; Taghinezhad, E.; Davari, M. Optimization of antifungal and insecticidal effects of garden thyme (*Thymus vulgaris* L.) essential oil through response surface methodology. *Biol. Cont. Pest Plant Dis.* **2018**, *7*, 9–19. [CrossRef]
54. Ebadollahi, A. The essential oil extracted from *Thymus kotschyanus* Boiss. & Hohen as a natural substance for management of the lesser grain borer, *Rhyzopertha dominica* F. *Agric. Forest* **2018**, *64*, 490–505. [CrossRef]
55. Nuruzzaman, M.; Rahman, M.M.; Liu, Y.; Naidu, R. Review nanoencapsulation, nano-guard for pesticides: A new window for safe application. *J. Agric. Food Chem.* **2016**, *64*, 1447–1483. [CrossRef] [PubMed]
56. Pavoni, L.; Pavela, R.; Cespi, M.; Bonacucina, G.; Maggi, F.; Zeni, V.; Canale, A.; Lucchi, A.; Bruschi, F.; Benelli, G. Green micro- and nanoemulsions for managing parasites, vectors and pests. *Nanomaterials* **2019**, *9*, 1285. [CrossRef]

MDPI
St. Alban-Anlage 66
4052 Basel
Switzerland
Tel. +41 61 683 77 34
Fax +41 61 302 89 18
www.mdpi.com

Processes Editorial Office
E-mail: processes@mdpi.com
www.mdpi.com/journal/processes

www.ingramcontent.com/pod-product-compliance
Lightning Source LLC
Chambersburg PA
CBHW061450210926
51850CB00008B/1670